# Agricultural
# Economics
# and
# Agribusiness

Seventh Edition

# AGRICULTURAL ECONOMICS AND AGRIBUSINESS

**Gail L. Cramer**
L. C. Carter Professor
*University of Arkansas—Fayetteville*

**Clarence W. Jensen**
Professor Emeritus
*Montana State University*

**Douglas D. Southgate, Jr.**
Professor
*The Ohio State University*

**John Wiley & Sons, Inc.**
New York   Chichester   Weinheim
Brisbane   Singapore   Toronto

ACQUISITIONS EDITOR Whitney Blake
PRODUCTION SERVICE Ingrao Associates
INTERIOR DESIGN Karin Kincheloe
ASST. MANUFACTURING MANAGER Mark Cirrillo
MARKETING MANAGER Wendy Goldner
SENIOR ILLUSTRATION COORDINATOR Anna Melhorn
PHOTO EDITOR Hilary Newman

This book was set in Palatino by Alexander Graphics and printed and
bound by Malloy Lithographers. The cover was printed by Phoenix Color.

*Library of Congress Cataloging in Publication Data:*

Cramer, Gail L.
    Agricultural economics and agribusiness/Gail L. Cramer, Clarence W.
Jensen, Douglas D. Southgate, Jr.—7th ed.
        p.    cm.
    Includes bibliographical references and index.
    ISBN 0-471-17376-2 (cloth : alk. paper)
    1. Agriculture—Economic aspects.    2. Agriculture—Economic
aspects—United States.    3. Agricultural industries.
4. Agricultural industries—United States.    I. Jensen, Clarence W.
II. Southgate, Douglas DeWitt.    III. Title.
HD1415.C7    1997
338.1—dc21                                                              93-29529
                                                                              CIP

Printed in the United States of America

10 9 8 7 6 5 4 3

Printed and bound by Malloy Lithographing, Inc.

## To Our Families

*Marilyn, Karilee, and Bruce Cramer*
*Elaine, Marcia Logan,*
*Daryll and Kathy Jensen*
*Myriam, Elizabeth, and Richard Southgate*

# About the Authors

**Gail L. Cramer** is currently L. C. Carter Chair Professor in the Department of Agricultural Economics and Rural Sociology at the University of Arkansas, a position he has held since 1987. He attained his Bachelor's degree in 1963 from Washington State University, his Master's degree from Michigan State University in 1964, and his Ph.D. in Agricultural Economics from Oregon State University in 1967.

In 1967, Gail Cramer was appointed Assistant Professor in the Department of Agricultural Economics and Economics at Montana State University. He was promoted to Associate Professor in 1972 and to Full Professor in 1976. While at Montana State University, he taught courses in advanced, intermediate, and beginning microeconomics, agricultural marketing and agricultural policy. He won four teaching awards at Montana State University including the Phi Kappa Phi University-Wide Award in 1980.

Primary research assignments are in wheat and rice marketing. He has published over 170 journal articles and other publications in the general area of grain marketing. Cramer won the E. G. Nourse Award for his outstanding Ph.D. dissertation on Cooperative Mergers. Other awards include the 1980 award for "Excellence in Quality of Communication" for

*Agricultural Economics and Agribusiness* with Clarence W. Jensen by the American Agricultural Economics Association. This book has been translated into Spanish, Malaysian, Chinese, and Russian. Also, Cramer received the Rice Technical Workers' Research Award for the outstanding rice research program in 1992 jointly with Dr. Eric J. Wailes at the University of Arkansas. In 1992, Cramer was selected for the Distinguished Faculty Award for Research and Public Service from the Arkansas Alumni Association. His research is domestic and international in scope. His rice research has taken him throughout the world, and he has presented seminars on his research in the Philippines, Japan, Indonesia, Mexico, Taiwan, Australia, Hong Kong, Singapore, Egypt, and England.

In 1990, Cramer was one of the principal founders of the International Agribusiness Management Association, which was incorporated in Arkansas in 1990. He was instrumental with Mike Woolverton in forming *Agribusiness: An International Journal* published by John Wiley and Sons, New York in 1985 and is one of the leading journals in agricultural economics. In addition, he served on the White House Agribusiness Commission, which was useful in expanding the number of undergraduate and graduate programs in agribusiness.

Cramer is married to the former Marilyn Jean Karlenberg and they have two grown children—Karilee and Bruce.

**Clarence W. Jensen** is currently Professor Emeritus in the Department of Agricultural Economics and Economics, an appointment he has enjoyed since 1980.

He attended Montana State College where he was granted Bachelors and Masters degrees in Agricultural Economics and Economics in 1951 and 1952. Further graduate work at Michigan State University was completed, where he received the Ph.D. degree in Agricultural Economics and Economics in 1958.

In 1955, Clarence Jensen was appointed Assistant Professor in the Department of Agricultural Economics and Economics at Montana State University, then was appointed to Associate Professor in 1959 and Full Professor in 1961. He later served as Department Head from 1966 to 1968.

At Montana State University, Jensen taught a wide variety of courses in Agricultural Economics and Economics (Economics of American Agriculture; Farm and Ranch Management; Economics of Natural Resources; Advanced Farm and Ranch Management; Agricultural Production Economics; Advanced Production Economics; General Economics; Microeconomic Principles; Intermediate Economics; Public Finance; Welfare Economics; and Managerial Economics) at both the undergraduate and graduate levels. He also was Advisor for many of the department's undergraduate majors, and served on numerous graduate committees as Chairman or committee member.

Research assignments were generally in the areas of agricultural and natural research management and productivity, production economics, costs and returns in farming and ranching, and the economics of water resource development and use. A sizable number of publications resulted from those research efforts.

During his tenure at Montana State University, Jensen served as Consultant to the Bureau of Land Management, and with visiting research appointments at Texas A&M University's Water Resources Institute, Winrock International, and L. C. Carter Rice Institute at the University of Arkansas.

Jointly with Gail Cramer, Jensen earned the 1980 Award for "Excellence in Quality of Communications" by the American Agricultural Economics Association for the first edition of our book *Agricultural Economics and Agribusiness*, now in its sixth edition.

Clarence Jensen married the former Elaine D. Johnson in September 1944. They have two children, each now with families of their own.

**Douglas D. Southgate, Jr.** is a professor of agricultural economics at Ohio State University, where he also directs the International studies Center and the Latin American Studies Program. He was awarded a bachelor's degree in economics by the University of Oregon in 1974 and was elected to Phi Beta Kappa. He completed his Ph.D. in agricultural economics at the University of Wisconsin, and the same year joined the faculty at Ohio State. He was promoted to associate professor in 1986 and to full professor in 1995.

A natural resource economist, Dr. Southgate specializes in the study of environmental problems in Latin America. He has published three books and a large number of journal articles and scholarly papers addressing the causes and consequences of tropical deforestation, the economics of watershed management in developing countries, and related topics. Since the early 1980s, he has consulted in twelve Latin American nations for the World Bank, the Inter-American Development Bank, the United States Agency for International Development, and the Ford Foundation.

In 1987, Dr. Southgate held a Fulbright Research Fellowship in Ecuador. He returned to that country in 1990 for a three-year Joint Career Corps assignment with the United States Agency for International Development. His primary duty was to provide advice on natural resource policy. From 1988 to 1990, he served on the Tropical Ecosystems Directorate of the United States "Man and the Biosphere Program."

Dr. Southgate is married to Myriam Anota Posso, and they have two children, Elizabeth and Richard.

# Preface

Agriculture is front-page news. On any given day, we may read of large grain or fiber sales, food shortages or surpluses, farm strikes, meat boycotts, trade embargoes, tariffs and import quotas, or export subsidies. Such events have significant impacts on human beings, both domestically and internationally, and can be understood and predicted better with sound training in agricultural economics and agribusiness, especially when supplemented by a good background in technical agriculture.

This book examines the structure and organization of the agricultural industry, then discusses basic micro- and macroeconomics principles as they apply to agriculture. Principles of economics are used to demonstrate to the student that theory actually makes reality more understandable. Theory allows us to abstract from the workaday world in order to simplify real-world problems so they can be examined, explained, and understood.

The material in this text is designed for a one-quarter or one-semester class in an introductory agricultural economics or agribusiness course. If all the applied chapters are covered with some outside reference work, however, it can be used for a two-quarter sequence.

Two different routes can be taken in going through the material in this book. For a freshman-level course, we recommend covering Chapters 1, 2, 3 (excluding indifference curve analysis), 4, 5 (excluding Chapter 5's appendix), 6, 7, 8, 9, and then selecting preferred applied chapters. In a sophomore-level course, we recommend full coverage of all theory chapters before dealing with the applied chapters.

We are interested in presenting the basic economic concepts appropriate to the agricultural industry in a clear and understandable manner. Many general economics textbooks have been able to accomplish this goal with respect to other industries. Agriculture, however, is a unique area of study with its own characteristics and opportunities. Moreover, we feel that students interested in agriculture should have a textbook directly relating to their profession. This book provides the necessary background for more advanced agricultural economics, agribusiness, and economics courses.

We have prepared the seventh edition with a number of special targets in mind: to produce a book whose size alone neither overwhelms the student nor inhibits the instructor making his or her own contribution to the class; to make the necessary topical revisions; to update relevant agricultural data; to clarify concepts that students have found to be especially difficult; and to maintain the coverage of macroeconomics and its linkages to agriculture. The reader can readily identify specialized words and terms, as each is set in boldface italics at its first mention. A section at the end of each chapter lists again those same words and phrases as a reminder of their importance.

Recent changes in U.S. domestic agricultural policy and this nation's trade relations with other countries have been included here. The financial crisis in agriculture and its credit institutions has resulted in a number of changes throughout the credit delivery system. Although these are likely not yet complete, legislated system changes as of mid-1996 have been incorporated in this edition.

Much more difficult to identify are the recent developments in command economies. Major institutional changes are underway in the Soviet Union and the People's Republic of China, and we track these as best we can where recent information permits.

Both the field and the literature of agricultural economics and agribusiness have grown because of the significant contributions of a great many people in the profession. At the end of each chapter, we give a brief biographical sketch of an individual who has made a major contribution to the literature in agricultural economics and agribusiness.

An *Instructor's Manual* accompanies this edition. For each chapter there is (1) a brief summary of the chapter; (2) a suggested lecture outline; (3) a list of economic terms that can serve both as key points for emphasis in going through the chapter material and as a source of essay questions requiring written answers; and (4) a core group of 15 true-false and 15 multiple-choice questions (all with correct answers provided) to help speed the task of preparing quizzes and exams. We also include duplicate transparencies of all graphs in the text that can be used with an over-head projector, as a lecture aid.

A *Student's Study Guide* has also been prepared to accompany this edition. Each chapter in the *Study Guide* is keyed to the text, briefly summarizing the basic economic concepts to expedite the student's grasp of agricultural economics and agribusiness. As with the *Instructor's Manual*, the *Study Guide* also contains chapter outlines; lists of economic and specialized words and terms for written-answer testing; and a number of test questions—fill-in, true–false, multiple choice, and problem exercise—to aid the student in reviewing and preparing for exams, with an answer key at the end of each chapter.

It is impossible to write a book without the help and contributions of many people. We owe a large debt to our instructors and to the authors of other textbooks. We cannot thank everyone who has been of help to us, but we would like to single out a few people who have been of direct assistance to us in the preparation of this book. Dr. John M. Marsh reviewed many of the original chapters, Dr. Dan Dunn gave freely of his assistance in expanding and updating Chapter 10, and Larry Simmons reviewed and made suggestions to improve the macroeconomics chapter. Marilyn Cramer reviewed and also typed the rough drafts of many of the chapters. Our colleagues at the University of Arkansas, Montana State University, and The Ohio State University have given us continued encouragement, and Dr. J. C. Headley, Head of Agricultural Economics and Rural Sociology at the University of Arkansas, has supported this project in many ways.

We are indebted for many valuable suggestions made by John Adrian of Auburn University; Kenneth Boggs of the University of Missouri—Columbia; James Garrett of the University of Nevada—Reno; Dean Linsenmeyer of the University of Nebraska—Lincoln; Lester Mandersheid of Michigan State University; Carl Pherson and John Shields of California State University—Fresno; Lawrence Bohl of Purdue University; Robert Emerson of the University of Florida—Gainesville; Ronald Deiter of Iowa State University; Josef M. Broder of the University of Georgia—Athens; Alvin Schupp at Louisiana State University; Roger D. Herriman at Illinois Central College; John Rogalla at California Polytechnic State University; and James Shataua at University of Wisconsin—River Falls.

A special note of appreciation also is due Dr. Ken Young for his skill in editing and updating the manuscript for this edition. Finally, although we appreciate all the excellent advice and counsel of others, errors and omissions that remain are the sole responsibility of the authors.

GAIL L. CRAMER
CLARENCE W. JENSEN
DOUGLAS D. SOUTHGATE

# Contents

# Agricultural
# Economics
# and
# Agribusiness

# CHAPTER 1

# Introduction

$\mathbf{A}$t least two important characteristics of the agricultural industry make it distinct from all others. One is the cyclical nature of production caused primarily by physical and biological factors. The other is price instability resulting from the effects of changes caused both within the market for agricultural products and physically.

The cyclical nature of production can be seen in the cattle market, for instance, where there is a long lag between the time when increasing beef prices motivate producers to increase production and when the results of their decisions show up at the meat counter of the local grocery store. It takes almost one and a half years from breeding a cow to weaning her heifer calf. Another year is normally required before that heifer can be bred. Yet another year is required before the heifer's calf can be started along the way toward the market for slaughter. An additional year and a half is needed to fatten and market the first offspring from the enlarged beef herd. Herd reduction can be a much more rapid process, but it still adds up to present cattle production cycles of about 10 years' duration.

Climatic conditions and diseases also affect production, causing price and income fluctuations. The corn blight in the United States in 1970 is an example. Only about 20 percent of the corn crop was affected, yet the price of corn increased from $1.07 to $1.43 per bushel as news of the disease spread. There also was concern that the blight would infect the next year's crop because of the manner in which hybrid seed is developed. To combat the problem, scientists worked on blight resistant varieties, and government policy was changed to encourage increased planting. Greater corn acreage and favorable growing conditions in 1971 resulted in a record crop and a more than 30 percent decline in the farm price of corn.

Encouraged by higher prices and the 1975 grain agreement with the former Soviet Union, American farmers increased their production of wheat and coarse grains. U.S. crop and livestock output both reached record levels in the early 1980s. Production and carry-over stocks of grains reached such levels that a 1983 acreage reduction program took 80 million acres out of production. While wheat stocks remained large, the 1983 Corn Belt drought helped reduce corn production by 50 percent, increasing corn prices by about $1.00 per bushel. Given normal weather in the years ahead, production of these grains can be expected to set new records in this country, unless effectively prevented by government acreage reduction programs.

Producer decisions, shifting emphasis between their major crop and livestock enterprises, also contribute to changing the prices of their products. Following the large sale of grains to the Soviet Union in 1972–73, and given no reason to expect higher livestock prices in the near future, many acres of forage and range land were converted from livestock to crop production.

Substantial cost outlays are required in preparing land for cropping, but returns can begin in a year, or less. Switching from crops to livestock takes a considerably longer time, however. Some production from hayland is possible within the first year of replanting. But earnings from land reseeded to grass are postponed over the several years it may take to get a productive stand of grass reestablished. Thus a sizable increase in the relative price of beef is needed before it becomes possible to increase livestock output from such land.

Certain market characteristics also have an impact on prices. Agriculture in general is relatively competitive, especially in the production of beef, hogs, and food and feed grains because there are many producers that, individually, do not produce a large enough volume to have any influence on market prices. And the products of these firms are very like the same commodity produced by other firms, either because of the characteristics of the product itself or through the grades and standards established by the market. In addition, entry barriers are low so that it is easy for producers to change to the most profitable enterprise.

These conditions cause the total supply and demand for each product to influence its price (as discussed in more detail in Chapter 6). The high proportion of fixed costs in agriculture frequently inhibits output reduction even though prices signal that such a change is desired. Consumers of agricultural products also change their buying habits as relative prices change. These factors cause wide fluctuations in agricultural prices, resulting in the boom or bust nature of agricultural incomes.

Price variability in the United States is further intensified by changes in world markets. Although U.S. production of wheat and coarse grains amounts to about 26 percent of total world output, more than 48 percent of those grains entering world trade channels in 1995–96 came from this country. By way of comparison, U.S. exports of wheat and coarse grains in the 1960s averaged 22 percent of total U.S. output, 32 percent in the 1970s, becoming nearly 46 percent by 1983–84. As the percentage of U.S. output being exported increases, domestic prices will be more greatly affected by world market changes.

Until 1972, U.S. agriculture was protected from the "free market" by production controls and an inventory policy that built up large stocks of food and feed grains. The 1972–73 drop in world output (about 10 percent) caused a 250 percent increase in many American commodity prices[1] and a sharp reduction in stocks of agricultural products. With world stocks at about 75 percent of the previous year's level, prices high, and

---

[1] Ray A. Goldberg, "U.S. Agribusiness Breaks Out of Isolation," *Harvard Business Review*, Boston, Mass.: Harvard Business School, May–June 1975, pp. 81–95.

government output restrictions removed, producers responded with increased output. Surpluses again increased, with stocks of wheat, coarse grains, and rice exceeding 464 million metric tons by the 1987–88 crop year and dropping to 232 million metric tons by 1996.

# CHOICE AND ITS ECONOMIC MEANING

Nature and mankind combine to cause the need to economize in deciding what things we will buy with our spendable incomes and to require decisions on how and for what purposes our resources will be used. Few people, if any, have enough money, nor do they have sufficient resources to produce the goods and services it would take to satisfy all their wants completely.

This basic fact is a problem, not just for individuals, but for nations as well. Whether the nation utilizes a free market or total government control, the problem remains. The relationship between the amounts of things available and the amounts desired causes a universal condition for any good or service that we call *scarcity*.

A good or service is scarce when we must give up (*sacrifice*) some amount of one thing to get some of another good or service. And the moment we realize that a valued good or service is being sacrificed so as to gain something else, we become aware of the economic meaning of the word *cost*. As we make our choices in the face of scarcity, costs are generated. These costs we call *opportunity costs* because they constitute the value of alternative opportunities foregone, or sacrificed.

As we choose between desired alternatives, with the value of forfeited opportunities being the true cost of any chosen alternatives, we can consider ourselves better off only if the good or service chosen is worth more to us than the one sacrificed. But what determines worth? It can't be the physical amount of something we give up, but how we value that something. To get a couple of fresh eggs to eat, you would probably give up rotten eggs by the hundreds of dozens, yet be unwilling to sacrifice a T-bone steak dinner for those two eggs. We value goods and services because of the satisfactions we can get for them. You're probably a jump ahead of us here: "Surely the amounts of a good we now have must influence how strongly we want another unit of that good," and you are right!

Let's discard the rotten eggs as useless and look at the fresh eggs and T-bone steak, for example. If you have a great many eggs and very little steak, you would place a high value on steak (willing to give quite a few eggs in exchange for one steak), but if you have no eggs at all and a freezer full of T-bone, you'd likely give quite a few of those steaks in exchange for only a few eggs.

The principle involved here, called the *law of diminishing marginal utility*,[2] is quite clear. The more of any good we have, the less we value another unit of it because the satisfaction derived from each additional unit declines as we consume more and more of that good. Because of this phenomenon, anything available in the market will be purchased by some people (it's "worth" more than the sacrificed goods), yet others pass it up as not worth the cost. This situation is fundamental in making rational choices. We buy (first) those things that will generate the greatest amount of satisfaction per dollar spent until we have reached the limit of our spendable incomes.

The same principle holds true in the area of resource use. Our gains have been maximized when we have so allocated our resources that no greater value of output can be derived from them in alternative uses.

As an example, suppose a rancher has 100 cows in the brood herd and only one (tired) bull. If the market prices of bulls and cows were such that one bull were equal in value to five cows, our rancher (with limited funds to invest in brood stock) might be willing to give as many as 10 cows to get just one more bull. On the other hand, if that rancher had 25 bulls and 100 cows, their values would be sharply different: A one-for-one exchange might now seem like a bargain. Diminishing returns limits what one more cow could add to total output in the first instance, or in the latter, what an additional bull could add to production. Their values to the owner are affected accordingly.

## ECONOMICS IN BRIEF _____

Economics is a social science because it deals with people in their daily activities where choices are required. The science of economics has developed over a long period of time in response to the need for appropriate criteria by which choices can properly be made. Essentially, economics is a reasoning method that compares the benefits (income or other desired outcome) resulting from an action with the sacrifices (costs) of that action.

In the overall, economics is concerned with overcoming the effects of scarcity by improving the efficiency with which scarce resources are allocated among their many competing uses, so as to best satisfy human wants. Economics is thus concerned with the future rather than the past, by attempting to predict what will happen if some action is taken.

---

[2] Note the use of the word *marginal* here. Marginal is an important concept in economics meaning added, additional, or extra that is measured as the change from one quantity or value to another.

The field of economics spans two major areas of economic activity. When the subject studied is a single decision-making entity—a consumer or producer—we call this *microeconomics*. The problem may be the kind a family faces in deciding how best to stretch its limited budget funds among all the possible ways that money could be spent. Or the problem may be that of a producing firm deciding which of the many possible different uses of its scarce resources will be the most profitable. The point of emphasis in such cases is the individual decision unit.

On the other hand, the economic system as a whole may be the point of interest—we call this *macroeconomics*. We may be studying changes in the money supply, its causes and effects in flows of goods and services, employment and unemployment, or national income, and the important analytic tool is macroeconomics.

## Model Building

In the complexity of the real world, we are forced to simplify a situation by making a number of assumptions that will permit a clear identification of cause-effect relationships. An economic model may be a simple "if, then" statement of the relationships between two variables. The model may be a word description, a diagram or graph relating two variables, or more complex mathematical equations describing the relationships within a system.[3] Whatever the situation, we are forced to consider only those facts relevant to the situation, varying some, holding other influential factors constant, while also ignoring an infinite number of unrelated real-world facts.

We may or may not recognize it, but we do essentially the same thing in many of our everyday activities. Step on the gas while driving down the highway and we expect our speed of travel to increase, assuming, however, that we're not already driving at full speed, that the wheels won't fall off or the motor blow apart, or the tank run dry just at the time of pushing the throttle down. Other possible happenings, if totally unrelated to the vehicle's increased speed, will be ignored: an accident on a Los Angeles freeway, a traffic jam in New York City, and so on, are unrelated (not caused by this decision) phenomena.

The farmer who is deciding whether to install a sprinkler irrigation system to irrigate a 160-acre field will ignore many facts about the farm and be concerned only with estimating the additional revenue generated by the greater crop output to compare with additional costs caused by in-

---

[3] Students frequently voice a fear of economics "because it is so mathematical," or "because it is graphs," yet there is little factual basis for this attitude. Neither mathematics nor graphics is economics; they are simply shorthand methods by which the analysis of economic problems is facilitated, and nothing more.

stalling and operating the system, exactly as we do later in this text. With other farm costs and returns unchanged, "What happens to total costs and returns as this irrigation system is added?" is the only question in need of an answer for a proper decision to be made.

## Agricultural Economics

*Agricultural economics* may be defined as *an applied social science dealing with how humans choose to use technical knowledge and scarce productive resources such as land, labor, capital, and management to produce food and fiber and to distribute it for consumption to various members of society over time.* Like economics, agricultural economics seeks to discover cause–effect relationships. It uses the scientific method and economic theory to find answers to problems in agriculture and agribusiness.

The application of economic theory to agricultural problems has gone through a process of slow acceptance. Like a tree's roots, the origins of the field now known as agricultural economics reach back in many directions and over a long period of time. We may briefly outline the growth of this field as coming from two separate sources: first in time, from the physical sciences, and later, from economic theorists.

The severity and length of the agricultural depression beginning in the 1880s caused increasing attention to be devoted to its causes and possible solutions. The most notable early efforts were made primarily by agronomists and horticulturists. They recognized that the ability to grow plants and animals was not sufficient to make farmers succeed. The scientific interests of some of these people shifted to the problems of managing the farm, with special emphasis on the selection and handling of crop and livestock enterprises within the farm firm.

The study of enterprise costs and returns was begun in 1902 by W. M. Hays and Andrews Boss at the University of Minnesota. These men, both agronomists, established a route system in which a number of farmers were contacted on a regular basis to collect detailed information on the costs of various enterprises maintained by those farms. Variations between farms were analyzed in terms of the types of enterprises, their labor and other input requirements, and the apparent differences in management capabilities of the participating farmers.

G. F. Warren, a horticulturist at Cornell University, took over and redirected a survey of farms in Tompkins County, New York, that had been initiated the year before by T. F. Hunt. Under Warren's direction, this and later studies became the primary source of costs and returns information for New York agriculture over the following 40 or more years. Furthermore, that method was the prototype of many such studies conducted by a large number of state Experiment Stations throughout the country.

The Cornell method was founded on the idea that the factors affecting a farmer's success or failure could be identified only by studying a large number of farms with similar enterprise organizations. With the conviction that averages tell the story, elements of cost were identified and tallied for each farm enterprise. Over the years, the survey method evolved into a set of performance standards that sought to determine the most profitable size and type of farm for each area studied.[4]

Ignored early by economic "purists" and suffering outright rejection by other agricultural sciences at first, the "economics of agriculture" gained grudging acceptance before finally winning its present stature. Early proponents of the usefulness of economic theory in solving agricultural problems drew on the principles developed and handed down through Adam Smith, Thomas R. Malthus, David Ricardo, Alfred Marshall, and many other well-known economic theorists.

Foremost among the early theorists in agricultural economics were H. C. Taylor and T. N. Carver. While at the University of Wisconsin, Taylor taught a course in agricultural economics during the 1902–03 school year. In that course, emphasis was put on the law of diminishing returns and its use in identifying the proper intensity of applying labor and capital to land. Taylor published a book in 1903, *Introduction to the Study of Agricultural Economics*, whose title alone suggests the broader use of economic concepts than the more restricted use of economics in the farm management books of his contemporaries.[5]

Carver taught his first agricultural economics course at Harvard, in 1904, under the title "The Economics of Agriculture With Special Reference to American Conditions." He, too, treated the problem of resource use-intensity as an economic problem subject to the law of diminishing returns (or, as he preferred, variable proportions). This and other topics in agricultural economics were contained in Carver's book, *The Distribution of Wealth*, published in 1904.[6]

A third theoretical treatment of problems in agriculture warrants special recognition. John D. Black's book *Production Economics*,[7] published in 1926, demonstrated the applicability of economic theory to a broad array of problems of the farm firm and the agricultural industry.

---

[4] Selecting only the Minnesota and Cornell approaches should not be taken to imply that they were the only important methods, or that their contributions were better than those of the many other respected workers in this period of time. The Minnesota and Cornell methods are only representative of two distinctly different approaches to the same problem; and they gained the widest acceptance throughout the country over the next two decades.

[5] Henry C. Taylor and Anne Dewees Taylor, *The Story of Agricultural Economics in the United States, 1840–1932*, Ames, Iowa: The Iowa State College Press, 1952, pp. 81, 127–129.

[6] Ibid.

[7] John D. Black, *Production Economics*, London: George G. Harrap and Co., Ltd., 1926.

The following rapid growth in the ranks of professional agricultural economists, and the ever-increasing public use of their special talents, bear testimony to the foresight of those early pioneering theorists.

Most beginning students probably have only a vague concept of agricultural economics. For the student, it is a blend of many subject areas. An agricultural economics curriculum ordinarily includes classes in technical agriculture, science, statistics, mathematics, business, general economics, and other social sciences. Students taking a curriculum in agricultural economics may major in such areas as farm management, production economics, agricultural marketing, agricultural policy, finance, economic development, natural resources, and community development or public affairs.

Many agricultural economics undergraduate degrees may be called agribusiness programs. This trend began in the early 1960s to more adequately describe the economic usefulness and job opportunities for certain majors in agricultural economics. Although there were fewer possibilities for students to return to production agriculture, the employment horizons were broadening in the total food and fiber sector that services agriculture. Agricultural economists are involved throughout our economic system in financial institutions, on farms and ranches, with oil companies, grain elevators, railroads, fertilizer companies, universities, feedlots, and so on.

Some interpret the word *agribusiness* narrowly, meaning only very large or conglomerate businesses within the agricultural industry. We favor the original meaning given by Davis and Goldberg, when they defined agribusiness to include *the sum total of all operations involved in the manufacture and distribution of farm supplies; production operations on the farm; and the storage, processing, and distribution of farm commodities and items made from them.*[8] For bachelor of science graduates, agribusiness is probably more descriptive of the type of positions in which most students will be employed. At this level of training, there is not much difference between traditional agricultural economics degrees and degrees in agribusiness at the land grant colleges and universities.

Agricultural economics is an important subject area because it is concerned with society's basic needs. Getting food and fiber to all people in the world in the right form at the right time is an extremely complex process. About 50 percent of the world's population is involved in the basic industry of providing food. Many think of the United States as a highly industrialized country with agriculture being a relatively small part, but approximately 50 percent of the total assets of all U.S. corporations and

---

[8] John H. Davis and Ray A. Goldberg, *A Concept of Agribusiness*, Boston, Mass.: Research Division, Harvard Business School, 1957.

farms combined is in agribusiness.[9] Also, around 17 percent of our labor force is employed in agribusiness operations and roughly 22 percent of our consumer expeditures are for food and clothing made from U.S. farm products.

Because of the rapid growth in world population, increasing food output will receive greater attention in future years. To date, three major world food conferences have been held. The first convened in Washington, D.C. in 1963, the second in The Hague in 1970, and the third in Rome in 1974. The main theme of these conferences was to attempt to get the world to make an effort to wipe out poverty and hunger. This, of course, requires concerted attention over a long period of time because of the complex nature of the problem. Progress is being made, and this book should assist in giving the reader a better understanding of the economic realities of this nagging problem of humankind. The dilemma involves the entire food sector, from farm supply firms to farming and ranching to food processing and distribution.

Agriculture is an integral part of the world food system. Crop and animal production are the foundation of that system. Agricultural economists must have a thorough understanding of that foundation because of the impact it has in the purchasing of inputs and in meeting the needs of consumers.

In order to understand the economic interrelationships that form the conceptual focus, the student must recognize the physical basis of all agricultural production. Each type of product has its own characteristics. The student should have an understanding of the influence of climatic environments in determining how and which commodities can be produced and distributed. In addition, the student should be aware of government policies of all types and their impact on production and distribution patterns.

## Choice and the Functions of Management

In its broadest sense, *management* is the process used to control or direct a situation. Each of us makes choices every day that affect our lives, thus we are managers. Whenever a choice of one action over another is made, there is an implication that the chosen alternative was (rationally) preferred over the one discarded. This decision process, a prerogative of management, is discussed more fully in Chapter 4, but it bears more than passing mention here.

Ordinarily, the information for decisions is available to the individual only after conscious effort to obtain it. How does one know what kinds of, how much, and when data are needed, or would be of use? Bradford

---

[9] Ray A. Goldberg, Harvard Business School, May 27, 1993.

and Johnson have identified and outlined five basic steps as the foundation for managerial decisions.[10]

1. Observing a problem and thinking about its solution.
2. Analyzing with further observations.
3. Making the decision.
4. Taking action.
5. Accepting responsibility.

Let us note each of these steps in a little more detail. Management must first have a set of objectives or goals. We often think of the goal of businesses in general as the maximization of profits. But personal and family satisfactions, participating in community activities or services, and increased leisure time might also be basic goals. Problems and opportunities will then be identified within this framework of objectives.

A problem exists when conditions are such that one would prefer them to be improved. The present situation thus deviates from one's ideal of that situation, giving rise to the need for information to help in its solution.

The assembling of facts may take many forms, depending on the seriousness and complexity of the problem. Observation may simply be in reading about the problem in a trade magazine, station bulletin, or other source of information, or noting the experiences of others in handling the same problem.

The type and intensity of analysis also depends on the particular problem. Careful analysis and consideration of several alternatives may be necessary before major decisions are attempted.

It is then time to make the decision and to take action, based on the analysis and within the framework of goals. Some managers find themselves unable to carry out this fourth step, a not uncommon failing. They may have good, well thought out ideas for improving a situation, yet never seem able to carry them out.

After a decision has been made and a project launched, the final step is to accept responsibility for action taken. One must be able to accept blame for the unwanted results of a wrong decision as well as credit for successes. Because of uncertainties over weather, disease, prices, and so on, expectations often do not materialize. Failure may cause serious financial consequences, which management must be in position to withstand.

The detail and attention given each of these steps will vary with different problems. In practice it might be difficult to clearly separate each

---

[10] Lawrence A. Bradford and Glenn L. Johnson, *Farm Management Analysis*, New York: John Wiley & Sons, 1953, pp. 3–12.

step from the others. However, if the process leading through carrying a decision to its conclusion is analyzed, you will find that all five of these points have been followed.

## Value Judgments

The type of economics referred to as *positive economics* deals with *what is*, or *what can be*. If a farmer were to ask "what can I do to increase my net income?" a careful appraisal might result in the following conclusion: "You could buy a nearby acreage and your net income will increase by $y$ dollars; rent that land and your net income will increase by $z$ dollars." Were these the only alternatives available the farmer could decide on the basis of positive economic information. Those conclusions can be tested and proved correct or incorrect.

Suppose, on the other hand, that you were to note another's financial situation and tell that person they should do thus and so to increase their net income. This is *normative economics*. You would have based your recommendation on the belief that the other individual *should* try to increase his/her income. And this is a value judgment. It may even rank low on the scale of things that other person holds to be important in life and be rejected because it conflicts with that individual's values.

As social scientists, economists find it difficult to remain impartially objective concerning economic matters of importance to people. We all have been exposed to economic incidents throughout our lives and have formed opinions of their causes and meanings. But our individual interpretations are formed within (and biased by) the framework of religious, ethical, moral, family and social attitudes, and beliefs that we have accepted during our lives. These attitudes and beliefs form the basis for ideas or convictions we hold about the way things *ought to be*, or what we *should* do. Such normative statements can neither be proved right nor wrong because they stem from basic values that we hold. They constitute value judgments.

Holding different value systems, economists frequently disagree over fundamental economic policies and what should be done to improve particular situations. They too have values and biases that influence their attitudes about where emphasis should be placed. Where value judgments cannot be avoided, the agricultural economist as a public policy advisor, finds it necessary to state the judgments involved so that people may understand that person's position.

## Using Graphs

A fundamental purpose of economic analysis is to explain people's responses to changes in their economic environment. If the causal relationships in this sphere can be correctly identified and specified, future behavior can be predicted with some degree of certainty.

As with any science, relationships between two or more variables can be stated in words alone. But words alone often lead to dry and tedious explanations. Whatever can be put in words can also be described in a graph. Simple graphs frequently make things much clearer because they help us visualize abstract economic ideas and relationships. We can see the situation so much better than with just words alone.

A graph can visually demonstrate relationships between variables. Because of the extensive use we make of graphs in economics, a well-grounded understanding of their methods and meaning is especially helpful both in learning economics and in comprehending economic solutions.

In Figure 1-1, we draw two intersecting lines, one horizontal and the other vertical. These lines divide the graph space into four quadrants that can be used to show how two variables are related. The vertical line we label $Y$, and the horizontal $X$, each showing positive and negative quantities in directions opposite from the zero point where the two lines intersect.

Four quadrants can be noted, which we label as quadrant I in the northeast part of the diagram, counterclockwise to quadrant II, and so on. Posi-

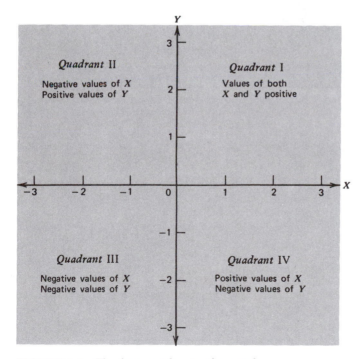

**FIGURE 1-1** The four quadrants of a graph.

tive values are measured along the X-axis to the right of zero, negative to the left; values upward from zero along the Y-axis are positive, negative downward from zero. Only in quadrant I do both $X$ and $Y$ have positive values. In quadrant II, values of $X$ are negative and $Y$ values are positive; quadrant III has negative values of both $X$ and $Y$; and in quadrant IV, $X$ values are positive while $Y$ values are negative. In this book we make almost exclusive use of quadrant I to demonstrate relationships, so we can now direct our attention to that quadrant only.

The four different graphs (charts A, B, C, and D) in Figure 1-2 generalize the major types of relationships between two variables X and Y that you can expect to use in this course of study. We use charts such as these to show how the quantity of one variable ($Y$) varies with quantities of the

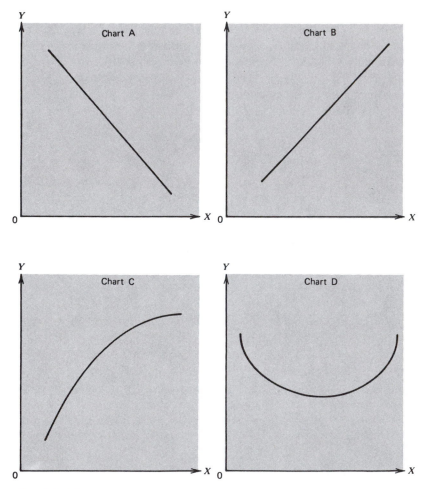

**FIGURE 1-2**   Some typical graphs used in economics.

other variable ($X$). More specifically, we chart how $Y$, the **dependent variable**, is changed because of changes in the **independent variable** $X$, that is, that $Y$ is a function of (or depends on) the quantity of $X$.[11]

In chart A, the line of relation has a negative (or downward) slope. When the quantity of $X$ is increased, the amount of $Y$ declines, similar to the typical demand curve (discussed in Chapter 3) and a number of other economic relationships discussed elsewhere in the text.

Chart B exhibits a positive relationship between $X$ and $Y$; as the quantity of $X$ is increased, the number of units of $Y$ also increases as a consequence. This chart demonstrates, in a general way, a topic such as market supply in which price of the product and quantities produced are positively related (as in Chapters 6 and 7).

Charts C and D demonstrate changing relationships between $X$ and $Y$. In chart C the effect of $X$ on $Y$ diminishes as larger quantities of $X$ are involved, similar to the functional relationships discussed in Chapter 4. The changing relationship shown in chart D is of a type similar to the cost curves discussed in Chapter 6. With increasing quantities on the $X$-axis, the $Y$-axis values first decline, then the relationship changes and the $Y$-axis values increase with further increases in $X$.

There are special requirements in the use we make of graphs, *implicit assumptions* if you will, that we wish to make especially evident here. We make frequent use of the Latin phrase *ceteris paribus*, which means "all other things remaining equal," or "constant." When making a graph, this assumption means that all things not measured along the two axes (that could affect the relationship shown in the graph) are held constant. For instance, time is an important variable, and must therefore be held constant as a single time period over which a relationship is specified or tested.

When you mark off a scale of units with a ruler along the two axes, you use a constant measure to indicate a given number of units. Each unit away from the zero point is exactly the same length as all other similar units along that scale. We must adhere rigidly to the same requirements when measuring physical units of resources or products along the axes. We make this clear with the requirement of **homogeneous units** of $X$ and $Y$: Each unit along the $X$-axis is an exact duplicate of every other unit, and similarly for $Y$-axis units.

---

[11] We will use the $Y$-axis quantity as the dependent (or resulting) variable and the $X$-axis for the independent (or causative) variable, except when discussing demand or supply separately. A demand curve question that asks "How much of this good will consumers buy at a number of different prices?" and a supply curve question that asks "How much of this good will firms produce at a number of different prices?" have reversed the causal sequence. Stated in this way the question has made the $Y$-axis variable (price) the independent or causal variable and the quantities measured along the $X$-axis dependent on those different prices.

Another assumption, *divisibility*, means that we assume we can divide units into small fractions of the units measured along the axes. This assumption permits a smooth curve to be drawn through identified points in the quadrant space, as in the charts in Figure 1–2.

In a number of graphs throughout this book, we have used straight lines (such as those in charts A and B) to help explain a principle. In some instances, there is no need to raise questions of degree of curvature and possibly divert our attention from the basic principle of the relationship between the two variables.

A caveat to you, the student: *Don't try to memorize* any of the graphic demonstrations sprinkled throughout this book. You will make economics far more difficult than it need be. Instead, we urge you to *learn the economic principles* displayed by the graphs. In doing this you will find it much easier to apply these principles to a wide variety of real-world problems. And this will help you better to reason your way to problem solutions, which is the name of the game.

## LOOKING AHEAD _____

The food and fiber system is very large and complex. A basic understanding of its structure and participants is needed to come to grips with the economic relationships within the industry.

Next, and more importantly, this book deals with the principles of consumer behavior as people make their choices; of producer behavior in choosing rates of resource use and resource combinations, and which goods to produce; and supply and price determination in the market, within the context of a free enterprise system. Considerable effort is made to move slowly through the theoretical concepts. Applications of the theory are used to enhance meaning and help familiarize you with the economic way of thinking and analyzing problems.

The final chapters present the use of the theoretical concepts in problems of agricultural marketing, policy, finance, natural resources, agricultural trade, and world food supplies. Students will find these applied areas relevant to fitting the agricultural industry into a connected whole. Any economic change has direct and indirect effects on the entire agricultural sector. The impact on some subsectors is more pronounced than on others. You should be able to analyze these interrelationships as they bear on decision making.

The Chapter Highlights section at the end of each chapter summarizes briefly the major topics of that chapter and serves as special reminders to the student. If the meaning and use of these ideas is not clear, it should prompt a rereading of the text.

At the end of each of the chapters in this book you will find a biography of an outstanding researcher in the general topic area of that chapter. These individuals were selected because of their valuable contributions in the field of agricultural economics or agribusiness. They were chosen from among many other equally qualified professionals to give you a better idea of what an agricultural or agribusiness economist does.

## CHAPTER HIGHLIGHTS

1. The cyclical nature of production and price instability are two important characteristics of agriculture.
2. Agriculture is a relatively competitive industry.
3. It is because of scarcity that we must economize in choosing between desired alternatives.
4. Sacrificing one thing for another gives rise to the true costs of making choices.
5. The economic method abstracts from the complex real world in order to discover causal relationships within the economy.
6. Graphs are used because they help in visualizing economic relationships.
7. Agricultural economics is a social science applied to agricultural problems.
8. Management is the decision maker, coordinating the uses of all the resources in a farm or business firm.
9. Positive economics results from a scientific analysis of facts relevant to a situation; normative economics also involves our personal values.
10. Agribusiness includes the functioning of the entire food and fiber system from the input industry to the farm or ranch and the ultimate consumer.

## KEY TERMS AND CONCEPTS TO REMEMBER

| | |
|---|---|
| Agribusiness | *Ceteris paribus* |
| Agricultural economics | Dependent variable |
| Independent variable | Microeconomics |
| Law of diminishing marginal utility | Normative economics |
| Macroeconomics | Opportunity cost |
| Management | Positive economics |
| Marginal | Scarcity |

## REVIEW QUESTIONS

1. What is the meaning of scarcity?
2. Is scarcity a problem that cannot be solved?
3. How is scarcity involved in improving your grades this term? What sacrifices would be required? Would you consider those sacrifices as "costs"?
4. The facts about most any situation surely are around somewhere. So why construct an economic "model" of that situation? Won't facts speak for themselves?
5. What is economics? Agricultural economics? What makes either of them a "social science"?
6. Suppose you are a farmer with a swampy area in one of your fields.
   a. Apply the five steps in management to draining the wet area.
   b. How would your decision be biased if you were opposed to land drainage of any kind?
   c. Would the economic appraisal in your decision be affected if you couldn't tolerate an untillable area in one of your fields?
7. In a play on words, some wag once said that "if you fill this room with economists they would all point in different directions." How is it possible that in a field that claims a "science" status its professionals can arrive at such widely different solutions to the same problem?

## SUGGESTED READINGS

Barkley, Paul W. *Economics: The Way We Choose*, New York: Harcourt, Brace Jovanovich, 1977, pp. 1–33.

Boehlje, Michael D., and Vernon Eidman. *Farm Management*, New York: John Wiley & Sons, 1984, pp. 4–32.

Bradford, Lawrence A., and Glenn L. Johnson. *Farm Management Analysis*, New York: John Wiley & Sons, 1953, pp. 1–37.

Calkins, Peter H., and Dennis D. DiPietre. *Farm Business Management*, New York: Macmillan, 1983, pp. 1–18.

Castle, Emery N., Manning H. Becker, and A. Gene Nelson. *Farm Business Management*, 3rd ed. New York: Macmillan, 1987, pp. 1–15.

Harsh, Stephen B. *Managing the Farm Business*, Englewood Cliffs, N.J.: Prentice Hall, 1981, pp. 1–12.

Kadlec, John E. *Farm Management*, Englewood Cliffs, N.J.: Prentice Hall, 1985, pp. 1–7.

Kay, Ronald D. *Farm Management*, New York: McGraw-Hill, 1986, pp. 3–18.

Taylor, Henry C., and Anne Dewees Taylor. *The Story of Agricultural Economics in the United States, 1840–1932*. Ames, Iowa: The Iowa State College Press, 1952.

# AN OUTSTANDING CONTRIBUTOR

**Bruce F. Gardner**   Bruce Gardner was born and raised in McHenry County, Illinois, where his brother still operates the family dairy farm. He received a B.S. degree in agricultural science from the University of Illinois in 1964 and a Ph.D. in economics from the University of Chicago in 1968.

Gardner was nominated by President Bush to be Assistant Secretary of Agriculture for Economics and was confirmed by the Senate in October 1989. As Assistant Secretary, Dr. Gardner was responsible for direction and oversight of the Department's economics and statistics agencies and served as the chief economist in the U.S. Department of Agriculture.

Gardner is a professor in the Department of Agricultural and Resource Economics at the University of Maryland, College Park. He has held this position since 1981. He previously was professor of Agricultural Economics at Texas A&M University (1977–80) and assistant and associate professor at North Carolina State University (1968–75). During 1975–77, Gardner was a senior staff economist on the President's Council of Economic Advisors.

Gardner has published many articles in a number of agricultural economics and general economics professional journals and had authored three books on agricultural policy issues. He has been an active member of the American Agricultural Economics Association, and in 1989 was elected a Fellow by the AAEA.

Dr. Gardner was on the Board of Directors of the American Agricultural Economics Association from 1984 to 1987, Associate Editor of the *American Journal of Agricultural Economics* from 1983 to 1989, and the AAEA representative on the Board of Directors of the National Bureau of Economic Research since 1988.

His research was selected for the Outstanding Article award by the *AJAE* in 1977 for a paper titled "The Farm Retail Price Spread in a Competitive Food Industry." For his book *Optimal Stockpiling of Grain* he received the AAEA award for Quality of Research Discovery in 1980. The College of Agriculture at the University of Maryland also presented him with their 1987 Award for Excellence in Research.

He is married to the former Mary Agacinski, and they have two children.

CHAPTER **2**

# The Farm and and Food System

CHAPTER **2**

# The Farm and Food System

$\mathbf{I}$t is impossible to understand the economics of American agriculture without a knowledge of the dimensions of the industry, the interrelationships within the industry, and how it is interconnected with the total economic system. This chapter will describe U.S. agriculture. How many farmers and ranchers are there in the United States? What is their financial structure? What is the role of corporations and cooperatives? Where do producers market their products? Are international markets important?

Agriculture is an integral part of the general economic system. We subdivide our national economy so that the fundamental structure can be seen. Producing firms and consumers are the central economic units in the system.

# AGRICULTURE'S ROLE IN THE U.S. ECONOMY

Agriculture is by no means a homogeneous industry. It is made up of small *family farms*;[1] large corporate organizations; credit and other input supply firms; marketing and processing firms; transportation networks; wholesalers; restaurants; and food and fiber retailers. The agriculture or agribusiness industry is composed of a complex series of firms. These firms supply inputs to farms, produce farm products, process agricultural products, and market these commodities to the final consumers. No longer do we think of agriculture as solely the physical and biological production of agricultural commodities.

The United States is one of the most productive countries in the world, and the efficiency of American agriculture is second to none. If the United States has a relative advantage in the production of any product at this time, it is in agricultural commodities.

By providing quality food and fiber at reasonable prices to all consumers, agriculture is vital to the U.S. economy. Two measures of the national importance of an industry are the number of people it employs and the value of its production. These measures are shown in Table 2-1.

The most important economic activities in the United States are services, manufacturing, and government. The total service sector of the economy includes such activities as medical care, legal and accounting services, wholesale and retail trade, and entertainment. The service in-

---

[1] A family farm is usually defined as one in which the farm family constitutes the basic labor force.

**TABLE 2-1**  Industry Distribution of National Output and Employed Labor Force, United States, 1995

| Type of Industry | National Output (%) | Employed Labor Force (%) |
|---|---|---|
| Farming | 2 | 2 |
| Mining and construction | 5 | 5 |
| Manufacturing | 18 | 16 |
| Transportation, communication, and utilities | 8 | 5 |
| Wholesale and retail trade | 14 | 23 |
| Finance, insurance, real estate | 17 | 6 |
| Services | 22 | 27 |
| Government (state, local, and federal) | 14 | 16 |

SOURCE: "Economic Report of the President," U.S. Government Printing Office, Washington, D. C., 1996; and "Survey of Current Business," U.S. Department of Commerce, March 1996.

dustry directly accounts for 27 percent of the labor force and 22 percent of the national output. If we also broaden the definition of services and add wholesale and retail trade, financial services, and government, the total service sector of the economy accounts for 67 percent of the national output. Consequently, the basic goods industries (agriculture, mining, and manufacturing) account for only one-third of the national output.

Next in importance is manufacturing. Manufacturing contributes approximately one-fifth of the national output and employs 16 percent of the labor force. Manufactured items include two broad categories of goods termed *durable* and *nondurable*. Durable goods or "hard goods" are such products as metals, machinery, automobiles, and household appliances. Nondurable goods are "soft goods," which include food products, textiles, and apparel.

In terms of employment, government is the third major industry because of large expenditures on education, military, and social programs.

These measures indicate that farming is one of the smaller industries, producing 2 percent of national output and directly employing 2 percent of the employed labor force. However, agriculture indirectly accounts for much employment in other industries such as manufacturing and processing, wholesaling, and retail trade. In total, agriculture employs nearly 3.1 million as farm proprietors and hired farm workers, 0.4 million in agricultural services, 0.4 million in agricultural input industries, 3.3 million in agricultural processing and marketing industries, 14 million in wholesale and retail establishments such as restaurants and supermarkets, and 0.5 million in indirect agribusiness such as chemical and fertilizer mining. Agriculture is thus responsible for providing about 21.6 million jobs, constituting 15.8 percent of total employment in the United States. In addition, the farm and food system contributes 14 percent of the nation's gross domestic product.

# THE BUSINESS STRUCTURE OF FARMS

Most farms in the United States are classified as family farms. In 1992, individual proprietorships or family farms accounted for 86 percent of all farms, partnerships accounted for 10 percent, corporations 4 percent, and others (estates and trusts) less than 1 percent. Because many of the partnerships and corporations are family organizations, about 96 percent of our farms are really family farms. Of the 2.1 million farms, 72,567 were operated under a corporate management. The number of corporate farms in agriculture has increased 8 percent since 1987. Of the corporations, a very large percentage had 10 or fewer shareholders. This small number of shareholders also tends to indicate a family-type of structure. These corporate units accounted for 13 percent of the farm land and 27 percent of product sales in 1992.

In 1992 there were 18,875 corporations in the livestock industry, 17,457 cash grain, 8,572 in horticultural specialty farms, and 6,227 in fruit and tree nut farms. These corporate farms sold 30 percent of all cattle and calves, 69 percent of the nursery and greenhouse products, 11 percent of all grain, and 41 percent of all fruits, nuts, and berries sold in the United States.

Family farms are incorporated for at least three reasons.[2] First, a corporate form of organization can be used to transfer farms to others at a lower cost than other forms of business organization. Second, employee benefits such as social security and unemployment insurance are tax deductible for the corporation, but not in an individual proprietorship. Third, a corporation can separate management from ownership, which can be beneficial, as well as to reduce the liability of both management and owners.

Before 1982, the maximum marginal tax rate on a corporation was 46 percent compared to 70 percent for individuals. Because of the lower corporate tax rate for taxable income in excess of $25,000, an incorporated producer could shelter more earnings and build equity more rapidly. This advantage to the corporation was all but eliminated when the Reagan administration reduced the marginal tax rates on individuals and corpora-

---

[2] Donn Reimund, "Form of Business Organization," *Structure Issues of American Agriculture*, Economics and Statistics Service, USDA, November 1979, pp. 128–133.

tions. Under the Tax Reform Act of 1986, the corporate tax rate is 15 percent for an income up to $50,000; 25 percent for incomes between $50,000 and $75,000; and 34 percent for incomes over $75,000. Revised corporate tax rates as of 1995 were as follows: (1) 15 percent for a taxable income up to $50,000; (2) 25 percent for taxable incomes between $50,000 and $75,000; (3) 34 percent for taxable incomes between $75,000 and $100,000; (4) 39 percent for taxable incomes between $100,000 and $335,000; (5) 34 percent for taxable incomes between $335,000 and $10 million; (6) 35 percent for taxable incomes between $10 million and $15 million; (7) 38 percent for taxable incomes between $15 million and $18,333,333; and (8) 35 percent for taxable incomes above $18,333,333.

The personal tax rates for individuals in 1996 were: (1) 15 percent for taxable incomes between 0 and $24,000; (2) 28 percent between $24,000 and $58,150; (3) 31 percent between $58,150 and $121,300; (4) 36 percent between $121,300 and $263,750; and (5) 39.6 percent on taxable incomes above $263,750. These rates should provide much incentive for farmers to incorporate for tax reasons.

There also are advantages in using a corporate organization when farms are to be transferred to heirs. It often is easier to transfer stocks than physical assets. Through a combination of stock gifts and sales the transfer usually can be made more smoothly, and at less cost.

In terms of their number, the corporation is not an important type of business organization in farming. However, the number of corporate (other than family-held) farms increased from 6198 in 1987 to 8039 in 1992. Many people are concerned about corporate activity in agriculture because of the economic consequences that could occur with concentrated resource control. As a result, some states have attempted to limit the growth of corporation farming. Laws passed in several states prohibit corporate farming. Statutes restricting corporate farming have been enacted in several other states. Some states have laws requiring corporations to report the land that they own in the state. More than one-half of the states also have laws restricting ownership of real property by aliens, with a great deal of variation in the restraints provided by these laws as they are applied to alien ownership of property.[3]

Many producers are concerned about farming corporations because they think corporations are more efficient and that their size gives them market advantages, which may put the family-farm operators at a competitive disadvantage. Farmers believe capital markets, volume buying of production inputs, and volume selling of output afford advantages to corporate farms that are not available to them. However, most studies show

---

[3] Dale C. Schian and David A. Seid, "State Laws Relating to the Ownership of U.S. Land by Aliens and Business Entities, October 31, 1986," Economic Research Service, USDA, December 1986.

that moderate-sized family farms are as efficient as most corporate farms. With this situation plus the generally low returns to agricultural investments, one would expect very little growth in corporate agriculture. This situation does not mean, however, that corporations will not buy agricultural land for speculative purposes and farm it for a while if the rate of return on a given piece of land is expected to be high. A large shift to corporate farming is unlikely to occur unless agricultural profits become more consistent and predictable and superior to investment alternatives.

# ECONOMIC SIZE CLASSES OF FARMS _____

Farms in the United States may also be grouped according to economic classes, or sectors. Because of changes in their numbers through time, we may refer to them as the expanding, declining, and noncommercial sectors.

By classifying farms according to the value of farm products sold, three classifications can be established: (1) those with sales over $100,000, (2) those with sales between $20,000 and $99,999, and (3) those with sales under $20,000. The number of farms falling into the first class is increasing, thus the "expanding" sector; the number falling into the second group is decreasing, thus the "declining" sector. The noncommercial sector includes farms with annual farm product sales of less than $20,000. This latter classification was the largest group in number of farms in 1980 and in 1994.

The expanding sector of agriculture numbered 271,000 farms in 1980, increasing to 346,000 by 1994 (Table 2-2). These farms accounted for 17

**TABLE 2-2**  Expanding Sector of Agriculture—Farms with Sales Greater Than $100,000 Annually, Number and Percent of Total, 1980 and 1994

|  | 1980 | Percent | 1994 | Percent |
|---|---|---|---|---|
| Number of farms | 271,000 | 11.1 | 346,000 | 16.8 |
| Cash receipts (millions)[a] | $ 95,479 | 68.3 | $152,320 | 77.4 |
| Net income (millions) | $ 27,808 | 81.3 | $ 43,919 | 88.3 |
| Off-farm income per farm[b] | $ 10,635 |  | $ 23,969 |  |
| Total all income per farm[c] | $113,247 |  | $150,903 |  |
| Direct government payments per farm | $ 2,096 | 44.2 | $ 12,920 | 56.7 |

[a] Includes other income.

[b] Off-farm income per family for 1994.

[c] Net disposable income per farm family 1994.

SOURCE: Economic Research Service, USDA.

percent of all farms, produced 77 percent of the value of agricultural output, and received 57 percent of the government support payments. Off-farm income averaged $23,969 per farm. Average total income per farm including off-farm income, nonmonetary income and government payments amounted to about $150,903.

The declining sector of agriculture includes those farms that sold between $20,000 and $99,999 worth of products in 1994 (Table 2-3). These farms decreased in number from 637,000 in 1980 to 472,000 by 1994. Farms of this size-class produced 17 percent of the agricultural output, about 18 percent of the net income in agriculture, and received 30 percent of the government payments. Total income for these farm operators averaged $53,324 per farm in 1994.

The noncommercial farms in the United States totaled 1,246,000 farms in 1994 (Table 2-4). They produce very little agricultural output and receive very little income from agriculture or from direct government payments. Off-farm income is relatively large. Out of a total income per farm of $36,440, over 100 percent ($38,842) is derived from nonfarm sources. It is apparent from these data that most of the noncommercial farms are only part-time or retirement operations that should not be included in commercial agriculture.

# VERTICAL COORDINATION IN AGRICULTURE _____

Many people are concerned with the amount of vertical coordination in agriculture, as it affects the decision-making ability of producers and also

**TABLE 2-3**  Declining Sector of Agriculture—Farms with Sales between $20,000 and $99,999 Annually, Number and Percent of Total, 1980 and 1994

|  | 1980 | Percent | 1994 | Percent |
|---|---|---|---|---|
| Number of farms | 637,000 | 26.1 | 472,000 | 23.0 |
| Cash receipts (millions)[a] | $ 35,351 | 25.3 | $ 33,076 | 16.8 |
| Net income (millions) | $ 8,276 | 24.3 | $ 8,835 | 17.8 |
| Off-farm income per farm[b] | $ 9,837 |  | $ 34,606 |  |
| Total all income per farm[c] | $ 22,829 |  | $ 53,324 |  |
| Direct government payments per farm | $ 879 | 45.3 | $ 4,940 | 29.6 |

[a] Includes other income.

[b] Off-farm income per family 1994.

[c] Net disposable income per farm family 1994.

SOURCE: Economic Research Service, USDA.

**TABLE 2-4**   Noncommercial Sector of Agriculture—Farms with Sales Less Than $20,000 Annually, Number and Percent of Total, 1980 and 1994

|  | 1980 | Percent | 1994 | Percent |
|---|---|---|---|---|
| Number of farms | 1,532,000 | 62.8 | 1,246,000 | 60.4 |
| Cash receipts (millions)[a] | $  8,906 | 6.4 | $ 11,309 | 5.8 |
| Net income (millions) | $ –1,884 | –5.5 | $ –2,993 | –6.0 |
| Off-farm income per farm[b] | $ 16,677 |  | $ 38,842 |  |
| Total all income per farm[c] | $ 15,447 |  | $ 36,440 |  |
| Direct government payments per farm | $      103 | 12.3 | $      860 | 13.7 |

SOURCE: Economic Research Service, USDA.

[a] Includes other income.

[b] Off-farm income per family 1994.

[c] Net disposable income per farm family 1994.

affects the control of resources. The terms *vertical coordination, contract production*, and *vertical integration* are often used interchangeably.

Vertical coordination is the term most generally used; it includes the linkage of successive stages in the marketing and production of a commodity in one decision entity. Vertical integration means that successive production stages and/or marketing stages are coordinated within one firm. An example of this type of integration might be a wheat farmer buying a flour mill or vice versa.

Contract production involves the use of production agreements between farmers or ranchers and processors, dealers, or others who are at the first stage before or after the farm. These agreements specify the type of crop to produce, how the crop is to be grown and harvested and perhaps the price to be paid the producer.

The proportion of total farm production under various forms of contracting and vertical integration increased from about 13 percent in 1960 to 18 percent in 1994. Contract production increased from 8 to 11 percent and vertical integration from 4 to 8 percent over the same span of years. Both contracting and vertical integration are more concentrated in livestock, poultry, and fruits and vegetables than in crops (Table 2-5). These total numbers reflect a moderate increase in vertical coordination.

# FARMER COOPERATIVES _____

Farmer cooperatives are an integral part of agriculture and the free enterprise economy. As an organizational form they are an alternative or addition to an individual proprietorship, partnership, or corporation. A

**TABLE 2-5**  Production Contracts and Vertical Integration

| Products | Production Contracts | | | | Vertical Integration | | | |
|---|---|---|---|---|---|---|---|---|
| | 1960 | 1970 | 1980 | 1994 | 1960 | 1970 | 1980 | 1994 |
| | (Percent) | | | | | | | |
| Crops | | | | | | | | |
| Feed grains | 0.1 | 0.1 | 1.2 | 1.2 | 0.4 | 0.5 | 0.5 | 0.5 |
| Hay and forage | 0.3 | 0.3 | 0.5 | 0.5 | 0.0 | 0.0 | 0.0 | 0.0 |
| Food grains | 1.0 | 2.0 | 1.0 | 0.1 | 0.3 | 0.5 | 0.5 | 0.5 |
| Fresh vegetables | 20.0 | 21.0 | 18.0 | 25.0 | 25.0 | 30.0 | 35.0 | 40.0 |
| Processing vegetables | 67.0 | 85.0 | 88.1 | 87.9 | 8.0 | 10.0 | 10.0 | 6.0 |
| Dry beans and peas | 1.5 | 1.0 | 2.0 | 2.0 | 1.0 | 1.0 | 1.0 | 1.0 |
| Potatoes | 40.0 | 45.0 | 60.0 | 55.0 | 30.0 | 25.0 | 35.0 | 40.0 |
| Citrus fruits | 0 | 0 | 0 | 0 | 8.9 | 9.4 | 11.2 | 6.9 |
| Other fruits and nuts | 0 | 0 | 0 | 0 | 15.0 | 20.0 | 25.0 | 25.0 |
| Sugar beets | 99.0 | 99.0 | 99.0 | 99.0 | 1.0 | 1.0 | 1.0 | 1.0 |
| Sugarcane | 24.4 | 31.5 | 29.3 | 27.3 | 75.6 | 68.5 | 70.7 | 72.7 |
| Cotton | 5.0 | 5.0 | 1.0 | 0.1 | 3.0 | 1.0 | 1.0 | 1.0 |
| Tobacco | 2.0 | 2.0 | 1.4 | 9.3 | 2.0 | 2.0 | 2.0 | 1.5 |
| Soybeans | 1.0 | 1.0 | 1.0 | 0 | 0.4 | 0.5 | 0.5 | 0.4 |
| Seed crops | 80.0 | 80.0 | 80.0 | 80.0 | 0.3 | 0.5 | 10.0 | 10.0 |
| Livestock | | | | | | | | |
| Fed cattle[a] | NA | NA | NA | NA | 6.7 | 6.7 | 3.6 | 4.5 |
| Sheep and lambs[a] | NA | NA | NA | NA | 5.1 | 11.7 | 9.2 | 29.0 |
| Market hogs | 0.7 | 1.0 | 1.5 | 10.5 | 0.7 | 1.0 | 1.5 | 8.0 |
| Fluid-grade milk | 0.1 | 0.1 | 0.3 | 0.1 | 0.0 | 0.0 | 0.0 | 0.0 |
| Manufacturing-grade milk | 0 | 0 | 0 | 0 | 2.0 | 1.0 | 1.0 | 1.0 |
| Eggs | 7.0 | 20.0 | 43.0 | 25.0 | 5.5 | 20.0 | 45.0 | 70.0 |
| Broilers | 90.0 | 90.0 | 91.0 | 92.0 | 5.4 | 7.0 | 8.0 | 8.0 |
| Market turkeys | 30.0 | 42.0 | 52.0 | 60.0 | 4.0 | 12.0 | 28.0 | 28.0 |
| Total farm output | 8.3 | 9.3 | 11.5 | 10.7 | 4.4 | 5.3 | 6.2 | 7.6 |

SOURCE: Economic Research Service, USDA.

[a] NA means not available.

*cooperative* is defined as a business that is organized, capitalized, and managed for its member-patrons, furnishing and/or marketing goods and services to the patrons at cost.[4] Farmer members sell their products through marketing cooperatives or buy their inputs through supply co-

---

[4] Ewell P. Roy, *Cooperatives Today and Tomorrow,* Danville, Ill.: The Interstate Printers and Publishers, 1964, p. 1.

operatives. In doing business with these cooperatives they derive a profit called *net savings*. These savings or patronage dividends are returned to the member-patrons in proportion to their business transactions with the cooperative. Therefore, in a cooperative the primary purpose is to make a "profit" for the patron-owners and not for investors as in a corporation.

The latest available data show that there were 2,173 farmer marketing cooperatives, down from 1960s total of 5727. Farm supply cooperatives declined to 1,496 from the 3222 that were in business in 1960. Estimated total membership of farmer cooperatives is around 4.0 million. Farmer cooperatives are important to agricultural producers, marketing about 31 percent of the agricultural products produced and providing about 29 percent of the major agricultural inputs used by American farmers. These inputs include feed, seed, fertilizer, petroleum, and livestock and poultry.

# FARM OUTPUT _____

Farm output increases over the past 40 years have essentially been accomplished with larger amounts of purchased inputs embodying a significant amount of technical change, improved management, and less labor. The total number of farms has been declining while the average size of farms has been increasing (Table 2-6). In 1995, there were 2.1 million farms with an average of 469 acres per farm, compared to 1920 when 6.5 million producers farmed an average of 147 acres per farm.

Between 1950 and 1995, the number of farm workers in agriculture dropped from 9.3 million to 2.3 million. Most of these laborers migrating out of agriculture were family workers. Of those remaining, family workers totaled 1.7 million and hired workers totaled 629,000.

Much of the off-farm migration has been caused by advances in agricultural science and technology. These improvements have increased output and lowered agricultural prices. As a result, some producers have been forced to find other employment where the probability of higher wages exists. On the other hand, some producers engage in agricultural pursuits at lower wage rates than could be earned in the nonfarm economy. They do so because of the added amenities of living in a rural area, the social impediments to mobility such as family and friends, or for other reasons that may be strictly personal in nature. Their reasons may stem back to what some refer to as *agricultural fundamentalism*.

Agricultural fundamentalism is a philosophy that draws deeply on the French Physiocratic school that dates from the mid-1700s. This school of thought held the basic belief that industry, trade, and the professions were useful, but unproductive. Only agriculture, forestry, fisheries, and mining were considered to be productive because they appeared to produce a "surplus." The Physiocrat's concept of production held that productive

**TABLE 2-6**  Number, Population, and Size of Farms in the United States

| Year | Number of Farms[a] | Farm Population (000) | Average Farm Size (Acres) | U.S. Population on Farms (%) |
|------|------|------|------|------|
| 1920 | 6,518,000 | 31,974 | 147 | 30.1 |
| 1930 | 6,546,000 | 30,529 | 151 | 24.9 |
| 1940 | 6,350,000 | 30,547 | 167 | 23.2 |
| 1950 | 5,648,000 | 23,048 | 213 | 15.3 |
| 1960 | 3,962,000 | 15,365 | 297 | 8.7 |
| 1970 | 2,924,000 | 9,712 | 383 | 4.8 |
| 1975 | 2,521,420 | 8,864 | 420 | 4.2 |
| 1980 | 2,439,510 | 6,051 | 426 | 2.7 |
| 1985 | 2,292,530 | 5,355 | 441 | 2.2 |
| 1990 | 2,140,420 | 4,591 | 461 | 1.9 |
| 1991 | 2,105,060 | 4,632 | 467 | 1.8 |
| 1992 | 2,095,740 | 4,665 | 468 | 1.8 |
| 1993 | 2,083,000 | 4,645 | 469 | 1.8 |
| 1994 | 2,065,000 | 4,605 | 471 | 1.8 |
| 1995 | 2,073,000 | 4,623 | 469 | 1.8 |

SOURCE: Economic Research Service, USDA.

[a] Over time the Bureau of the Census has used varying definitions of a farm. In 1959, 1964, and 1969, places of less than 10 acres were counted as farms if estimated sales of agricultural products for the year amounted to at least $250, and places of 10 acres or more were counted as farms if their sales amounted to at least $50 per year. The Census definition of a farm was changed in 1974 to an establishment that had or normally would have had sales of agricultural products of $1000 or more.

effort in these industries caused useful physical material to appear that previously did not exist. Agricultural fundamentalism holds that there is something special and unique about the farm way of life. Today, however, we realize that all industry is productive and agriculture is a business like any other economic activity.

Farms with sales over $5,000 are many and heterogeneous. Table 2-7 indicates the relative importance of various classifications of farms, as shown graphically in Figure 2-1. Obviously, type-of-farming areas do not follow state lines, as shown in the figure, even though generalizing such areas by states would make it appear so.

Cash grain farms totaled 352,000 and represented 28 percent of the farms in 1992. This type is comprised primarily of farms producing wheat, corn, and soybeans. These grain farms accounted for more than 70 percent of the total cropland from which crops were harvested. All cash grain crops are important all across the United States; however, the greatest con-

**TABLE 2-7**   Types of Farms in the United States with Annual Sales of $5000 or More

| | Number | | | | |
|---|---|---|---|---|---|
| | 1974 | 1978 | 1982 | 1987 | 1992 |
| Total U.S. | 1,405,064 | 1,532,813 | 1,426,441 | 1,334,545 | 1,262,875 |
| Cash grain | 509,701 | 467,998 | 478,992 | 377,640 | 352,245 |
| Tobacco | 68,937 | 78,362 | 81,699 | 46,451 | 63,572 |
| Cotton | 25,333 | 28,559 | 19,258 | 25,988 | 19,331 |
| Other field crops | 65,391 | 61,715 | 51,857 | 62,129 | 67,312 |
| Vegetables | 16,538 | 20,284 | 19,549 | 20,828 | 21,486 |
| Fruit and nuts | 42,941 | 48,341 | 46,251 | 50,613 | 48,210 |
| Poultry | 41,281 | 40,379 | 34,236 | 32,666 | 29,855 |
| Dairy | 191,523 | 164,260 | 162,143 | 136,892 | 112,647 |
| Livestock | 359,704 | 525,273 | 447,321 | 500,221 | 473,276 |
| General | 52,623 | 61,178 | 44,775 | 36,395 | 22,631 |
| Miscellaneous[a] | 31,092 | 36,464 | 40,360 | 44,722 | 52,310 |

SOURCE: U.S. Department of Commerce. *Census of Agriculture*, Bureau of the Census, Washington, D.C.

[a] Includes nursery, greenhouse products, and such things as mink production.

centration of cash grain farms was in the North Central Region. Soybeans and corn were the principal cash-grain crops in 12 states. Cash grain farms had an average size of 605 acres, including an average of 394 acres of cropland.

Tobacco farms use relatively few acres of land and much labor. They made up 5.0 percent of all farms while containing only 0.3 percent of total harvested cropland. These farms are located predominantly in the South where more than 87 percent of them are concentrated in four states. The average size of a tobacco farm was 126 acres, and the average amount of land cropped was 35 acres. Ninety-five percent of the tobacco farms harvested less than 25 acres of tobacco.

The total number of cotton farms declined 33 percent between 1978 and 1982, increased 35 percent between 1982 and 1987, and decreased 26 percent between 1987 and 1992. The longrun decline in numbers of cotton farms has been associated with an increase in the size of the remaining farms, averaging about 939 acres. Cotton farms comprised 4.2 percent of total cropland. The leading states in cotton production are Texas, California, Mississippi, Arkansas, Louisiana, and Tennessee.

Approximately 5.3 percent of the farms are "other field crop" farms. They cultivate 5.9 percent of the total harvested cropland, with an average size of 308 acres. Hay and potatoes are the principal crops in this classification.

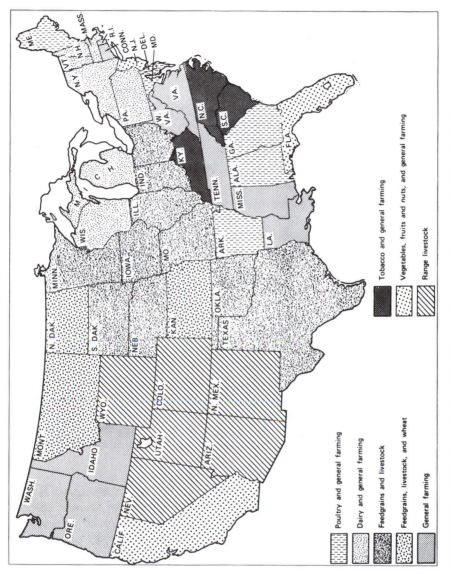

**FIGURE 2-1** Major types of agricultural enterprises by cash receipts received by farmers and ranchers, in the contiguous United States.

Vegetable farms are widely scattered, but California, Wisconsin, Florida, Minnesota, Texas and four other states are most important in producing fresh and processed vegetables. Production of vegetables is concentrated on a small number of highly specialized farms. Much of the cropland used for vegetable production is irrigated. The average size of vegetable farms is 256 acres, accounting for 1.7 percent of all farms with sales over $5,000, and 1.2 percent of the total cropland harvested.

Fruit and nut farms are also highly specialized. A wide variety of fruits and nuts is grown in California, Florida, Washington, Texas, Michigan, Georgia, and New York. These states had 67 percent of all fruit and nut farms. Owing to climate, different fruits are predominant in each of these states. Grapes, oranges, peaches, prunes, apricots, avocados, olives, strawberries, dates, walnuts, grapefruit, almonds, kiwifruit, nectarines, cherries, lemons and pears are important in California while oranges and grapefruit are the major fruits in Florida. In Washington, the principal crops are apples, pears, cherries, grapes, and peaches; Georgia produces pecans and peaches; in Michigan the main fruit crops are apples and cherries. New York produces a large amount of apples and grapes, and Texas produces substantial amounts of grapefruit and pecans. Cranberries are produced in Massachusetts and Wisconsin; and California and South Carolina are major peach producers. Fruit and nut farms comprise 3.8 percent of all farms and 0.2 percent of the total cropland, averaging 69 acres per farm.

Most of the poultry broiler operations are located in Arkansas, Georgia, North Carolina, Mississippi, Maryland, Texas, Delaware, and California. All poultry farms account for 2.4 percent of the farms with sales over $5,000.

The greatest concentration in dairying is in the Northeast, California, and the six states bordering the Great Lakes. Dairy farms numbered 113,000 in 1992, with an average size of 336 acres.

Livestock farms constitute the largest groups of farms, comprising 55 percent of all farms, and containing 64 percent of all land in farms. These farms, however, account for only 37 percent of all farms with sales over $5,000. In 1992, livestock farms sold 83 percent of all cattle and calves, 84 percent of all hogs and pigs, and 88 percent of all sheep and lambs. These farms are rather large and average 630 acres per farm. Livestock ranches, a classification used primarily in the Western states, are very large specialized cow-calf or sheep operations.

General farms are those with 50 percent or more of the total value of all farm products sold from seed crops, hay, and silage, and are farms on which no one product provided 50 percent or more of the total value of all farm products sold. These farms comprised 1.8 percent of all farms. The average size of these farms was 387 acres.

Even though many farms are still rather small, specialized agriculture continues to grow. Whether or not farms will continue to become larger, more specialized organizations depends on factors such as capital limitations, management ability, technological developments, and the risk and uncertainy facing farm producers.

# THE AGRIBUSINESS COMPLEX _____

The farming industry is closely related to the marketing industries that are essential to transform, transport, and transfer food and fiber to the consumer. In addition, farming is served by a large number of industries that manufacture and distribute durable goods and other farm supplies used in agriculture (Figure 2-2).

Farmers are buying more of their inputs rather than using farm produced inputs. These inputs include feed, fertilizer, petroleum products, farm machinery, chemicals, and other farm supplies and services.

Food processors are the link between farmers and food wholesalers and retailers. There are about 21,000 processing firms that add form utility to the raw farm product, for example, the transformation of sugar beets into sugar and wheat into bread.

Processors may sell to wholesalers and retailers through food brokers. The broker is an independent sales agent for the processor who neither takes possession nor title to grocery products.

Food wholesalers link food processors with retailers who sell directly to consumers and institutional outlets. There are about 43,000 of these wholesalers.

A retail food chain is a group of 11 or more stores owned and operated by the same firm. Chain stores began in 1859 with the Great Atlantic and Pacific Tea Company. Between 1910 and 1930, Atlantic and Pacific expanded, and by 1930 they had over 15,000 stores.

The major impact of the retail chain store movement was on wholesale operations. Chain stores combined wholesaling with retailing, and to a lesser extent, processing, and were able to lower costs. They were able to cut prices, forcing higher-priced independents out of business. Then after World War II, the chains began adopting the supermarket type of operation.

Because the independent grocery retailers and wholesalers lost business to the chains, they responded with their own vertically integrated organizations. In some cases, groups of independent retailers created cooperative wholesale systems to supply them with merchandise. In other cases, wholesalers assembled "voluntary groups" of retailers into chain-type organizations, adopting most of the methods used by the chains. The number of grocery stores in the United States declined from about 280,000

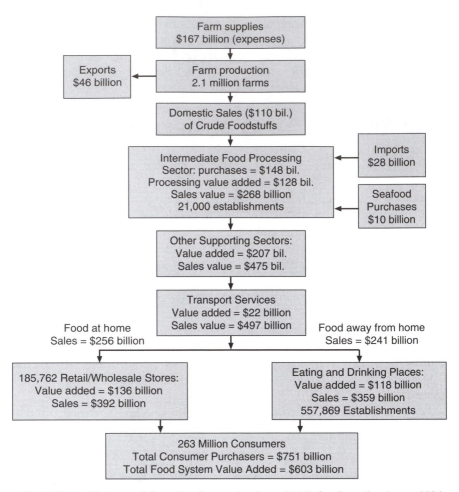

**FIGURE 2-2**   Estimated functional organization of U.S. food agribusiness, 1994.

in 1954 to less than 162,000 in 1994. The 20 largest grocery store chains sold about 41 percent of all grocery items.

Consumers spent $751 billion for food at the retail level in 1995. The farm value of that food was $148 billion, and the total value added by marketing was $603 billion.[5] One must remember that the marketing job is extremely important because farm products are raw materials and the marketing system adds time, form, place, and ownership utility or satisfaction costing more than three times the farm value of the original output.

_____

[5] Farm value and value added by marketing do not add to total consumer spending for food because such values as imported items, tips, and taxes are not included.

The major factors increasing the marketing bill over time have been rising labor costs, the demand for services such as special packaging and convenience foods, plus increased volume of food products handled by the system. Affluence has permitted the replacement of domestic workers in the kitchen with mechanical devices and prepared foods.

The American consumer has a wide selection of food products from which to choose at a relatively low proportion of disposable personal income. U.S. consumers spend about 11 percent of their disposable personal income for food, one of the lowest of any country in the world, and this percentage has been relatively stable in the last five years.

Consumer demand for food, however, changes over time. In general, per capita consumption of poultry, corn sweetener, rice, skim milk, fresh fruit, fats and oils, cheese, fruit, and vegetables has been increasing while consumption of tea, whole milk, buttermilk, butter, cottage cheese, and eggs has been declining. Red meat consumption per capita has declined in recent years, while flour and cereal products have been increasing.

# INTERNATIONAL TRADE _____

Since the "Russian wheat deal" of 1972, international trade in agricultural commodities has received much national attention from consumers and producers. Crop shortages, rising incomes, and population growth have increased the value of exports of agricultural commodities to $55.8 billion in the calendar year 1995, $12.9 billion more than 1992 (Table 2-8). The increase in U.S. exports of farm commodities kept the U.S. agricultural trade balance (exports minus imports) at about $25.8 billion. Imports in 1995 were $30.0 billion, up from $24.6 billion in 1992 (Table 2-9). The increase in commodity prices and export volumes, particularly for meat, feed grains, soybeans, fruits, vegetables, and cotton, accounted for most of the increase in the value of agricultural exports.

Although international trade is relatively unimportant to the total U.S. economy, it is very important to agriculture and to our level of living. Although the United States exports and imports only about 7 to 9 percent of its gross domestic product, about 23 percent of all farm income is derived from agricultural product exports. In 1995-96 U.S. agriculture exported 58 percent of its wheat and 31 percent of its corn output. In that year, the nation used about 224 million acres to produce its corn, wheat, barley, oat, sorghum, soybean, rice, and cotton crops. The output from 40 percent of those acres (90 million) was exported to other nations. Thus, out of every 3 crop acres used for production, over one acre was used to produce exports of those crops.

**TABLE 2-8**   U.S. Agricultural Exports: Value by Commodity, Calendar Years

| | 1970 | 1982 | 1986 | 1989 | 1992 | 1995 |
|---|---|---|---|---|---|---|
| Commodity Exports | (Millions of Dollars) | | | | | |
| Animals and animal products | | | | | | |
| Dairy products | 127 | 347 | 438 | 405 | 726 | 771 |
| Fats, oils and greases | 247 | 663 | 411 | 512 | 525 | 827 |
| Hides and skins | 187 | 1,022 | 1,521 | 1,717 | 1,346 | 1,748 |
| Meat and meat products | 132 | 978 | 1,113 | 2,346 | 3,339 | 4,522 |
| Poultry and poultry products | 56 | 515 | 496 | 720 | 1,211 | 2,345 |
| Other | 101 | 410 | 530 | 673 | 718 | 780 |
| Total animals and animal products | 850 | 3,935 | 4,509 | 6,373 | 7,925 | 10,933 |
| Grains and preparations | | | | | | |
| Feed grains | 1,064 | 6,444 | 4,330 | 7,874 | 5,737 | 8,341 |
| Rice | 314 | 997 | 621 | 970 | 725 | 996 |
| Wheat | 1,111 | 6,927 | 3,279 | 6,150 | 4,675 | 5,681 |
| Other | 107 | 273 | 398 | 2,208 | 3,035 | 3,519 |
| Total grains and preparations | 2,596 | 14,641 | 8,628 | 17,202 | 14,172 | 18,537 |
| Oilseeds and products | | | | | | |
| Cottonseed and soybean oil | 244 | 692 | 343 | 461 | 432 | 786 |
| Soybeans | 1,228 | 6,218 | 4,321 | 3,944 | 4,380 | 5,400 |
| Protein meal | 358 | 1,447 | 1,302 | 1,176 | 1,398 | 1,140 |
| Other | 91 | 784 | 493 | 735 | 980 | 1,597 |
| Total oilseeds and products | 1,921 | 9,141 | 6,459 | 6,316 | 7,190 | 8,923 |
| Other products and preparations | | | | | | |
| Cotton | 372 | 1,955 | 786 | 2,268 | 1,999 | 3,714 |
| Tobacco (unmanufactured) | 517 | 1,547 | 1,209 | 1,341 | 1,651 | 1,400 |
| Fruit and nut preparations | 334 | 1,917 | 2,040 | 2,413 | 2,732 | 4,650 |
| Vegetables and preparations | 206 | 1,174 | 1,024 | 1,604 | 2,871 | 3,889 |
| Other | 463 | 2,312 | 1,349 | 2,394 | 4,389 | 3,768 |
| Total other products and preparations | 1,892 | 8,905 | 6,468 | 10,020 | 13,642 | 17,421 |
| All commodities | 7,259 | 36,622 | 26,064 | 39,911 | 42,929 | 55,814 |

SOURCE: Economic Research Service, USDA. "Foreign Agricultural Trade of the United States," January/February 1996.

# OVERVIEW OF OUR ECONOMIC SYSTEM _____

Two major economic entities are involved in our free enterprise economic system. These are households and business firms (including farms

**TABLE 2-9** U.S. Agricultural Imports: Value by Commodity, Calendar Years

| | 1970 | 1982 | 1986 | 1989 | 1992 | 1995 |
|---|---|---|---|---|---|---|
| Commodity Imports | (Millions of Dollars) | | | | | |
| Animals and animal products | | | | | | |
|   Cattle | 111 | 469 | 665 | 662 | 1,245 | 1,413 |
|   Dairy products | 125 | 612 | 809 | 857 | 857 | 1,089 |
|   Hides and skins | 110 | 198 | 210 | 216 | 185 | 199 |
|   Meat and meat products | 1,011 | 2,037 | 2,444 | 2,548 | 2,638 | 2,478 |
|   Other | 204 | 304 | 326 | 784 | 755 | 811 |
|   Total animals and animal products | 1,561 | 3,620 | 4,454 | 5,067 | 5,680 | 5,990 |
| Grains and preparations | 70 | 375 | 678 | 1,217 | 1,586 | 2,362 |
| Oilseeds and products | 202 | 475 | 586 | 916 | 1,219 | 1,815 |
| Other products and preparations | | | | | | |
|   Sugar, cane and beet | 725 | 789 | 625 | 593 | 642 | 683 |
|   Tobacco | 139 | 342 | 602 | 551 | 1,353 | 718 |
|   Fruits and preparations | 146 | 975 | 1,557 | 1,093 | 1,408 | 1,631 |
|   Nuts and preparations | 100 | 225 | 370 | 344 | 467 | 485 |
|   Vegetables and preparations | 298 | 1,134 | 1,579 | 2,049 | 2,184 | 3,103 |
|   Wine | 145 | 783 | 1,006 | 930 | 1,087 | 1,153 |
|   Malted beverages | 32 | 466 | 784 | 855 | 864 | 1,166 |
|   Bananas | 188 | 581 | 749 | 871 | 1,097 | 1,140 |
|   Cocoa beans | 201 | 704 | 1,111 | 977 | 1,080 | 1,134 |
|   Coffee | 1,160 | 2,903 | 4,544 | 2,432 | 1,706 | 3,263 |
|   Crude rubber | 231 | 530 | 615 | 958 | 770 | 1,629 |
|   Spices | 55 | 152 | 353 | 275 | 262 | 334 |
|   Tea | 53 | 129 | 133 | 139 | 179 | 190 |
|   Carpet wool | 31 | 30 | 38 | 58 | 29 | 43 |
|   Other | 390 | 898 | 1,183 | 2,368 | 3,011 | 3,154 |
|   Total other products and preparations | 3,937 | 10,753 | 15,333 | 14,552 | 16,139 | 19,826 |
|   All commodities | 5,770 | 15,223 | 21,051 | 21,752 | 24,624 | 29,993 |

SOURCE: Economic Research Service, USDA. "Foreign Agricultural Trade of the United States," January/February 1996.

and ranches). Households are our dwelling places made up of families or individuals. They are consuming units that purchase the nation's goods and services. In addition they own our economic resources. Business firms, on the other hand, are the economic actors that produce the nation's output of goods and services. In order to accomplish this process they must purchase or hire economic resources. Business firms in the

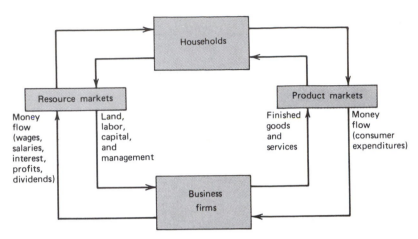

**FIGURE 2-3** Circular flow diagram—households exchanging resources for money income to spend on goods and services.

United States are organized as either single proprietorships, partnerships, corporations, or cooperatives.

A circular flow diagram is used in Figure 2-3 to illustrate how households and business firms interact in our economy. The right side of the model shows the flow of finished goods and services from business firms to consumers. This movement of products represents all markets for consumer goods and services such as food, clothing, furniture, and television sets. The reverse money flow shown in the figure is the payments made by householders for those goods and services. The product market for goods and services establishes prices that regulate the quantity and quality of goods produced and consumed.

The left side of the circular flow diagram shows the movement of economic resources (land, labor, capital, and management) from households to business firms. The return money flow is from business firms paying for these resources in the form of wages and salaries for labor and management, dividends and interest for capital, and rents on land. These flows make up the resource market. This market determines resource prices that regulate the flow of resources from consumers to producers.

Now if we visualize the entire diagram, you can see that households are selling their resources to firms in order to purchase finished goods and services. Also, business firms are selling goods and services to consumers in order to purchase factors of production to continue the production process.

Households are the selling side of the resource market and the buying side of the product market. Business firms are on the buying side of the resource market and the selling side of the product market. The money flowing through the product market is determined by the "dollar votes" of consumers in the market and the prices at which goods and services sell. The amount of money flowing through the resource market depends on the amounts of resources put on the market by households and the price of resources that is determined in the resource market.

The transactions between households and business firms are limited by scarcity. Consumers have limited incomes, but unlimited wants. Business firms are also constrained in production by limited resources to produce final goods and services,

A farm or ranch is both a household and a business firm. The farm family provides some of the economic resources to produce food and fiber that it sells to other households. With that money flow it can purchase final goods and services from other producers and purchase other economic resources from other households. Thus, this simplified circular flow model applies to the individual farm or ranch as it does to the entire economic system.

A circular flow diagram presents a macroeconomic view of an economy by dealing with the economy in the aggregate. In order to measure the flows of money and goods among households, businesses, and government, a system of national accounts is used. These accounts provide the basis for much economic planning and government policy.

Basic to such a national accounting system is the concept of *gross national product* (GNP). Gross national product is the total value of all finished goods and services produced in the economy within a given time period, such as a year whether these goods and services were produced in the United States or overseas. Therefore, American output produced in Argentina is included in U.S. GNP. Also, foreign-produced output in the United States is excluded from U.S. GNP.

In 1992, the United States changed from GNP accounting to gross domestic product (GDP) accounting. *Gross domestic product* is the total value of goods and services produced within the United States by either foreign or domestic resources. One can calculate GDP as the total expenditures of households, business firms of all kinds, governmental units, and net foreign transactions (exports minus imports). In this case, we are adding up the values of purchases made by final consumers in the *product market* (the right-hand loop of the circular flow diagram). With the expenditure method, we avoid double counting by including only sales to *final* consumers, and exclude the sales of intermediate goods and services.

A simplified example of the several stages of production from the original contributor to the final purchase of a cotton shirt will help make this more clear. The tabulation below begins with a farmer using resources to produce the cotton used to make that shirt. Each later stage along the way buys from the previous stage at a market-determined price for the good

| Stages of Production | Product Value at Point of Sale | Value Added |
|---|---|---|
| Cotton farmer | $2 | $2 |
| Cotton mill | 4 | 2 |
| Textile mill | 9 | 5 |
| Shirt manufacturer | 14 | 5 |
| Wholesaler | 16 | 2 |
| Retail outlet | 20 | 4 |
| Totals | $65 | $20 |

through that producer's contribution to value. Suppose that each stage is paid the full value of its product. When all steps have been completed, a shirt worth $20 is available for sale to a final user.

These stages have coordinated their functions in such a way that an economic good worth $20 has been produced, which is what GDP is supposed to account for. If we had tallied sales between all economic units in the system, $65 would have been counted. This total would have greatly overstated GDP since we have, at the end of it all, one shirt worth $20 to the consumer.

Rather than calculate GDP from the expenditure approach, one can total up the incomes earned in the *factor markets* (transactions in the left-hand loop of the circular flow diagram). In this case, GDP is the total value of all resource earnings such as wages and salaries, interest, rents and profits, minus business taxes and capital consumption (wear and tear on machines and equipment), less United States income earned overseas. With the income approach, we avoid double counting by adding only the resource earnings at each stage of production throughout the economy.

An economy is said to be in *equilibrium* when opposing forces within the system just offset one another and there is no incentive or pressure to change. So when expenditures by consumers and investors (the product flow loop) are just equal to the payments to the factors of production (the income flow loop), the system is in equilibrium. Thus, an economic equilibrium or balance has been achieved when factor payments have generated enough income to purchase the amount of goods and services that have been produced by the system.

If expenditures by consumers and investors are less than the value of goods and services produced, business firms cannot sell all their output at current prices, so inventories rise and future output and employment will decline. On the other hand, when expenditures are greater than the value of goods and services produced, inventories decline and future output and employment should increase.

The conscious direction of economic activity involves three primary areas of influence by which the government attempts to achieve its national goals of full employment, price stability, and economic growth: (1) monetary policy, (2) fiscal policy, and (3) administrative regulation.

*Monetary policy* influences economic activity in the system through the government's actions in managing the money supply and interest rates, *Fiscal policy* relates to the actions of the government in exercising its spending and taxing powers. And the effects of *administrative regulation* are evident when the government establishes or changes rules and regulations relating to the terms and conditions for loans and installment purchases, controlling wages and prices, restricting environmental pollution, embargoing or otherwise regulating exports and imports, and many similar regulatory actions elsewhere in the economy,

If it is deemed necessary to increase spending in the economy, the government may choose, among many other such actions, to buy government bonds in the market (expansionary monetary policy) or cut federal tax rates (also expansionary). Such actions as these "inject" money into the economy by leaving more spendable money in the hands of consumers and business firms than would have existed in the absence of those policies, thus stimulating aggregate economic activity with increased output and employment.

If, on the other hand, reduced economic activity is necessary, the government can sell bonds in the market or increase taxes. These actions will reduce the amount of money that people and business firms have to spend and will reduce aggregate output and employment.

Specific application in the three general policy areas is dealt with in the chapters on macroeconomics, market regulation, agricultural finance, price and income policies, and international trade, where their relevance to the agricultural sector is more direct.

# SUMMARY

Agriculture is a very large, heterogeneous industry. It involves many types of businesses in producing and distributing food and fiber to consumers. Agriculture includes farms; credit and supply firms; marketing, processing, and distribution firms; restaurants, and retailers.

American agriculture has about 2.1 million farms and ranches. These farms vary widely in size, but the average farm size is 469 acres. Most farms are cash grain and livestock operations, scattered throughout the country.

The American consumer has a wide selection of nutritious foods available at "reasonable" prices. American consumers spend about 11 percent of their disposable income for food, one of the lowest in the world. International consumers are purchasing more American farm products. Food and feed grains, meat products, soybeans, fruits, and vegetables, and cotton are the major export commodities. At present the United States is exporting the products from one out of every three acres of cropland harvested. Coffee, meat, fruits, and vegetables are our major imported items.

Households and business firms are the two economic actors in our free enterprise economic system. Households sell resources to business firms for money and in turn spend this money on final goods and services produced by business firms.

GDP measures the total value of final goods and services produced in the economy during a given period of time whether by foreign or domestic resources. GDP can be measured in either the product market or the factor market. Government policies are used to influence the level of economic activity in the system.

## CHAPTER HIGHLIGHTS

1. Agriculture is composed of the complete food and fiber system. It does not consist solely of the physical and biological production of agricultural commodities.

2. Farms produce 2 percent of the U.S. gross national product (GNP) and directly employ 2 percent of the labor force. The total agribusiness system, however, employs 16 percent of the U.S. labor force.

3. Contrary to popular opinion, most farms and ranches are family enterprises. Corporations account for only 4 percent of U.S. farms, 13 percent of the farm land, and 27 percent of agricultural sales.

4. Most family farms are as efficient as corporation farms.

5. Farms that sold more than $100,000 worth of products in 1994 make up about 17 percent of all farms, produce 77 percent of U.S. farm output, and have a per family income of $150,903 per year.

6. The average noncommercial farmer has an off-farm income of $36,440 per year and a negative net income from farming.

7. About 18 percent of U.S. agricultural production is contracted or produced under vertically integrated arrangements.

8. Farmer cooperatives market 31 percent of U.S. agricultural products and handle 29 percent of all agricultural inputs.

9. Most American farms are small, on the average. In 1995 the mean farm size was 469 acres.

10. Agriculture includes 21,000 food processors, 43,000 wholesalers, and over 740,000 retail and institutional (eating and drinking places, etc.) food outlets.

11. Consumers spend about $751 billion on food, or about 11 percent of their disposable incomes.
12. International trade is very important to agriculture and to U.S. consumers.
13. Households and business firms are the two major economic entities in our economic system.
14. Households buy products and sell resources.
15. Business firms buy resources and sell products.
16. Government serves important functions in the uses of its monetary, fiscal, and regulatory powers.

## KEY TERMS AND CONCEPTS TO REMEMBER

| | |
|---|---|
| Administrative regulation | Fiscal policy |
| Agricultural fundamentalism | Gross domestic product (GDP) |
| Contract production | Gross national product (GNP) |
| Cooperative | Monetary policy |
| Equilibrium | Product market |
| Factor market | Vertical coordination |
| Family farm | Vertical integration |

## REVIEW QUESTIONS

1. What is a farm? Why has the definition of a farm been changed in recent years?
2. The number of American farms has been declining. What are some factors influencing this trend?
3. How does a cooperative differ from a "for profit" corporation?
4. Corporate farms have been increasing in number. What impact could this development have on family farms?
5. Agricultural producers are responsible for producing the final product that the consumer purchases. Thus, the agricultural marketing system is unproductive. Discuss.
6. Farmers and ranchers should adjust their production to the food and fiber needs of consumers. Discuss.
7. "The U.S. government should restrict imports of agricultural commodities, since we can produce all commodities more efficiently than any other country." Discuss.

**8.** What is gross domestic product (GDP)? What does it measure?

**9.** How do the government's regulatory powers, monetary policy, and fiscal policy influence economic activity in the system?

# SUGGESTED READINGS

Abrahamsen, Martin A. *Cooperative Business Enterprise, New* York: McGraw-Hill, 1976, Chapters 1–3.

*Agricultural Outlook,* a monthly magazine. Washington, D.C.: Economic Research Service, U.S. Department of Agriculture.

*Handbook of Agricultural Charts,* Washington, D.C.: Economic Research Service, U.S. Department of Agriculture.

Goldberg, Ray A. *Agribusiness Coordination,* Boston, Mass.: Division of Research, Harvard Business School, 1968, Chapter 1.

McConnell, Campbell R. and Stanley Brue. *Economics,* 13th ed. New York: McGraw-Hill, 1996, Chapter 34.

Schertz, Lyle P. et al. *Another Revolution in U.S. Farming?* Agricultural Economic Report No. 441, Washington, D.C.: Economics, Statistics, and Cooperatives Service, USDA, December, 1979.

Economics, Statistics, and Cooperatives Service. *A Time to Choose: Summary Report on the Structure of Agriculture,* Washington, D,C.: USDA, January 1981.

Economics, Statistics, and Cooperatives Service, *Structure Issues of American Agriculture,* Agricultural Economic Report No. 438, Washington, D.C.: USDA, November 1979.

# AN OUTSTANDING CONTRIBUTOR

**Ray A. Goldberg**   Ray Goldberg is George M. Moffett Professor of Agriculture and Business at the Harvard Business School, a chair he has held since 1970.

Born and raised in Fargo, North Dakota, he has bachelor's and M.B.A. degrees from Harvard University which he received in 1948 and 1950. He earned his Ph.D. degree in 1952 at the University of Minnesota. Following about three years of private business at Moorhead, Minnesota, Dr. Goldberg joined the faculty of the Harvard Business School in 1955.

Dr. Goldberg has served as an officer and director of numerous agribusiness firms in the food and financial sectors of agriculture, and has been consultant to a diverse group of public and private agencies including the Commodity

Futures Trading Commission, Agency for International Development, President's Food and Fiber Commission, the National Marine Fisheries Service, U.S. Comptroller of the Currency, National Academy of Sciences, World Food System, Inc., Ford Foundation, Winrock International, institutes in Nicaragua and Mexico, and many other agriculturally oriented organizations. In addition, he is on the Editorial Board of the *Food Policy Journal* and similarly served for the *American Journal of Agricultural Economics.*

About 150 Harvard M.B.A.s each year concentrate on an agribusiness major under Professor Goldberg's supervision. He directs the national and international agribusiness Continuing Education programs at the Harvard Business School. These programs bring together about 150 of the world's agribusiness leaders each year from private industry to analyze case studies of agribusiness production, marketing, and financial situations.

Dr. Goldberg has written or edited 13 books and a great many articles on agribusiness and related topics. He has built an outstanding agribusiness program at Harvard and has had a major influence on domestic and international agribusiness firms.

He is married to the former Thelma Englander, and they have three children.

# CHAPTER 3

# Consumer Behavior and Demand

$\mathbf{I}$n the last chapter, it was noted that households and business firms were the major economic actors in our economic system. This chapter concentrates only on households, or consumers, and the behavior of people in meeting their desire for goods and services. It is in the observed market behavior of people that the concept of demand rests. While demand will be specifically discussed later in the chapter, suffice it to say at this point that demand means the quantities of a product bought at alternative prices holding everything else constant.

## THE UTILITY BASIS OF DEMAND _____

When consumer behavior is studied, certain characteristics can be noted. One feature is that consumers spend everything they earn on goods and services, including savings. Another is that consumers never seem to get enough of most things. We can infer from this characteristic of consumer behavior that human wants are insatiable and that more is preferred to less.

One of the reasons consumers do not buy infinite quantities of everything is that they have a limited amount of money income. In economics, we assume that consumers, with a given money income, will purchase clothing, housing, food, haircuts, and all the other things that they want in amounts that will maximize utility or satisfaction for them. The utility of a product or service is derived from the inherent characteristics or qualities that cause them to be desired. These may be objective or subjective qualities. But it is unlikely that two individuals would attain the same utility or satisfaction from the consumption of the same amount of a product.

Another noticeable feature of consumer behavior is that income is not spent on a single item; a variety of goods and services is purchased. The reason for this behavior is contained in what is called the *law of diminishing marginal utility.* This law means that when an individual consumes additional units of a specific commodity, consumption of other goods and services unchanged, the amount of satisfaction derived from each additional unit of that good decreases. (Remember that marginal means additional.)

### An Example of Diminishing Marginal Utility

Suppose a traveler has been in Death Valley without water for three days. Under these conditions that person may be dead, but let us assume our traveler is alive (and still rational). Let's also assume that this individual has some money and encounters an entrepreneur who has water to sell. For the first glass of water, our traveler may be willing to pay a great deal of money because the utility derived from that first glass of water is very high. A second glass of water will also add utility, but a lesser amount than the first, so the traveler would be willing to pay less for this second glass of water as it adds less utility than the first. If you carry this example to its logical conclusion, the traveler would not be willing to pay anything for, say, the thirtieth glass because that person would be full, with no desire for more.

Note that the concept of utility requires a specified time period. If, for instance, the supplier of water were to leave in one hour, the amount of water our thirsty traveler would be willing to buy and drink would be quite different than if the supplier were to travel along with him for a day, a week, or more. And if our buyer could purchase canteens full to carry along, the amount purchased would also be quite different.

Were even these few conditions to change, we would be unable to determine what effect a glass of water has on utility, or satisfaction. The same problem would exist for all other goods, and economists might as well use a roulette wheel to determine the answers to economic questions. But *ceteris paribus* (all other things remaining unchanged) saves it for us. With all other things held constant, the effect of one variable on another is determinable.

The concept of utility, fundamental as it is to understanding demand, is just that—a theoretical concept. It gives us no measurable basis by which to determine the demand for a good. Neither can a physicist specify the comfort one would feel when the thermometer registers 70 degrees Fahrenheit.

If we could assign values to the traveler's units of utility, the problem would be much simpler. That person would buy more water until the value of the last glass of water bought was worth just what it cost. The next glassful would not be purchased because it would not be worth its cost.[1]

## Another Example of Diminishing Marginal Utility

The law of diminishing utility can be made more explicit by assuming the following utility schedule for doughnuts consumed by Tom.

The first column in Table 3-1 shows Tom's consumption of doughnuts per day. The second column indicates the total amount of satisfaction

---

[1] How much water to buy for the utility it can produce is identical to the producer's problem of using quantities of a valuable resource to produce something else of value, as will be discussed in the next chapter.

**TABLE 3-1**   Tom's Utility Schedule for Doughnuts

| (1)<br>Number per Day | (2)<br>Total Utility | (3)<br>Marginal Utility |
|---|---|---|
| 0 | 0 | |
| | | 6 |
| 1 | 6 | |
| | | 5 |
| 2 | 11 | |
| | | 4 |
| 3 | 15 | |
| | | 3 |
| 4 | 18 | |
| | | 2 |
| 5 | 20 | |
| | | 1 |
| 6 | 21 | |
| | | 0 |
| 7 | 21 | |

measured in terms of some unit of satisfaction, such as "utils," from consuming various amounts of doughnuts per day. For instance, if Tom consumes two doughnuts per day his total utility is 11 utils. If he eats seven doughnuts in a day his total utility is 21 utils. Thus Tom's total utility curve for doughnuts rises and then levels off, as more doughnuts are consumed.

Column 3 shows added or *marginal* utility. Notice that Tom's added utility from consuming additional doughnuts decreases, which is consistent with the law of diminishing marginal utility. Tom's marginal utility from consuming the first doughnut is six utils; from the fourth, three utils; and from the seventh, zero utils. In comparing Tom's total and marginal utility schedules for doughnuts, be aware of the fact that when marginal utility is positive, total utility is rising. When marginal utility is zero, total utility is constant. Also, it is possible for Tom's marginal utility to be negative at higher levels of doughnut consumption. If that was the case, total utility would decrease. A negative marginal utility suggests that additional doughnuts are a "surplus" problem. He has eaten all his stomach can hold, and consuming more doughnuts would actually create disutility.

## CONSUMER CHOICE _____

The fact that people buy a certain combination of goods with their spendable money, and not some other combination, implies that the choice made

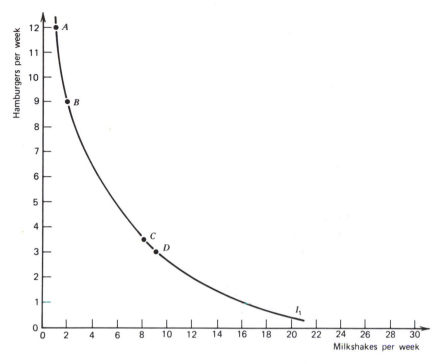

**FIGURE 3-1**  Susan's indifference curve.

was more satisfying to the consumer than some other allocation of those funds might have been.[2]

The theoretical basis of such choices can be demonstrated graphically as in Figure 3-1, where the problem is limited to choosing between two goods only. We measure units of one good along the horizontal axis and units of the other good along the vertical axis. Some generalizations about the amounts of utility derived from consuming different quantities and proportions of those goods can be made.

## Indifference Curves

Consider a student, Susan, choosing between different combinations of hamburgers and milkshakes. Her food preferences are shown by the *indifference curve*, labeled $I_1$. That curving line is called an indifference curve because it shows all the combinations of those two goods that will give her the same amount of satisfaction or utility. All points *along* the indifference curve are equally satisfying; utility is constant at all points on that curve.

---

[2] We assume that the consumer is rational and that the person maximizes the satisfaction to be obtained from his or her spendable income.

With many different combinations of hamburgers and milkshakes being able to produce identical amounts of utility, as indicated by the curve $I_i$, there can be no reason for preferring one combination over another. Thus we can say that Susan is indifferent regarding the combinations shown (12:1, 9:2, 3.5:8, and 3:9 hamburgers and milkshakes). Connecting those four identified points traces out the curve of constant utility. The indifference curve shows that 12 hamburgers and one milkshake (point A) give Susan the same satisfaction as three hamburgers and nine milkshakes (point D); the proportions indicated by points B and C also produce the same utility.

The shape of the indifference curve shows the willingness to substitute one good for another. At point A, our student has, relatively speaking, many hamburgers and few milkshakes. Because of diminishing marginal utility, the twelfth hamburger adds only a little utility; another milkshake would add quite a bit. Consequently, Susan would be willing to exchange three hamburgers for another milkshake. But as we move downward on the indifference curve, the marginal utility of another milkshake is falling, while the marginal utility of another hamburger is rising. By the time we get to point D, we can see that Susan would be willing to give up the ninth milkshake in exchange for only one-half of a hamburger. Their *values* to her depend on how much more utility she would get from consuming one more unit of either good, rather than on their prices. Between points A and B a milkshake is worth three hamburgers, or six times more valuable than between points C and D where a milkshake is worth only one-half a hamburger.

As we move from point A to point B, compared with moving from point C to point D, we can see a change in the *rate* at which milkshakes will substitute for hamburgers, a characteristic called the **diminishing marginal rate of substitution.** We use the symbolism $MRS_{mh}$ to mean the "marginal rate of substitution of milkshakes for hamburgers," and $\Delta h$ and $\Delta m$ as "change in hamburgers," and "change in milkshakes," respectively. Thus $MRS_{mh} = \Delta h / \Delta m$ shows the number of hamburgers that a milkshake will replace in Susan's weekly diet without changing her total satisfaction. The marginal rate of substitution is said to be diminishing because, as we move downward along $I_1$, each additional milkshake will replace fewer hamburgers than the previous one did.[3]

---

[3] Since utility along the indifference curve is constant, we may rewrite this relationship as $\Delta m \cdot MU_m = \Delta h \cdot MU_h$, where $MU_m$ means the marginal utility per milkshake and $MU_h$ the marginal utility per hamburger. Along $I_1$ from point A to point B the units of change in hamburger times their *MU* must equal the units of change in milkshakes times their *MU*. The marginal rate of substitution is negative because the indifference curve slopes downward to the right. Ignoring sign, the equality $\Delta m \cdot MU_m = \Delta h \cdot MU_h$ can be rewritten as $\Delta h / \Delta m = MU_m / MU_h$. Thus we can note that the relative marginal utilities are the determinants of the rate at which two goods substitute for one another.

The shape of the indifference curve tells a great deal about Susan's preferences. For instance, if the indifference curve is a right angle it means she would be unwilling to substitute one commodity for the other. She would purchase only one combination of goods, such as six hamburgers and four milkshakes. However, if she perceives them to be good substitutes the indifference curve would be closer to a straight line. In this case she would be willing to interchange the goods at a nearly constant rate.

Indifference curves have three major characteristics. The first feature is that they are downward sloping to the right. This characteristic means that if a consumer gives up one commodity, the loss in satisfaction must be compensated for by additional units of the other commodity if utility is to remain constant along the indifference curve.

The second characteristic is that indifference curves are convex to the origin. A requirement such as this is fulfilled if the marginal rate of substitution decreases as one moves along the indifference curve. If the indifference curve was not convex but concave (bending away from the origin), it would suggest that our consumer is getting equal satisfaction from giving up more and more of one commodity to obtain given amounts of another. This fact does not coincide with observed consumer behavior.

A third feature of indifference curves is that they cannot intersect. Higher indifference curves are larger combinations of both goods and hence represent higher levels of satisfaction (Figure 3-2). Because we assume that the consumer is not satiated with goods, $I_1$, $I_2$, $I_3$ show progressively higher levels of satisfaction. Each consumer has an *indifference map* (family of indifference curves as demonstrated by Figure 3-2) showing that person's own tastes and preferences.

## The Budget Line

What Susan can consume is determined by her money income. If she has $7.50 per week to spend on these two goods, and the price of hamburgers is $1.00 each and milkshakes are $.50 each, then she could buy 7½ hamburgers ($7.50/$1) if her entire $7.50 budget were spent on hamburgers. If the total weekly income is spent on milkshakes, she could buy 15 ($7.50/$.50). It is also possible to buy other combinations of hamburgers and milkshakes with $7.50. You can see that if two-thirds of her income were spent on hamburgers and one-third on milkshakes, five hamburgers and five milkshakes could be purchased. These three possible combinations of $7.50 expenditures are shown in Figure 3-3. Connecting points *A*, *B*, and *C*, derives the **budget line,** which shows all the possible combinations of hamburgers and milkshakes that the consumer can buy for $7.50. The slope of the budget line is a negative 7.5/15, or - 1/2. Thus the slope of the budget line is

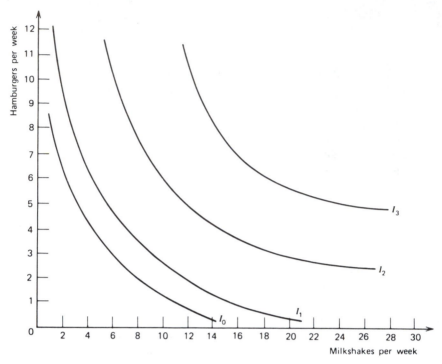

**FIGURE 3-2**   Susan's indifference map.

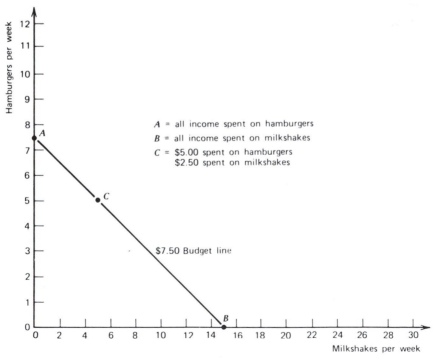

$A$ = all income spent on hamburgers
$B$ = all income spent on milkshakes
$C$ = $5.00 spent on hamburgers
      $2.50 spent on milkshakes

$7.50 Budget line

**FIGURE 3-3**   The consumer's budget line.

equal to the price of milkshakes $(P_m)$ divided by the price of hamburgers $(P_h)$, $- .50/1.00 = -1/2$.

## Consumer Equilibrium

We now have both the consumer's preference system and the budget line. Putting them together permits an analysis of the combination of goods that will be purchased to maximize satisfaction or utility and result in efficiency in consumption.

The rational consumer wants to get to the highest indifference curve, given the budget constraint. From Figure 3-4, one can see that the optimal position of the consumer is at point $B$ where the budget line is just tangent to indifference curve $I_1$. At that preferred point, the consumer is maximizing satisfaction (given $7.50 of income) by purchasing three hamburgers and nine milkshakes per week. This consumer would not be in equilibrium at point $A$ or $C$, because by moving her purchases to point $B$, she could consume on indifference curve $I_1$ rather than attain less satisfaction on $I_0$. This consumer cannot consume at point $D$ on $I_2$, because

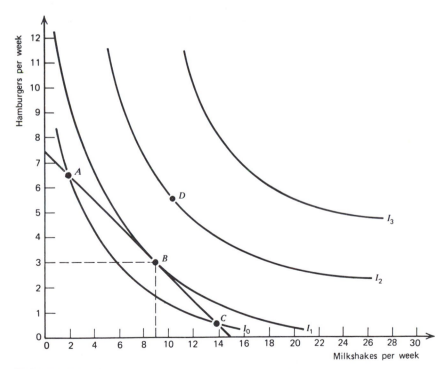

**FIGURE 3-4** Consumer equilibrium in consumption.

she does not have the income to reach that utility level. Your dictionary defines *equilibrium* as a state of balance between opposing forces. We use that same meaning in economics, with the opposing forces being Susan's desires for these two goods on the one hand and their market prices on the other. Once having found an equilibrium at point *B, ceteris paribus,* there is no incentive to change her spending from three hamburgers and nine milkshakes. But change Susan's relative desires for these goods, her spendable money income, or their relative prices, and a new equilibrium will be determined.

The slope of the indifference curve (*MRS*) is equal to the ratio of the marginal utility of milkshakes to the marginal utility of hamburgers. Also, we know that the slope of the budget line is equal to $P_m/P_h$. Hence the point of consumer equilibrium at point *B* is the same as saying that in equilibrium,

$$MU_m / MU_h = P_m / P_h^{\,4}$$

Rearranging this equation, we have:

$$MU_m / P_m = MU_h / P_h$$

In this form one can see that this consumer's optimum is where the marginal utility from each good purchased is proportional to its price. Another way of stating the same condition is to say that the consumer is in equilibrium when the marginal utility per dollar spent on milkshakes is equal to the marginal utility per dollar spent on hamburgers. This condition must hold for all goods purchased for the consumer to be in equilibrium.

## The Effect of Price Changes

Let us return to our example with Susan and her fixed income of $7.50 per week, with the price of hamburgers at $1.00 each and the price of a milkshake at $0.50. Now let's assume the price of a milkshake drops to $0.25. We can plot a new budget line representing the same amount of money income ($7.50). These two budget lines are shown in Figure 3-5.

As was mentioned previously, Susan will be in equilibrium consuming at the point where the budget line is tangent to the highest possible indifference curve. This situation can be illustrated by adding Susan's indifference map to Figure 3-5, resulting in Figure 3-6. This figure shows that before the price of milkshakes dropped from $0.50 to $0.25 she was

---

[4] In this form, the equation shows that the rate at which these goods substitute *in consumption* ($MU_m/MU_h$) is identical to their rate of substitution *in the market* ($P_m/P_h$), a condition for equilibrium.

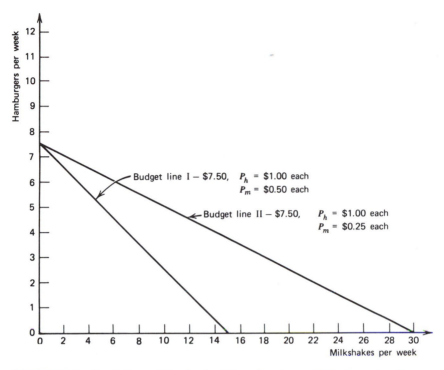

**FIGURE 3-5** A rotation of the budget line due to a milkshake price decrease.

purchasing nine milkshakes per week. After the price of milkshakes fell to $0.25, Susan is in equilibrium purchasing 15 milkshakes. She buys more milkshakes when the price decreases. The increase in the quantity of milkshakes purchased is due to the *substitution effect* and *income effect* of the price change. The substitution effect occurs when the price of milkshakes declines relative to hamburgers. Susan will substitute milkshakes for hamburgers because the price of milkshakes has fallen. When the price of milkshakes fell, this increased Susan's real income. This relationship is called the income effect. The *real income effect* of a decline in the price of milkshakes means that Susan is now able to buy more of either or both of these goods even though her money income has not changed.

## The Demand Curve

The effect of a price change of a commodity, *ceteris paribus*, normally is to change the quantity demanded in the opposite direction. A *demand curve* shows the quantities of a good that consumers will buy at different prices for that good at a point in time, everything else unchanged.

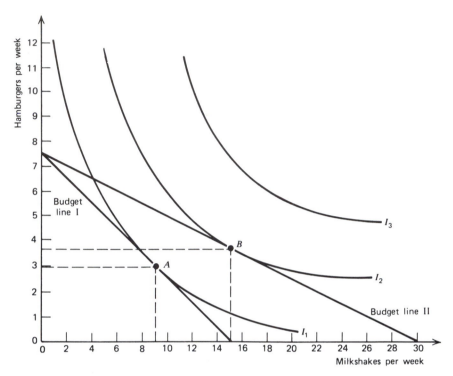

**FIGURE 3-6**   The effect of a price change for milkshakes.

We can derive two points on Susan's demand curve for milkshakes from Figure 3-6. Budget line I shows a price of milkshakes at $0.50 and nine milkshakes purchased (point *A*). Budget line II shows the price of milkshakes at $0.25 with 15 milkshakes purchased (point *B*). These points are plotted in Figure 3-7, making up two points on Susan's demand curve for milkshakes. Other points on the demand curve could be derived by varying the price of milkshakes (as in Figure 3-6) and deriving an equilibrium quantity purchased at each different price.

The price of an item must fall for the consumer to be willing to purchase more of it, because each additional unit consumed adds less utility than the previous one yielded. Because the utility derived from additional units of a good declines, those additional units are worth less to the consumer, resulting in the downward slope of the demand curve.

The market demand curve for milkshakes is the horizontal summation of all individual demand curves for that good. The **market demand curve** is defined as the quantities of a commodity all consumers are willing to purchase at different prices, in a given period of time, holding all other factors constant.

An example of adding two individual demand curves for milkshakes to obtain the market demand curve for milkshakes is shown in Figure 3-8.

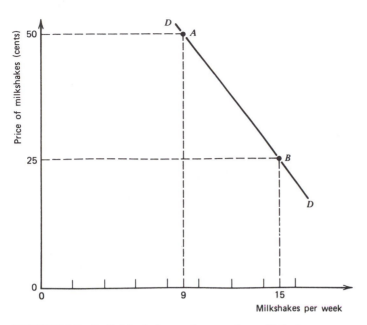

**FIGURE 3-7** Individual demand curve for milkshakes.

At a price for milkshakes of $0.50 each, Susan purchases nine shakes; Joe buys none. Thus, one point on the market demand curve is represented by Susan's purchase of nine milkshakes at $0.50. At $0.25 per shake, Joe buys 10 and Susan buys 15 milkshakes. Thus, another point on the market demand curve at $0.25 is a total of 25 milkshakes taken by the two consumers. Additional points are derived in a like manner.

## Price Elasticity of Demand

The downward sloping demand curve gives rise to another phenomenon of considerable importance—a price-quantity relationship referred to as the *price elasticity of demand.* Price elasticity of demand measures the responsiveness of quantity demanded to a change in price, *ceteris paribus.*

Price elasticity of demand is computed as the percentage change in quantity demanded divided by the percentage change in price. A simple method to derive the price elasticity of demand over a small segment of the demand curve may be expressed by the following formula:[5]

$$E_d = \frac{(Q_1 - Q_2)/(Q_1 + Q_2)}{(P_1 - P_2)/(P_1 + P_2)}$$

---

[5] This formula measures "average" price elasticity between two points on the demand curve, and is called *arc elasticity.* The formula could have been written as

$$E_d = \frac{(Q_1 - Q_2)}{(Q_1 + Q_2)/2} \div \frac{(P_1 - P_2)}{(P_1 + P_2)/2}$$

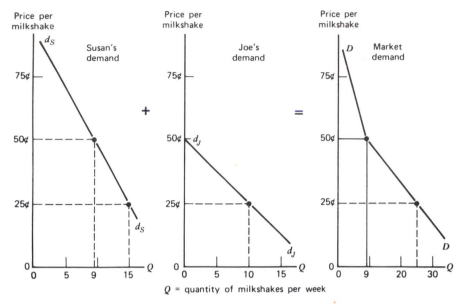

**FIGURE 3-8**   Market demand curve for milkshakes.

Note carefully the price and quantity notations, and their meanings. In this formulation, $Q_1$ stands for the first quantity observation, and $Q_2$ for the second quantity. Likewise, $P_1$ is the first observed price, and $P_2$ is the second price. The percentage change in quantity demanded is represented by $(Q_1 - Q_2) \div (Q_1 + Q_2)$, and the percentage change in price by $(P_1 - P_2) \div (P_1 + P_2)$.

To understand the use of this method in calculating elasticity, refer to Figure 3-9, where $Q_1$ = 4 billion bushels and $Q_2$ = 5 billion bushels, $P_1$ = $3.00 per bushel and $P_2$ = $2.25 per bushel. By substituting these values into the formula,

$$E_d = \dfrac{\dfrac{4-5}{4+5}}{\dfrac{3.00-2.25}{3.00+2.25}} = \dfrac{-\dfrac{1}{9}}{0.75 \big/ 5.25} = -\dfrac{1}{9} \cdot \dfrac{5.25}{0.75} = -0.78$$

the price elasticity is computed to be − 0.78, or inelastic.[6] When the coefficient is less than one (ignoring the negative sign), demand is said to be

---

But since the 2's cancel one another, the coefficient is unaffected. Differential calculus would permit the determination of price elasticity at a specific point on the demand curve, called *point elasticity*.
[6] Because the demand curve slopes downward to the right, the coefficient of price elasticity will always be negative. Note, too, a characteristic of straight-line demand curves that slope

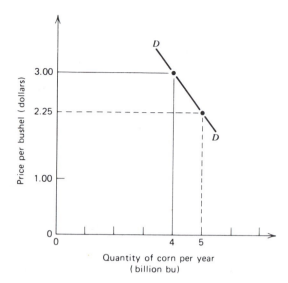

**FIGURE 3-9**   An inelastic segment of a demand curve.

inelastic, and when elasticity is greater than one (ignoring the negative sign), demand is elastic. When price elasticity equals one, demand is said to be of unitary elasticity.

If demand is *inelastic*, the quantity demanded changes relatively little compared to the change in price. A decrease of 75 cents in corn price, from $3.00 to $2.25, as shown in Figure 3-9, increases the quantity demanded from 4 billion bushels to 5 billion bushels. Despite the apparently large increase in corn purchases, total expenditures still drop from $12 billion to $11.25 billion because of the inelastic demand. Were the price of corn to increase, on the other hand, total expenditures would increase because of the relative unresponsiveness of quantity demanded to these price changes.

Price elasticity of demand is called *elastic* when the change in quantity demanded is large relative to the change in price, as shown in Figure 3-10. At a price of $0.80 per dozen eggs, a consumer may be willing and able to purchase 20 dozen eggs per year for a total expenditure of $16.00. With a drop in the price of eggs to $0.70 per dozen, our consumer buys 40 dozen eggs for an expenditure of $28.00.

Likewise, if the price of eggs were to increase rather than decrease, total expenditures would fall because of the responsiveness of quantity demanded to price changes as indicated by the elasticity coefficient of –5.0.

Price elasticity of demand is termed *unitary* when the percentage changes in price and quantity demanded are the same. With unitary elas-

_____

downward to the right: All such curves will have ranges over which they are elastic, unitary elastic, and inelastic.

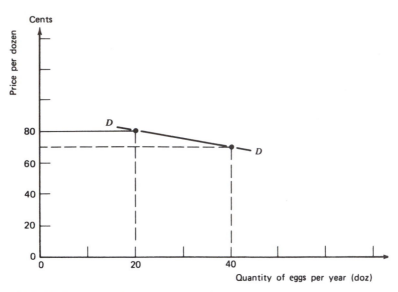

**FIGURE 3-10**   An elastic segment of a demand curve.

ticity of demand, total expenditures do not change as price changes, be
cause changes in quantity bought just offset the expenditures that would
be gained (or lost) as the price increases (or decreases).

Elasticities are usually measured for only small price movements along
a demand curve, with different elasticities at each point along any straight
line demand curve. Demand is generally more elastic at higher prices and
more inelastic at lower prices.

Price elasticities of demand for farm commodities vary widely. For example, the price elasticity of demand at the farm level for cattle is − 0.68;
for calves −1.08; for eggs − 0.23; for vegetables − 0.10; and for wheat
− 0.03.[7]

## Factors That Influence Demand Elasticities

Three primary factors influence the elasticity of demand:

1. Whether good substitutes for the product are available.
2. Whether or not there are many alternative uses for a product.
3. Whether the product is an important expenditure in a consumer's total budget.

---

[7] George E. Brandow, *Interrelationships Among Demands for Farm Products and Implications for
Controls of Market Supply*, University Park, Pa.: Agricultural Experiment Station Bulletin 680,
1961.

The elasticity of demand for a product will tend to be greater the more substitutes that are available, the wider the range of uses of the product, and the more important the product is in the consumer's budget.

Most raw agricultural products have inelastic demands because there are few good substitute products for them. For example, different wheat varieties are used to make flour for bread, rolls, cakes, noodles, macaroni, and spaghetti. Some of these products could be made from flours ground from rye, corn, or barley, but in many cases the quality of such products is poor. Because of the poor substitutability of other flours for wheat flour, the price of wheat flour can be decreased or increased, prices of substitute flour unchanged, without causing consumers to shift rapidly to or from the use of wheat flour.

The greater the number of alternative uses for a commodity, the greater is its price elasticity. The demand for ground beef is relatively elastic because it can be used for such things as hamburgers, as a steak, mixed with extenders such a soybeans, or as an ingredient in a casserole. If the price of hamburger changes, large variations can occur in the quantity purchased.

The demand for automobiles, homes, furniture, televisions, and so on is elastic because they are large expenditure items and take a large portion of the family budget. When expenditures are large it tends to make consumers budget more carefully and shop for the best deal, taking into account the quality and price of substitute products. Therefore, a small change in price may be noticed by many consumers, causing a large change in the quantity taken.

So far, the discussion has been limited to a single demand curve and movements along that curve. Any movement along a given demand curve indicates a *change in quantity demanded* in response to a change in price. When the entire demand curve shifts for some reason, it is termed a *change* or shift *in demand*. (Notice the different phraseology here.)

## Changes in Demand

A change in demand means that at any given price a larger or smaller quantity will be demanded. In Figure 3-11, the demand for butter has increased as indicated by the shift from demand schedule *DD* to demand schedule *D'D'*. Notice that along the original demand curve *DD*, and a price of $0.80 per pound for butter, consumers purchased 10 pounds per week. After the demand curve shifted to demand curve *D'D'*, consumers purchased 20 pounds rather than 10 pounds per week even though the price remained unchanged at $0.80 per pound. Thus, when the demand increases, more units of a product will be sold at each price (or what amounts to the same thing, paying a higher price for the same quantity).

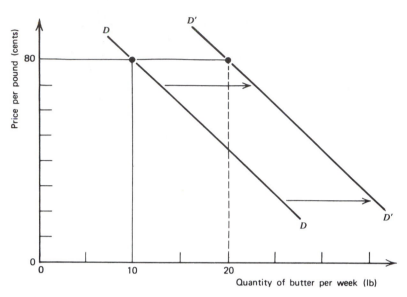

**FIGURE 3-11**   An increase in demand for butter.

Conversely, if demand decreases, or shifts to the left, fewer units of a product will be demanded at each price as shown in Figure 3-12. If consumers originally were purchasing 10 pounds per week at $.80 per pound, a decrease in demand to $D'D'$ causes them to buy only 5 pounds per week at that price. Again note that consumers would purchase 10 pounds, but only at a lower price.

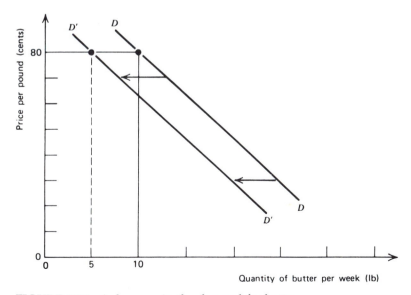

**FIGURE 3-12**   A decrease in the demand for butter.

Whenever a demand curve is drawn, it is assumed that many factors are held constant. When any of these factors change, a shift in the demand curve results. These factors generally include changes in consumer incomes, population, tastes and preferences, related product prices, and peoples' expectations.

As disposable incomes increase, consumers have more money to spend on goods and services, resulting in increased demand for these items, such as illustrated in Figure 3-11. A decrease in incomes due to such things as increased unemployment will shift the demand curve for most consumers to the left, as in Figure 3-12.

As the number of people increases, more housing, food, clothing, and services are needed. Therefore, as population grows so does the demand for these items (Figure 3-11). When population drops in an area, the demand for these same items decreases (Figure 3-12).

Related products are classified as being **substitutes** or **complements**. The effect of relative product price changes may be seen, using feed grains as an example. Barley and corn are substitute feed grains. If an increase in the price of barley occurs for some reason, the quantity of barley consumed will decrease and the demand for corn will increase as consumers of barley shift to the relatively cheaper feed grain.

If two products are used together, such as bread and butter, we call these products complements. An increase in the price of bread will reduce the quantity of bread demanded and consequently reduce the demand for butter because less butter is used as bread consumption falls.

Tastes and preferences change slowly. Many of our tastes are developed as a result of our cultural environment and may change over time because of experiences and education. Advertising and promotional efforts of business firms are an attempt to change our tastes and increase the demand for their products, and to make the demand curve more inelastic by developing brand loyalties. As people acquire less of a taste for a product, its demand curve will shift to the left. An example is the declining per capita consumption of fresh potatoes. At the same time, however, the per capita consumption of frozen potatoes is increasing.

Consumer expectations are difficult to handle in economics, but they play a significant role in market behavior. If the general price level is increasing and is expected to continue to increase, people will increase their current demand for products. They will buy more now in the expectation that prices will rise further. This response is especially true during periods of rapid inflation. On the other hand, if consumers expect prices to drop, they may postpone purchases until some future date.

## Income Elasticity of Demand

The relationship between changes in consumer income and quantity of an item purchased is called an *Engel curve*. As income increases more or less of a commodity may be bought. A normal good is one in which con-sumers buy more of it as income increases. An inferior good is one that consumers buy less of as income increases.

A different Engel curve exists for each commodity and for each indi-vidual. Let's use food as an illustration. The quantity of food purchased increases as income rises, but at a decreasing rate. Thus, the proportion of income spent for food decreases as income increases (Figure 3-13). Other items such as clothing can be characterized by an Engel curve rep-resented in Figure 3-14. The steepening curve shows that the quantity of clothing purchased changes substantially as money income rises.

*Income elasticity of demand* is defined as a measure of the responsive-ness of quantity of a good purchased with changes in income, holding all other factors constant. It can be expressed as the percentage change in quantity bought divided by the percentage change in income at any point along an Engel curve. When the income change is small, a rough esti-mate of the income elasticity of demand can be made using the follow-ing arc income elasticity formula:

$$E_1 = \frac{(Q_1 - Q_2)/(Q_1 + Q_2)}{(I_1 - I_2)/(I_1 + I_2)}$$

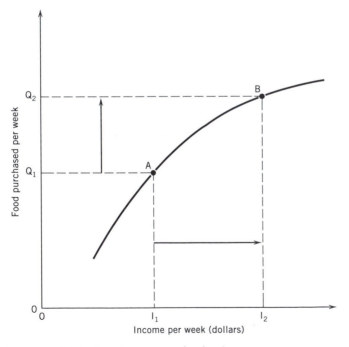

**FIGURE 3-13**  An Engel curve for food.

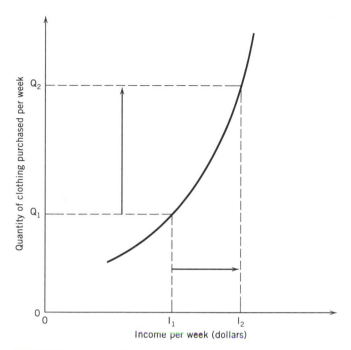

**FIGURE 3-14** An Engel curve for clothing.

The arc income elasticity coefficient can be calculated from the Engel's function shown in Figure 3-15 over segment *AB*.

$$E_1 = \frac{(10-30)\big/(10+30)}{(200-400)\big/(200+400)} = \frac{20\big/40}{200\big/600} = 2\big/4 \cdot 6\big/2 = 1.5$$

The income elasticity coefficient of steak is 1.5, meaning that a 1 percent increase in income results in a 1.5 percent increase in the quantity of steak purchased. Analyses such as income elasticity of demand are important in determining the impact of income changes on the purchases of farm food items. "The income elasticity for food in the aggregate, as well as for many individual food products, is thought to decrease as incomes increase."[8] Therefore, income elasticities will usually change over various income levels and can be positive or negative. Positive income elasticities indicate normal goods, and negative income elasticities indicate inferior goods.

_____

[8] William G. Tomek and Kenneth L. Robinson, *Agricultural Product Prices*, Ithaca, N.Y.: Cornell University Press, 1972, p. 31.

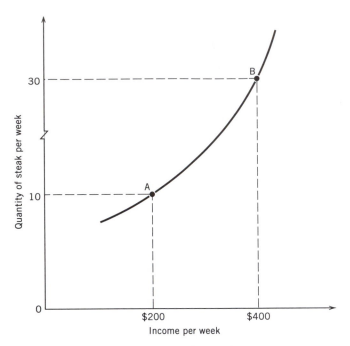

**FIGURE 3-15**   An Engel curve for steak.

## Cross Elasticity of Demand

Cross price elasticity of demand is another elasticity concept similar to price and income elasticities of demand. *Cross price elasticity* is a measure of the responsiveness of the quantity demanded of good X to a change in the price of good Y, holding all other factors constant.

Arc cross price elasticity of demand for commodity X with respect to a small change in the price of commodity Y can be illustrated with the following algebraic expression:

$$E_{xy} = \frac{(Q_{x1} - Q_{x2})/(Q_{x1} + Q_{x2})}{(P_{y1} - P_{y2})/(P_{y1} + P_{y2})}$$

If the cross elasticity coefficient from the calculated equation is positive, the two commodities (X and Y) are called substitutes. An example of substitute commodities are peanut and soybean oil. An increase in the price of soybean oil will decrease its quantity demanded and increase the demand and price of peanut oil. Therefore the increase in price of soybean oil and the increase in consumption of peanut oil are both positive, making $E_{xy}$ positive.

Commodities that have negative cross elasticities of demand are classified as being complementary commodities. Bread and butter provide an everyday illustration. If the price of butter increases, the quantity of butter demanded decreases and the demand for bread would also decrease. Hence, as the price of butter increased, the consumption of bread would decrease, giving a negative sign to the cross price elasticity between these two complementary commodities.

The cross price elasticity of pork with respect to the price of beef is around + 0.15. The interpretation of this coefficient is that the quantity of pork purchased will increase 0.15 percent for each 1 percent increase in the price of beef, *ceteris paribus*.

Commodities with high cross elasticities are very close substitute commodities, whereas coefficients close to zero show commodities that are unrelated. High negative coefficients represent strong complementary commodities.

# SUMMARY

The utility or satisfaction that a consumer receives from consuming goods and services is partly subjective in nature. We assume that human wants are insatiable. However, consumer's incomes are limited. Therefore, the consumer is attempting to purchase goods and services that will maximize satisfaction given that person's limited money income. Once the consumer is in equilibrium, it is possible to determine the quantities of a good purchased. By changing the price of a good, and hence the consumers budget constraint, additional price–quantity relationships can be found (review Figure 3-6). This price–quantity data can be plotted to derive the individual's demand curve for this product.

Individual demand curves for a product are added horizontally to obtain the market demand curve for a product.

Price elasticity of demand is a measure of the responsiveness of quantity demanded to a change in price, *ceteris paribus*. It is calculated over a small segment of a given demand curve. When the elasticity coefficient is less than one in absolute value, demand is inelastic; when greater than one in absolute value, demand is elastic; and when price elasticity is equal to one, demand is unitary.

If demand for a product is inelastic, price and consumer expenditures vary directly. An increase in price increases consumer spending. A decrease in price decreases consumer expenditures on the product. On the other hand, if demand is elastic, price and consumer expenditures on the product vary inversely.

Income elasticity of demand is a measure of the responsiveness of quantity of a good purchased to changes in income.

How products are related can be determined by the cross price elasticity of demand. Commodities with negative cross elasticities are complementary. Commodities with positive cross elasticities of demand are substitutes.

# CHAPTER HIGHLIGHTS

1. Consumers derive utility or satisfaction from the consumption of goods and services.
2. The concept of diminishing marginal utility is based on observations of consumers' market behavior. It states that as an individual consumes additional units of a commodity the amount of utility attained from each additional unit decreases, other things remaining equal.
3. An indifference curve shows all the different combinations of two goods that will give a consumer the same amount of satisfaction or utility.
4. The marginal rate of substitution is the rate at which one good can be substituted for another without changing the consumer's total utility.
5. Goods substitute for one another at a diminishing marginal rate because of diminishing marginal utility for each; it takes progressively more of one good to replace the satisfaction lost as successive increments of the other good are given up.
6. The budget line shows all the combinations of two goods that the consumer can buy with a given amount of money.
7. The consumer is at an equilibrium when an additional dollar spent on each good would return the same marginal utility per dollar, or $MU_a / P_a = MU_b / P_b$.
8. A change in relative prices will change the slope of the budget line and change the proportion of goods bought at equilibrium for the consumer.
9. Each tangency point between the budget line and an indifference curve identifies a point along the demand curve.
10. A demand curve is a schedule that shows, *ceteris paribus*, how many units of a good the consumer will buy at different prices for that good.
11. The market demand curve for a good is the horizontal summation of the demand curves of all individuals in the market for that good.
12. An individual's downward sloping demand curve for a commodity is caused by diminishing marginal utility.
13. A movement along a given demand curve is a change in quantity demanded.

14. Price elasticity of demand is a measure of the responsiveness of quantity demanded to a change in price, *ceteris paribus*.
15. The elasticity of a given demand curve for a product will depend on the availability of good substitutes for that product, alternative uses for the product, and the importance of the product in the consumer's budget.
16. A change in demand is a shift in the entire demand schedule.
17. Factors that will shift the demand curve are changes in consumer incomes, population, tastes and preferences, related product prices, and expectations.
18. Income elasticity of demand is a measure of the responsiveness of quantity purchased of a good to changes in income, *ceteris paribus*.
19. Cross elasticity of demand is a measure of the responsiveness of the quantity of one commodity to a change in price of another, *ceteris paribus*.
20. Substitute commodities have positive cross elasticities of demand.
21. Complementary commodities have negative cross elasticity of demand.

# KEY TERMS AND CONCEPTS TO REMEMBER

Budget line
Change in demand
Change in quantity demanded
Complements
Cross price elasticity
Demand curve
Diminishing marginal rate of
   substitution
Engel curve
Equilibrium

Income effect
Income elasticity of demand
Indifference curve
Indifference map
Law of diminishing marginal
   utility
Market demand curve
Price elasticity of demand
Substitutes
Substitution effect

# REVIEW QUESTIONS

1. Define utility. Is temperature measurable? Is utility measurable?
2. Indifference curves represent the tastes and preferences of a consumer. Can you add up indifference curves to show the tastes and preferences of a nation? Why, or why not?

3. Demand curves normally slope downward and to the right. Can you think of abnormal demand curves that have different slopes?
4. Given your knowledge of elasticity, why would legislators continue to raise the tax on products such as cigarettes and beer?
5. What is the difference between a change in demand and a change in quantity demanded?
6. Explain the factors that would shift a given demand curve.
7. Explain the meaning of a cross elasticity coefficient between bread and butter of −1.8.

## SUGGESTED READINGS

Awh, Robert Y. *Microeconomics: Theory and Applications*, Santa Barbara, Calif.: John Wiley & Sons, 1976, Chapter 6.

Goodwin, John W. *Agricultural Economics*, 2nd ed. Reston, Va.: Reston, 1982, Chapter 12.

Heyne, Paul T. *The Economic Way of Thinking*, 7th ed. New York: Macmillan, 1993, Chapter 2.

Leftwich, Richard H., and Ross D. Eckert. *The Price System and Resource Allocation*, 10th ed. Hinsdale, Ill.: The Dryden Press, 1987, Chapters 5 and 6.

Samuelson, Paul A., William D. Nordhaus and Michael J. Mandel. *Economics*, 15th ed. New York: McGraw-Hill, 1995, Chapter 22.

## AN OUTSTANDING CONTRIBUTOR

**Jean Kinsey**   Jean Kinsey is currently professor in the Department of Agricultural and Applied Economics and Director of the Food Retail Industry Center at the University of Minnesota.

Jean's agricultural background started in a creamery in Nelsonville, Wisconsin. Her father managed cooperative dairy plants in Wisconsin throughout his career. Jean taught senior high school home economics classes for several years in California after graduating with a B.A at St. Olaf College, Minnesota, in 1963. She received an M.S. degree in consumer economics from the University of California, Davis in 1965 and a Ph.D. in agricultural economics in 1976 at Davis.

Dr. Kinsey joined the Department of Agricultural Economics at the University of Minnesota after graduating with a Ph.D and was promoted to professor in 1982. She was a Resident Fellow at the National Center for Food and Agricultural Policy at Resources for the Future in 1986–87. She was appointed Director of Graduate Studies in the Department of Agricultural and Applied Economics from 1989–1992.

Dr. Kinsey's major research interests have been about consumer credit behavior, changing food consumption patterns, the economics of information and consumers' welfare loss, and the marginal propensity to consume food away from home when spouses work part-time and full-time. Some of her recent research projects include: examining the implications of food safety regulations (particularly pesticides) for consumer welfare and their potential to create non-tariff barriers to international trade, intergenerational transfers of time and money, farm and food lobbying impacts on farm incomes, and impacts of Indian gambling casinos on local county per capita income.

Dr. Kinsey's publications include 23 articles in professional journals such as the *American Journal of Agricultural Economics*, the *European Journal of Agricultural Economics, Food Policy*, the *Journal of Consumer Affairs*, and the *Journal of Consumer Research*, plus three books and ten chapters in books. Her book entitled *Food Trends and the Changing Consumer* (coauthored with Ben Senauer and Elaine Asp) received an honorable mention for quality of communication from the AAEA in 1992.

Most of Dr. Kinsey's teaching has been in consumer economics. At the University of Minnesota, she currently teaches Human Capital and Household Economics, The Economics of Food and Consumer Policy, and Intermediate Microeconomics of Consumer Behavior.

Dr. Kinsey has consulted with the U.S. Department of Agriculture, USAID, AT&T, the Minnesota Attorney General's Office, and various law firms and food companies. She is currently on the Executive Board of Directors and the Foundation Board of the American Agricultural Economics Association, and is Chair of the Board of Directors of the Federal Reserve Bank of Minneapolis. She was formerly President and Board Member of the American Council on Consumer Interests.

# CHAPTER 4

# Producer Decision Making: Single-Variable Input Functions

$\mathbf{T}$ he ultimate objective of all economic activity is the satisfaction of human wants, so any activity or process that satisfies a human desire (either directly or indirectly, presently or in the future) can be considered *production*. Viewed in this light, production is a process by which resources[1] are transformed into products or services that are usable by consumers.

Producers face a threefold decision problem: (1) what to produce; (2) how much to produce; and (3) how to produce. The subject matter of this chapter relates most directly to the producer's second question. But we concentrate on its other side by considering what happens when the quantity of a resource used in production is changed. Our approach here is to simplify the rate of resource use problem by directing our attention to the input–output relationship between a single resource and its product.

Production may be a many staged series of products, with the output from one process or stage being used as an input in a following stage until it reaches the form desired for final consumption. Such complexity is the rule rather than the exception. This complexity may be made clear by recognizing the many processes carried on before a slice of bread can find its way to your dinner table.

The farmer uses a wide variety of resources as raw materials to produce the wheat which is an input for the miller, whose product is an input to the baker, whose product is an input for the retail grocer, whose product is the consumable good, bread, an input in the act of consumption, valued for the utility or satisfaction it yields to you the final consumer. And this description has not bothered to identify the roles played by brokers and dealers, wholesalers, the transportation and financial industries, and many others involved in transferring products from one stage to another all along the route from the farmer to the final consumer.

It was stated earlier that scarcity creates the need to economize. Because there are too few resources to produce enough goods and services to satisfy all of our wants, we must economize on our resource commitments by choosing which of many alternative uses and combinations of resources will do the best job for us in meeting our wants.

Economists have defined a *resource* (an *input*) as a factor that can be used to produce a product that can satisfy a human want or desire. Because the number and variety of resources and the complexity of resource interrelationships defies mental comprehension, we are forced to

---

[1] These are also called "inputs" or "factors of production."

classify those resources and their relationships into generalized groupings.[2]

In the simplified example just mentioned of how resource services get transformed into consumable goods and services, each stage of production used a mixture of resources. At each step along the way decisions were made about which resources to use, quantities of each, and how much of the product to produce. At this point, we are unable to determine just how much each resource has contributed to the output and to the costs of obtaining the output. We need to establish the cause-effect relationships between the resources used and their product. This relationship is most clearly accomplished by grouping resources on the basis of similarities in their special characteristics.

## PHYSICAL RELATIONSHIPS ———————————

To make resource and product relationships as explicit and clear as possible, let's be a little extra careful for the moment. We can identify four basic categories of resources—land, labor, capital, and management[3]— each of which must be used in some combination with the other three before any product can be produced.

In the *land* group we put everything you ordinarily see in viewing the earth's surface. But there's more than this to our economic concept of land. We include not just the soil itself, but all its physical characteristics and all the natural environment that may influence the ability of land to yield a product.

Despite the fact that we frequently find labor and management in one and the same person, particularly in the single-proprietor firm, we reserve for *labor* the strictly physical act of performing a task; and for *management* the sole responsibility of decision making. Decision making includes the entrepreneurial functions of risk bearing, organizing resources into productive sets, deciding on which resources to use, their forms, and when and how much of each will be used in production.

---

[2] Because the universe is infinitely complex, and the human mind is finite, there is no way to study and understand the real world other than by classifying objects and things into groups that exhibit certain similarities. This abstraction from reality is fundamental to all science, not only to economics. By abstracting we reduce real-world problems to manageable proportions, making possible meaningful predictions—the basic goal of all science. For a more rigorous treatment of the scientific method, see, for example, George A. Stigler, *The Theory of Price*, New York: Macmillan, 1946.

[3] This grouping accords with the view of resource earnings held by firm operators, with the payment to *land* called "rent," the earnings of *labor* its "wage," the earnings of capital its "interest," and rewards to *management* being "profit."

In the one remaining group, *capital*, we toss every manufactured thing that can be used to aid or enhance production. Capital is inclusive of such physical things as buildings, machinery, brood stock, seed, equipment and tools, physically improved resources (e.g., land clearing, drainage, and leveling) that are made more productive as a consequence of that improvement, and any action by which current consumption of production is deferred so as to make resources more productive in the future.[4]

Common observation tells us that different quantities and combinations of these four resources will produce different amounts of the product. Despite their versatility, some resources are totally incapable of producing certain things: Given present technology, we don't find cotton being produced on the polar ice caps (land), nor would we expect a cement mixer (capital) to be of much use in polishing magnifying glass lenses. Within limits, however, most resources can be used to produce a variety of products and, in addition, many resources can be substituted for one another in production.

These characteristics, of relationships between resources and their products and between the resources themselves, are easily verified. That resources are productive can be demonstrated by changing the quantity of a resource used and observing that the quantity of product also changes. Further, whenever one resource is reduced in quantity and an increase in another resource prevents a decrease in the quantity of output, these two resources are substitutes for one another. These properties are basic to the theory of production and the economic decisions required as a consequence.

What these characteristics indicate is that output results from the particular set of resources used in some "functional" way. We call this

---

[4] Not all economists agree with such a rigid specification of types or classes of resources as this. Some will go only so far as a three-way classification (land, labor, and capital). The further we go in attempting to define specific attributes of individual resource categories, the more aware we must become of how one resource characteristic overlaps another. How, for instance, can we separate the original and indestructible gift of nature (land) from the improvements (capital) that have been made to the land resource over time by the succession of those who have managed and made use of it? The distinction between land and capital, in that resource, is made less useful, for some purposes, by the fact that the relative contribution of capital has been increasing through time. Similarly, the distinction between labor and management appears a bit artificial—even insulting to labor, one of the most important resources in any economy. Even in the simple act of digging a ditch, labor really is a combination of these two resource services, performing the physical work while also giving thought to where next to put the shovel point, how deeply, and so on. No special inference is intended here, we wish only to separate as clearly as possible the basic functions of each resource type so that later identification of resource relationships may be more easily understood.

the ***production function***.[5] We can represent the relationship symboli-
cally as

$$Y = f(X_1, X_2, X_3 \ldots X_n)$$

where $Y$ stands for the physical quantity of product, or output, the symbolism $f(\ )$ means "results from," "depends on," or "is a function of"; and
the $X$'s identify the different resources (inputs) used to produce that $Y$,
where $X_n$ refers to the last different input in the production function.

As we increase the amounts of the resources $X_1 \ldots X_n$, we find two general choices in their proportions, leading to two different results. Either
we increase them in the same proportion[6] and experience one kind of production response, or we change the resource proportions and get a completely different response.

## Constant Returns

If we consider increasing all inputs in a constant ratio, we could just as
well restate the function as $Y = f(X)$. Thus, if one unit of this input is composed of, say, 100 acres of land, 1 year of labor, $5000 of capital, and 1
month of management time, then two units must be *exactly* twice as much
of these *identical* resources, three units being three times as much, and so
on. And if one unit of this input $X$ yields 100 units of $Y$, then two units
of input $X$ can yield nothing other than 200 units of $Y$, three units of $X$
will produce 300 units of $Y$, and so on, resulting in a constant relationship between $X$ and $Y$, or ***constant returns***.[7] Figure 4-1A shows this relationship graphically, with the line of relationship (or function) labeled

---

[5] Much like demand and supply curves, the production function is also a "schedule"—it
shows how much output will be produced by a specific set of resources in a given period
of time and state of the arts (or technology). See C. E. Ferguson, *Microeconomic Theory*, rev.
ed., Homewood, Ill.: Richard D. Irwin, 1969, p. 116.

[6] If we don't do this, we can't even meet the implicit assumptions of graphics.

[7] Don't get trapped by "loose" thinking here. Each unit of our composite input $X$ is an exact duplicate (in *every* way) of every other unit of that resource. For instance, the $5000 capi-
tal component cannot be machine services in the first unit of $X$, fertilizer in the second, different types of seed in the third, pesticides or herbicides in the fourth. The land part of one
unit of $X$ cannot be "Grade A" land, for instance, with that in another unit being of a different quality, and so on.

At this level of abstraction, we haven't even the liberty of saying "But if you continue
doubling both $X$ and $Y$ you must eventually run out of space, if nothing else." Our definition of land includes the totality of the natural environment, including space that cannot be
conceived of as being finite and limiting as land is being increased. Thus depleting space
would set up a different situation in which our resource proportions are being changed, failing thereby to fulfill our requirement that resource proportions are not changed.

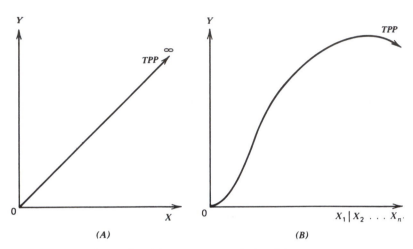

**FIGURE 4-1**   Generalized input-output relationships.

*TPP*, meaning ***total physical product***.[8] The relationship between the re-source and its product is also called a production function.

Now consider the other, more realistic choice open to us. As a produc-ing firm changes the quantity of its output, it will change one or more of the resources, but it is not very likely that the firm will (or even can) change all of them. If a farmer decides to increase the output of a par-ticular crop or livestock product, that may be accomplished by increas-ing the amount of labor used, acres of land devoted to that product, fer-tilizers, machinery, or some combination of these. But it is unlikely that all resources can be increased in the same proportion. Total land acres, buildings, or major machines, among other resources, would not neces-sarily be changed. If so, we then have some inputs that are *variable* inputs and others that are held constant, or *fixed*.

## Diminishing Returns

In the first instance, with everything varied and resource proportions constant, we are forced to conclude that the only result obtainable is a constant ratio of input to output. Here, because we are changing re-

---

[8] We use the label *TPP* to emphasize the fact that, at this point, we are dealing only with physical relationships, and not values. Further, we draw *TPP* as a continuous (smooth) curve by assuming that the input X is divisible into as small an increment as desired. This may seem a bit unreasonable at first, but consider the resource labor. One 8-hour day may be re-garded as a unit of labor, but that doesn't prevent the use of only 1 hour if that small an amount is desired. All other resource units, or their services, can be regarded as similarly divisible.

source proportions, another natural phenomenon that we cannot escape becomes obvious. We cannot avoid a *changing* input-output relationship that seems so absolute economists call it a "law," a consequence only of changing input proportions. This relationship can be represented by the function

$$Y = f(X_1 \mid X_2, X_3 \ldots X_n)$$

where the vertical bar is used to indicate that the inputs to its left are *variable* and those to the right of the vertical bar are held constant or *fixed*. This function is demonstrated graphically in Figure 4-1B, with the horizontal axis label indicating the general function.

The shape of the total product curve tells us what happens when we change the input proportions, making clear the physical basis of the **law of diminishing returns**[9]—the producer's exact counterpart of diminishing marginal utility in consumption. Our law states that as successive amounts of a variable input are combined with a fixed input in a production process, the total product will increase, reach a maximum, and eventually decline.

Let's use a hypothetical example to demonstrate this important principle. Assume a firm producing the product $Y$, using a resource $X_1$ that is variable, plus another set of resources $X_2 \ldots X_n$, all of which are held constant. Our production function then is of the form $Y = f(X_1 \mid X_2 \ldots X_n)$, with the data given in Table 4-1 and plotted in Figure 4-2.

Given rigid adherence to the specific requirements of the law of diminishing returns, we can consider our data as typical of real-world physical resource and product relationships. Plotting these data will result in a function that exhibits the characteristics stated in the law of diminishing returns.[10]

## Marginal Physical Product

Beginning with no $X_1$ used, as we add successive increments of the variable input, its physical productivity is low at first because there is too little

---

[9] Also referred to, with possibly a bit more clarity, as the "law of variable proportions," or "proportionality," strongly implying that changes in the shape of the *TPP* curve are the result of changes in the proportions between the resources used to produce that product.

[10] You might engage in an interesting bit of futility by trying to think of some real-life situation where the law of diminishing returns proves to be wrong, but don't forget its strict, limiting requirements.

**TABLE 4-1**   Hypothetical Data from the Function $Y = f(X_1 \mid X_2 \ldots X_n)$

| (1)<br>Variable<br>Input<br>($X_1$) | (2)<br>Total Physical<br>Product<br>(TPP) | (3)<br>Average Physical<br>Product<br>(APP) | (4)<br>Marginal Physical<br>Product<br>(MPP) |
|---|---|---|---|
| 0 | 0 | 0 | |
| | | | 7.5 |
| 10 | 75 | 7.5 | |
| | | | 17.0 |
| 20 | 245 | 12.3 | |
| | | | 19.0 |
| 30 | 435 | 14.5 | |
| | | | 12.5 |
| 40 | 560 | 14.0 | |
| | | | 8.8 |
| 50 | 648 | 13.0 | |
| | | | 6.2 |
| 60 | 710 | 11.8 | |
| | | | 4.3 |
| 70 | 753 | 10.8 | |
| | | | 2.9 |
| 80 | 782 | 9.8 | |
| | | | 1.8 |
| 90 | 800 | 8.9 | |
| | | | 1.0 |
| 100 | 810 | 8.1 | |
| | | | -0.2 |
| 110 | 808 | 7.3 | |

of the variable input relative to the fixed inputs.[11] Consequently, as we use greater amounts of $X_1$, we should expect its productivity to increase, causing the total product curve to bend upward. In this part of the function, total product is increasing at an increasing rate, as the data in Table 4-1 show.

At some point along the curve,[12] TPP will begin bending away from the Y-axis—increasing at a decreasing rate—because the variable resource has become more plentiful in relation to the fixed factors, and its

---

[11] There are few land-using examples in agriculture where the total product curve will actually begin at zero output when none of the variable input is present, because past management practices, resource substitutability, and the nature of the soil resource itself cause some of the varied input already to be in the soil. Agronomic experiments therefore make use of check plots with zero $X_1$, measuring "yield over check" to adjust for this.

[12] The point of steepest slope on the TPP curve (where TPP stops increasing at an increasing rate and begins increasing at a decreasing rate) is called the "inflection point"—labeled A in Figure 4-2.

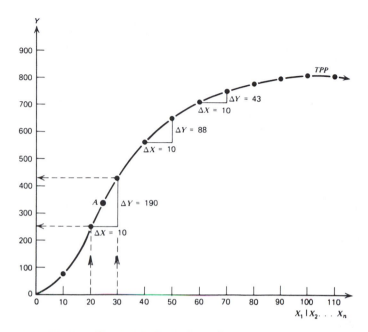

**FIGURE 4-2**  The total physical product curve.

productivity is relatively high.[13] Continued additions of $X_1$ will cause it to become excessive relative to the fixed inputs, losing more and more of its physical productivity because the fixed inputs become more and more limiting. Finally, using more $X_1$ can actually reduce output, as can be noted in the data in Table 4-1.[14]

Because economic decisions are based on additions or increments of the variable input, a more meaningful definition of diminishing returns is based on the marginal product of the variable input $X_1$, because it is an *incremental* measure of resource productivity. We define *marginal physical product* (MPP) as the amount added to total product when another unit of the variable input is used.

---

[13] Let's postpone until the section on two-variable input analysis where one of the fixed factors, $X_2 \ldots X_n$, may be used in different amounts from that set which generated the function we are using here.

[14] Excessive fertilizer application will burn up a crop, too much irrigation water will drown it out, or too many laborers doing a job will get in one another's way.

Note carefully that the law of diminishing returns is not a case of step by step sequential additions of $X_1$ but *either/or* choices in its application where we may use one unit of $X_1$, *or* two units, *or* three units. We cannot use one unit of $X_1$, then add a second, then a third, and so on.

The *MPP* curve is derived from the production function [column (4) in Table 4-1, and plotted graphically as in Figure 4-3] and is a measure of the *slope* of the *TPP* curve. The shape of the *MPP* curve more clearly describes the changing nature of the total curve. Total product increases at an increasing rate when *MPP* is increasing, increases at a decreasing rate when *MPP* is falling, reaches a maximum when *MPP* is zero, and falls absolutely when *MPP* is negative.

The formula by which *MPP* is derived can be shown as

$$MPP = \frac{\text{Change in output}}{\text{Change in input}} = \frac{\Delta TPP}{\Delta X_1} = \frac{\Delta Y}{\Delta X_1}$$

where the $\Delta$ symbol means change.[15] When *MPP* is computed in this manner we are able to determine by how much *TPP* changes whenever a change (either an increase or decrease) is made in resource use, at whatever level of $X_1$ use from which we measure that change. Our increments of $X_1$ (in Table 4-1) are in groups of 10, but the computed *MPP* is *per unit* of $X_1$. As a result, the figures in column (4) of Table 4-1 are placed midway between the $X_1$ observations because they are an average of those 10 units' MPPs; likewise, the plottings in Figure 4-3 are at the midpoints of the observations along the horizontal axis because the 10-unit groupings (and not just the tenth one in each increment of $X_1$) produced the change in output.

## Average Physical Product

One further useful derivation from the production function is the *average physical product* (*APP*), the formula being

$$APP = \frac{\text{Output}}{\text{Input}} = \frac{Y}{X_1}$$

The computed values [Column (3) of Table 4-1, and plotted in Figure 4-3] tell us how productive the variable resource is *on the average*, or *per unit* of $X_1$. If 20 units of $X_1$ produce 245 units of the product $Y$, the *APP* per unit of $X_1$ is 12.3, at 30 $X_1$ and 435 $Y$, *APP* is 14.5, and so on. As *TPP* increases, *APP* also increases, but only to the point along the *TPP* curve where *MPP* and *APP* are equal. From that point on, as $X_1$ is increased, *APP* falls, and becomes zero only when *TPP* becomes zero.

---

[15] We have used 10-unit rather than single-unit increments of $X_1$ in our production examples here (and 100-unit increments in our later cost analyses) to help the student remember the process of marginal computations. With single-unit changes in $X_1$, it is so easy to look only to $\Delta Y$ as the marginal product, forgetting that $\Delta Y$ is divided by $\Delta X_1$ to obtain *MPP*.

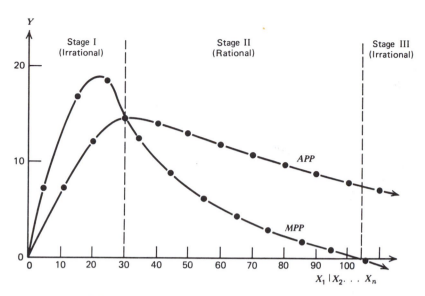

**FIGURE 4-3** Marginal and average product curves and the stages of production.

These two curves, *MPP* and *APP*, define three zones or stages of production, in only one of which (Stage II) can we find an optimum use of the input $X_1$. The two other stages (Stage I and Stage III) are both labeled *irrational* because, given our assumption that the firm's objective is to maximize its net earnings over the cost of $X_1$ used, the firm can increase its net revenue by moving into Stage II from either of the other stages.[16] It is a necessary condition that we know of the existence of our two irrational zones, but we can eliminate them from any further consideration since an answer to the economic question of how much $X_1$ to use, and how much product to produce, is to be found only in Stage II.

## VALUE RELATIONSHIPS _____

Up to this point our description has had a purely physical basis only. But economics is not concerned simply with the physical activities of

[16] We can make a preliminary (but positive) statement about Stage I that if it pays at all to produce in this stage it will pay even more to move into Stage II because the increment to output, whatever its value, is greater than the increment to cost. And in Stage III, the producer is incurring additional resource costs while at the same time reducing total output; valuable product is being thrown away while variable costs are increased. Thus, given profit maximization as the firm's objective, both Stages I and III are irrational.

producing something—that is a question of technical relationships. The economic concern is with economic feasibility—the manner in which human wants may be satisfied most efficiently, with economic criteria as the foundation for judging what is efficient. We need to be able to answer questions such as "What does it cost?" "Is it worth it?" "Is there a more profitable way of doing this?" We then are directed to comparing the values of products with the values of inputs used up in their production.

As a first step, a couple of assumptions will be useful and even a bit realistic: (1) There are so many firms producing this product that the actions of any one firm will have no influence whatsoever on either input or product prices; and (2) that the market does not differentiate one firm's product from that of another, that is, the firms produce a homogeneous product. Thus if a corn producer were to shut down completely or, alternatively, to produce the last possible extra bushel of corn, the market price of corn would not be affected. And provided that the corn meets certain quality standards, one producer's corn will not be discriminated against or offered a premium over that of other firms producing corn.

## Value Product

Now let's inject some prices into our production information. Given the data in Table 4-1, and assuming that the market price of the resource $X_1$ is \$5.00 per unit, and the market price per unit of $Y$ is \$1.00, what is the proper amount of $X_1$ to use so as to maximize profits? Table 4-2 provides the cost and value information to answer this question.

When we multiply the units of output (the *TPP* schedule) by the price of the product ($P_y$ = \$1.00), we obtain *total value product* ($TPP \times P_y = TVP$) in column (2). The function now shows dollars worth of output produced by the different amounts of the input $X_1$ used.

Dividing *TVP* by the units of $X_1$ used derives *average value product* (*AVP*), the average value of output per unit of $X_1$ at each level of use of that input—$AVP = TVP/X_1$.

*Marginal value product* (*MVP*) is derived by dividing the change in *TVP* by the change in the variable input, so we now are able to show how much additional value of output is produced by each additional amount of the variable input used by using the formula $MVP = \Delta TVP/\Delta X_1$.

## Marginal Factor Cost

We require a concept that measures costs at the margin in the same way that *MVP* measures output value at the margin. That measure is *marginal factor cost* (*MFC*). *MFC* is the amount that is added to total cost

**TABLE 4-2**  Functional Relationships (from Table 4-1) in Value Terms

| (1) Variable Input $(X_1)$ | (2) Total Value Product (TVP) | (3) Average Value Product (AVP) | (4) Marginal Value Product (MVP) | (5) Marginal Factor Cost (MFC) |
|---|---|---|---|---|
| 0 | $ 0 | $ 0 | | |
| | | | $ 7.50 | $5.00 |
| 10 | 75 | 7.50 | | |
| | | | 17.00 | 5.00 |
| 20 | 245 | 12.25 | | |
| | | | 19.00 | 5.00 |
| 30 | 435 | 14.50 | | |
| | | | 12.50 | 5.00 |
| 40 | 560 | 14.00 | | |
| | | | 8.80 | 5.00 |
| 50 | 648 | 12.96 | | |
| | | | 6.20 | 5.00 |
| 60 | 710 | 11.83 | $MVP = MFC$ | |
| | | | 4.30 | 5.00 |
| 70 | 753 | 10.76 | | |
| | | | 2.90 | 5.00 |
| 80 | 782 | 9.78 | | |
| | | | 1.80 | 5.00 |
| 90 | 800 | 8.89 | | |
| | | | 1.00 | 5.00 |
| 100 | 810 | 8.10 | | |
| | | | −0.20 | 5.00 |
| 110 | 808 | 7.35 | | |

when one more unit of the variable input $X_1$ is used. Because the market price of $X_1$ is $5.00 per unit, using another unit of that resource will add $5.00 to total costs. Therefore $MFC_{x1} = P_{x1}$.

An *optimum*[17] can now be determined by locating the point on the production function where one more unit of the variable input adds to revenue just what it adds to costs, that is, where $MVP = MFC$. Because $MFC$ is $5.00, we must find a $5.00 point in the $MVP$ schedule. That occurs at approximately 60 units of $X_1$.

If we were to use 70 $X_1$ instead, our value product per unit for the 10 additional units of $X_1$ would be $4.30. But they cost $5.00 each, so we would be losing $0.70 on each of them, a loss that can be avoided simply by using 60 rather than 70 units of $X_1$. And if we use only 50 units of $X_1$, the last 10 units of that resource added $6.20 each, which is $1.20 more than they cost, meaning that profit can be increased by using more $X_1$.

---

[17] By *optimum* we mean that one rate of $X_1$ use which yields the highest net return over the cost of $X_1$, given the fixed resource set.

**TABLE 4-3** Checking the Net Revenue Productivity of the Variable Input $X_1$[a]

| Variable Input ($X_1$) | Total Revenue (TR) | Total Cost of $X_1$ ($TCx_1$) | Net Revenue ($TR - TCx_1$) |
|---|---|---|---|
| — | — | — | — |
| — | — | — | — |
| — | — | — | — |
| 50 | $648 | $250 | $398 |
| 60 | 710 | 300 | 410  max. NR |
| 70 | 753 | 350 | 403 |
| — | — | — | — |
| — | — | — | — |
| — | — | — | — |

[a] This solution, using discrete numbers, will only accidentally determine exactly the same optimum as with calculus, or as with graphic analysis, such as in Figure 4-4.

Our optimizing rule, of finding the point where $MVP = MFC$, said that by using 60 units of $X_1$ we would maximize the net difference between revenue and costs. Let's check that in another way, as in Table 4-3.

Relabel our TVP schedule from Table 4-2 as TR (total revenue), develop a $TC_{x1}$ column (meaning the total cost of $X_1$) by multiplying the amount of $X_1$ used by $5.00, the price of a unit of $X_1$. Then subtract these cost amounts from the corresponding revenue figures to get net revenue (NR),[18] the amount by which revenue exceeds cost incurred for the variable input. Net revenue is maximized when 60 units of $X_1$ are used, which yields a TR of $710. If you compute the other values at both lesser and greater amounts of $X_1$ you will find that NR continues to fall in both directions away from the $410 maximum.

To graph our information, as in Figure 4-4, multiply the Y-axis physical unit scale by $1.00, which gives dollars worth of output. We still maintain physical units of $X_1$ on the X-axis because we want to know what happens to costs and returns as different physical units of that resource are used.

The TVP curve maintains the same shape as the TPP curve we started with because it has been multiplied by a constant factor, $1.00, the price of a unit of output. The market price of Y does not change as a result of decisions made by this firm because of the competitive market condition

---

[18] This is nothing more than a net over the cost of the variable input, but it still is the only relevant indicator in deciding how much $X_1$ to use, as will be discussed later. We can't call this figure profit because we have not identified the costs of the other (fixed) resources used to produce that income. However, at the point where net revenue is maximized, profit is also maximized, because fixed costs will not affect the optimal output and rate of resource use.

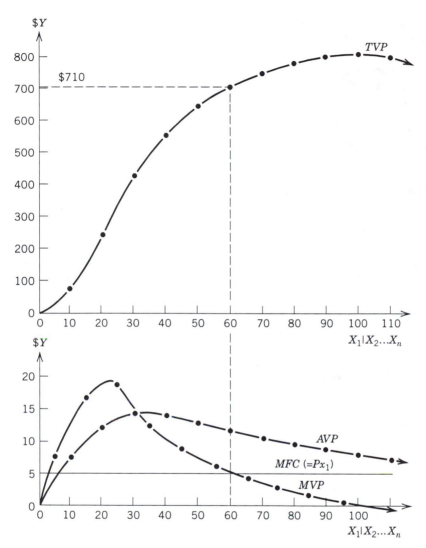

**FIGURE 4-4** Graphic determination of an optimum.

we have assumed. And the same holds true for the shapes of the *MVP* and *AVP* curves because they are derived from the *TVP* curve.

Likewise, the price of the variable input $X_1$ does not change as a result of the producer's decisions either to use none of the resource, the maximum amount, or some amount between these extremes. Because of this market condition, we can graph *MFC* as a horizontal line with $P_{x1} = \$5.00$.[19]

———————————————————————

[19] We can say that $MFC = P_{x1}$ because $MFC = (\Delta TC/\Delta X_1) = (\Delta X_1 \times P_{x1})/(\Delta X_1) = P_{x1}$.

We now have a graphic picture of the search for the optimizing point where $MVP = MFC$. Given our assumed prices, $MVP = MFC$ wherever the two lines cross. But this is true at two points along the $MVP$ curve, so let's dispose of the one at less than 10 units of $X_1$ by adding the requirement that $MVP$ must be declining and less than $AVP$, which forces us into Stage II. Drop a vertical line (from the point where $MVP$ and $MFC$ are equal) to the horizontal axis, and it shows 60 $X_1$, carry that vertical line up to the $TVP$ curve and over to the $Y$-axis, and it shows \$710. This is the optimum quantity of $X_1$ to use.[20]

## Adjusting to Price Changes

Our assumption about unchanging prices was valid regarding individual firm decisions, but that doesn't mean that we are unable to cope with changing resource and product prices, or that we even expect them to remain unchanged. Change is a fact of life, in markets as in anything else. Changing economic conditions, in total market supply and demand cause frequent price adjustments in both resource and product markets. So our optimum is correct, not for all time, but only until another price change occurs, then an adjustment in $X_1$ use must again be made to find a new optimum.

It would seem obvious that as the price of the product increases, producers should increase their output to take advantage of the price change, and to reduce output when product price falls. Possibly not so clear, however, are the necessary changes in resource use that must be made as the price of the resource changes. Let's look at these two types of price changes, first to see why a product price change should cause the volume of output to be changed.

We need to go back to Table 4-2 (and Figure 4-4) and ask what happens to the $TVP$ curve if $P_y$ is \$2.00, say, rather than \$1.00. With product price doubled it should be clear that $TVP$ will be twice as large at all levels of $X_1$ used. With that new $TVP$ curve, $MVP$ must also have increased. Its zero points (at $X_1 = 0$, and near 100 $X_1$) cannot change, but all other points will be twice as high as previously. Given $P_{x1} = \$5.00$, $MVP$ now must equal $MFC$ at a higher level of $X_1$ used, that is, at about 80 units. The optimal output is now \$1564 worth of $Y$, caused both by the product price increase and the greater amount of $X_1$ now being used. The price rise to \$2.00 per unit of $Y$ would increase $NR$ to \$1120 with no change in the quantity of $X_1$ used. But when $X_1$ is increased to 80 units, $NR$ is increased to \$1164.

---

[20] The increments in our production function do not permit a more accurate estimate of $X_1$ use than to say about 60 units, despite the fact that a mathematical derivation might result in exactly 62.5 $X_1$ as optimal. $MVP$ is \$6.20 between 50 and 60 $X_1$, and \$4.30 between 60 and 70. An $MVP$ of \$5.00 falls somewhere between these two observations, closer to \$4.30 than to \$6.20, and therefore something more than 60 units of $X_1$, but that degree of accuracy is beyond our needs here.

If $P_y$ were to fall to 50 cents, with $P_{x1}$ unchanged, optimal input use would drop to about 40 units with NR at \$155. And if $P_y$ were to fall even lower (about \$.34) so that maximum $AVP$ falls to less than \$5.00, the firm would cease production altogether. At maximum $AVP$ less than \$5.00, the firm would find no point in Stage II where $MVP$ = $MFC$, so no matter what its fixed resource losses might be, there would be an additional loss on the variable resource that just makes a bad situation worse.[21]

## The Firm's Demand for Resources

The effects of resource price changes on the rate at which resources are used are more easily noted than are the effects of product price changes. Begin with a price for $X_1$ of \$19.00 per unit. Equating $MFC$ and $MVP$ results in approximately 25 units of $X_1$ used. However, 25 units of $X_1$ cost \$475 and produces a $TVP$ of about \$325,[22] leaving a net loss of \$150. Obviously this situation is not feasible. Given $P_y$ = \$1.00, we can pay no more than \$14.50 for input $X_1$ (its maximum $AVP$) and even that price will leave a zero net return over the cost of $X_1$. As the price of input $X_1$ falls below \$14.50, the proper amount of $X_1$ to use is still found by equating $MVP$ and $MFC$, with the results shown for selected price–quantity observations in Table 4-4 and plotted in Figure 4-5.

What we now have is a set of data that shows how the use of a resource must be changed as its price changes. We, therefore, have price–quantity data just as explained in the chapter dealing with the consumer's demand curve, and that is exactly what it is—the firm's *demand* for the variable resource. The firm's demand curve for $X_1$ then is the $MVP$ of $X_1$ within Stage II. There is a maximum price of $X_1$ (\$14.50 and 30 $X_1$ used) above which the firm will quit using the resource entirely, and a maximum amount of $X_1$ (100 units) that can profitably be used even when the resource is free[23] to the user, with an optimum use of $X_1$ to be found at all possible prices between these two extremes. The firm's response to price changes thus fits the specifications of the law of demand—"quantity demanded varies inversely with price,"[24] as shown by the curve in Figure 4-5 labeled $D$.

---

[21] We will defer, until the chapter on costs, any consideration of what resource market price changes can be expected, and what alternatives may still be open to the owner of the fixed resources, should the product price fall so low.

[22] At 25 units of $X_1$ used, $AVP$ is \$13, so $TVP$ = \$13 × 25 $X_1$ = \$325.

[23] Even if $X_1$ were to be a free good, we could afford to use no more than about 100 units because nothing would be added to $TVP$ by employing a greater amount of $X_1$. Using more than 100 $X_1$ actually reduces $TVP$ (a phenomenon of Stage Ill) and is irrational.

[24] With the marginal productivity of a resource related to its price as the basis for the proper rate at which to use that factor as an input, what does this suggest to you regarding the consumer's use of a good to produce satisfaction (utility), and the comparable marginality of such use?

**TABLE 4-4**  Deriving the Firm's
Demand for an Input

| Price of Input $X_1$ | Quantity of Input $X_1$ Used |
|---|---|
| — | — |
| — | — |
| $14.50 | 30 |
| — | — |
| — | — |
| 10.00 | 40 |
| — | — |
| — | — |
| 5.00 | 60 |
| — | — |
| — | — |
| 2.00 | 80 |
| — | — |
| — | — |

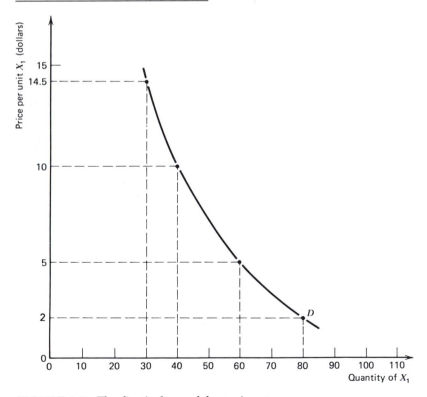

**FIGURE 4-5**  The firm's demand for an input.

To this point we have been discussing the simplest of functions, specifically: (1) a single-input function with nothing fixed, and its constant returns; and (2) a single-variable input function with fixed factors, and its diminishing returns.

Our analysis of the economics of production has necessarily been somewhat abstract, but this abstractness does no damage whatsoever to the validity or applicability of the concepts presented here. Note the simplifying (and correct) logic of the method—if this analysis is true using one exemplifying input as the variable, with all others held constant, it must also be true for all other productive resources.

You should recognize that, no matter what separately identifiable resource each different $X$ might represent, any one of those $X$'s could be used as the variable input. The farmer who may be thinking about buying a nearby tract of land as an add-on, but hoping to get by without increasing other resources too (e.g., more or larger machinery, or more labor, etc.), really is looking at that tract of land as the variable resource $X_1$ with all other inputs held constant. A rancher who is considering developing a tract of irrigated hay land may be dealing with a complex variable input by virtue of varying not only the capital investment, but more haying equipment may also be needed, with even an increase in the livestock herd size to take best advantage of the greater amount of forage to be available, with yet other resources held constant.

# SUMMARY

The major concern in this chapter has been the basic principles underlying (economically) proper choices. We have approached the question of how much of any variable resource to use within the framework of a single-variable input function.

Given the set of resources a firm uses to produce a good, it is convenient to classify these resources into four categories: land, labor, capital, and management. Then as we change any one of these resources we expect a corresponding change in the quantity of the product. The law of diminishing returns describes the effect on total output when the amount of a variable input is changed while all other resources are held constant. The production function is descriptive of this relationship.

As the price of a variable input falls, or its product price rises, it would appear logical that greater amounts of that resource would be used to increase the quantity of product produced; opposite changes should cause a reduction in the rate of resource use.

Such changes are based on the relationship between costs and values of an additional—marginal—unit of a variable resource. When the value output ($MVP$) of a marginal unit of that resource is greater than its cost

(*MFC*), it pays more to use that unit than to not use it. Profit is increased (or losses decreased) by using that unit because more is added to revenue than is added to cost. If, however, *MVP* is less than *MFC*, it pays to cut back the use of that resource. Profit is greater (or losses less) when that marginal unit is not used because costs are reduced more than is revenue by that reduction in resource use.

As we relate the quantities of a variable resource to different prices for that resource, we can note the special conditions in the demand for any good or resource. The higher its price, the less of any good or resource that will be used, and conversely. Plotting the pairs of price–quantity observations identifies the firm's demand for a variable resource. The Stage II portion (only) of the *MVP* curve of any variable resource is the firm's demand curve for that resource.

# CHAPTER HIGHLIGHTS

1. Resources, production, and goods and services are words or terms with economic meaning. In the process of using resources to produce goods and services, utility is created because human wants can be satisfied with them.

2. The universe of things and their relations to one another are far too complex for the human mind to grasp without a systematic abstraction and classification.

3. The relationship between resources and their product is called a production function.

4. A constant input–output relationship results when we do not change the makeup of the input unit formed from the four basic resources, land, labor, capital, and management. This relationship may be summarized with the general function $Y = f(X_1 \ldots X_n)$, or simply $Y = f(X)$.

5. Diminishing returns are the inevitable consequence of changing proportions among the inputs used to produce something, a relationship described by the general function $Y = f(X_1 \mid X_2, X_3, X_4)$.

6. The law of diminishing returns states that "as successive amounts of a variable input are combined with a fixed input, the total product will increase, reach a maximum, and eventually decline."

7. Marginal physical product is the amount that is added to total product when another unit of the variable input is used. *MPP* measures the rate of change in the input–output relationship and is obtained by dividing the change in total physical product (*TPP*) by the causal change in the variable input.

8. Average physical product is a measure of the average productivity of the variable input. *APP* is obtained by simply dividing total output by the number of units of the variable that produced the product.

9. Taken together, *MPP* and *APP* define three stages of production in only one of which (Stage II) will an economic optimum be found.

10. When output is multiplied by the price of the product we have total value product—$TPP \times P_y = TVP$. We then derive *MVP* and *AVP* in the same way as their physical counterparts were derived:

$$MVP = \frac{\Delta TVP}{\Delta X_1} \quad \text{and} \quad AVP = \frac{TVP}{X_1}$$

11. The amount that is added to total cost, when another unit of the variable input is used, is called marginal factor cost—*MFC*—which is the market price of that resource, so $MFC = P_{x1}$.

12. Knowing both returns and costs at the margin permits us to determine the optimal rate at which to use the variable input. The optimum is determined by equating *MVP* and *MFC*. Too little of the resource is being used if *MVP* is greater than *MFC*; the opportunity to capture that excess of value over costs is needlessly forfeited. If *MVP* is less than *MFC*, the last unit of output costs more to produce than it is worth, and too much of the variable input is being used.

13. The *MVP* curve within Stage II is the firm's demand curve for the variable input, observed by noting the optimal quantity of the input used at each different price for that resource.

# KEY TERMS AND CONCEPTS TO REMEMBER

Average physical product
Average value product
Capital
Constant returns
Input
Labor
Land
Law of diminishing returns
Management

Marginal factor cost
Marginal physical product
Marginal value product
Production
Production function
Resource
Total physical product
Total value product

# REVIEW QUESTIONS

1. We could list separately identifiable productive factors in each firm by the hundreds, and more. What good reasons could there be for grouping them into four categories?
2. Discuss the meaning of a production function.
3. What causes conditions to be so absolute that we summarize them as the law of diminishing returns?
4. Is the labor used by a cotton farmer during growing season tillage operations the same resource as labor for harvesting?
5. "Obviously, a farmer should handle all the land within the farm in such a way that crop yields per acre are maximized." Comment.
6. On page 92 the effect of changing the price of the product was described. With Table 4-2 as a starting point, recompute *TVP* and *MVP* using the two additional product prices discussed ($2/unit, and $.50/unit). Construct a graph and plot these *MVP* curves, as well as the *MVP* curve when $P_y$ = $1.00/unit. What is the optimal rate to use $X_1$, at each of these prices, given $P_{x1}$ = $5.00/unit? Do these adjustments in response to product price changes make sense?

# SUGGESTED READINGS

Bishop, C. E., and W. D. Toussaint. *Introduction to Agricultural Economic Analysis*, New York: John Wiley & Sons, 1958, Chapters 4–6.

Bradford, Lawrence A., and Glenn L. Johnson. *Farm Management Analysis*, New York: John Wiley & Sons, 1953, Chapter 8.

Brehm, Carl. *Introduction to Economics*, New York: Random House, 1970, Chapter 5.

Ferguson, C. E. *Microeconomic Theory*, rev. ed. Homewood, Ill. : Richard D. Irwin, 1969, Chapter 5.

Leftwich, Richard H. *Introduction to Microeconomics*, New York: Holt, Rinehart & Winston, 1970, Chapter 15.

Peterson, Willis L. *Principles of Economics: Micro*, 8th ed. Homewood, Ill.: Richard D. Irwin, 1991, Chapter 4.

# AN OUTSTANDING CONTRIBUTOR

 **Richard E. Just**   Richard Just is currently professor of agricultural economics in the Department of Agricultural Economics and Resource Economics at the University of Maryland. He was born and raised on a 1000-acre wheat and cattle farm near Tulsa, in northeastern Oklahoma. He managed the family farm through his high school years and continues to maintain a financial interest in the farm and related real estate development associated with urbanization around the farm. He earned his B.S. degree in agricultural economics at Oklahoma State University in 1969. Just received his M.A. degree in statistics in 1971 and his Ph.D. in agricultural economics in 1972 at the University of California, Berkeley.

Just's Ph.D. dissertation was published as a Giannini monograph that was selected for the Outstanding Published Research Award in 1975 by the Western Agricultural Economics Association. He was awarded this same honor again in 1983 for his research on multicrop production functions.

His research has focused on the economics of risk in agriculture, welfare economics of government policy, international trade, and futures markets. His research on international trade received the Quality of Research Discovery Award in 1977 from the American Agricultural Economics Association, and his research on the futures markets was selected for their 1981 Outstanding Journal Article Award.

He has published numerous journal articles in more than 35 refereed journals and contributed chapters to more than 20 books. His expertise has been utilized by many national and international agencies, including the World Bank, U.S. General Accounting Office, Winrock International, the Electric Power Institute, Safeway Stores, Inc., and Development and Resources, Inc.

Just served as editor of the *American Journal of Agricultural Economics*, on the Editorial Council of both the *American Journal of Agricultural Economics* and the *Western Journal of Agricultural Economics*, and on the Editorial Board of the *Journal of Development Planning Literature*.

He is married to the former Janet Lee Humphries.

# Producer Decision Making: Two-Variable Inputs and Enterprise Selection

$I$n the previous chapter we discussed the simplest production functions, specifically: (1) a single, composite, variable input function with nothing fixed, and its constant returns; and (2) a single-variable input function with other resources held constant. This functional relationship between a variable factor and its product is sometimes referred to as the *factor-product relationship.*

In deciding what to produce and how to produce it, the decision maker is faced with two other choices. Few resources, if any, are so limited in their adaptability that they can only be used to produce one thing. This flexibility means that a choice has to be made as to the proportions in which resources should be combined to do the best job of producing the product. Such a choice deals with the relationships among resources, or the *factor-factor relationship.*

Another choice facing the producer is what enterprise or combination of enterprises will be the most profitable for the firm.[1] Agricultural production decisions are seldom so simple that the operator can choose a single crop or single livestock enterprise as the firm's only product. Crop farms typically produce two or more different types of grain crops; stock ranches frequently produce both livestock and grain or forage crops, or even two or more types of livestock. The relationship between enterprises is referred to as the *product-product relationship.*

These two general problem areas—*factor–factor* and *product–product*—and the economic criteria for making these choices, are dealt with in turn in this chapter.

## TWO-VARIABLE INPUT FUNCTIONS _____

The simplified production function discussed in Chapter 4, using only one variable input, served as a useful beginning in developing an understanding of the economics of resource use. Numerous decision questions can be found where the problem is that simple. But what happens when two or more variable inputs are used? We still are able to classify specific resources into groups as we did earlier, developing two distinctly different types of functions: a two-variable input function with nothing fixed, of the general form $Y = f(X_1\ X_2)$; and a function with two or more inputs varied and one or more inputs fixed, of the general form $Y = f(X_1\ X_2|X_3\ X_4)$. In both these general cases, we face a factor–factor choice of the pro-

---

[1] An enterprise is a specific crop or type of livestock from which products are obtained, for example, cotton, wheat, beef cattle, hogs, or onions.

portions in which to use the variable resources. Physical relationships and economic factors cause producers to choose one particular set of resources and not some other combination with which to produce their products. Basic in this choice is the interaction between inputs and their effect on the productivity of the employed resources as the input mix is changed.

## Isoproduct Contours

Of great practical and economic importance to producers is the fact that different resource combinations are capable of producing a given quantity of output. With a two-variable input function, reducing the quantity of one resource will reduce output and also change the *MPP*s of those inputs, *ceteris paribus*. But the producer doesn't have to accept a reduction of output as the only possibility, because that loss of output may be regained by a compensating increase in the quantity of the other input. The wide variety of such substitution possibilities, for just one level of output, is illustrated in Figure 5-1. Different proportions of the inputs $X_1$ and $X_2$ can be used to produce a given amount of output, such as $Y_n$. Point $A$ in the diagram shows the quantity $Y_n$ being produced by a small amount of input $X_1$ and a large amount of input $X_2$. That same quantity of output may also be produced with smaller and smaller quantities of $X_2$, but

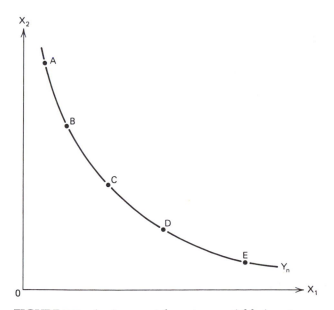

**FIGURE 5-1** An isoquant from two variable inputs.

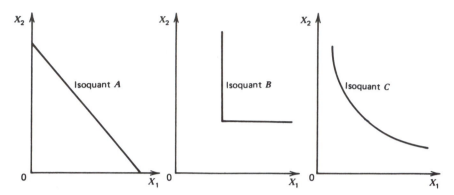

**FIGURE 5-2**   Isoquants for perfect substitutes (*A*), perfect complements (*B*), and imperfect substitutes (*C*).

only if compensating increases in $X_1$ are used, as shown by the points *B*, *C*, *D*, and *E*. A curve drawn through these points shows all the combinations of $X_1$ and $X_2$ that can be used to produce the given output, $Y_n$. This curve, called an *isoquant*,[2] identifies all points in the graph space with the same quantity of output. An isoquant thus shows all the combinations of these inputs that can be used to produce a given quantity of output.

## Input Substitution

A characteristic of resource relationships with special economic meaning is *resource substitution*. Resources are able to substitute for one another when the use of one resource can be increased as a replacement for another reduced in amount and still yield a given amount of product. The ease or difficulty of substituting one resource for another is made apparent by the shape of the isoquant, with its shape determined by the rate at which resources substitute for one another. Three basic types of resource relationships are discernible: (1) perfect substitutes, (2) perfect complements, and (3) imperfect substitutes.

*Perfect substitutes* are just that, perfectly able to replace one another without affecting output. The isoquant for perfect substitutes must be a straight line, as in Figure 5-2, Isoquant *A*. An example of perfect substitutes would be in choosing between 20 percent nitrogen as $X_1$ and 40 percent nitrogen as $X_2$. Disregarding the effect that may exist because of the

---

[2] Combined from the Greek *isos* (meaning equal) and *quant* (from quantity) to mean "equal quantity." Also variously referred to as an isoproduct contour or curve, equal output contour, product-indifference curve, all having the same meaning. Note the similarity between this curve and the one in Figure 3-1. It is the producer's exact counterpart of the consumer's indifference curve.

greater bulk of inert carrier in 20 percent nitrogen, it will substitute for 40 percent nitrogen at a constant 2:1 ratio.

Assuming no quality differences, irrigation water from one well ($X_1$) would substitute for irrigation water from another well ($X_2$) at a constant rate. Many other examples could be used. The firm's decision as to which resource to use with such a substitution relationship is easily made: use the cheapest one.

Resources that are *perfect complements* leave no room for choice in the proportion of their use. Perfect complements (Isoquant B) must be used in a technically prescribed ratio. An additional amount of one resource or the other will add nothing to total product, so the decision problem is one of whether or not to use the *pair* of resources rather than their *ratio* of use.

An example of (nearly) perfect complements is a tractor and plow. A tractor with plow can turn so much soil in a given period of time. Adding more tractors will not increase output, nor will output be increased if more plows are added to your tractor.[3]

The third resource relationship is probably more typical of the kind of choice problem faced by the agricultural producer—the relationship categorized as *imperfect substitutes*. When resources are imperfect substitutes for one another, successive incremental reductions of $X_2$ will cause output to fall unless increasingly larger offsetting amounts of $X_1$ are added, as indicated by the curved isoquant (C) in Figure 5-2. Such isoquants are concave from above (convex to the origin), indicating that it takes larger and larger amounts of one resource to replace equal incremental reductions in the other resource, maintaining output at some given level.

For instance, one can use less and less land to produce a product, if the sacrifice of land is offset with more intensively applied capital, but continuous substitutions of capital for land would eventually be unable to maintain output.

Poultry producers use capital as a substitute for land when they build confinement facilities that reduce the need for land area. The Western wheat farmer substitutes land for capital by using a larger land area rather than develop irrigation (or other capital intensive) facilities to produce that product.

Other substitution possibilities exist among the four basic resources, or among innumerable types or kinds of resources within each category. The limit of such substitution is where the *MPP* of a resource becomes zero. Thus no matter how much effort one might make, further additions of that resource could not prevent a reduction in output.

---

[3] Providing the tractor/plow technical unit was already proper as to load, speed of travel, quality of plowing done, and so on. Or maybe you'd prefer this example (no special mechanical adjustments, please!): right ($X_1$) and left ($X_2$) drive wheels on a standard two-drive-wheel tractor.

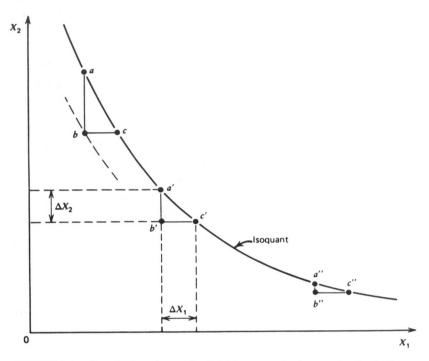

**FIGURE 5-3**   Graph showing a diminishing marginal rate of substitution.

We are describing a specific characteristic of the *rate* at which resources substitute for one another, or the ***marginal rate of substitution***.[4] The marginal rate of substitution of $X_1$ for $X_2$, ignoring sign, is indicated by the formula

$$MRS_{x1x2} = \Delta X_2 / \Delta X_1$$

which can be read to be the number of units of $X_2$ that a unit of $X_1$ can replace without changing output,[5] and is, therefore, a measure of the *slope* of the isoquant. Since the isoquant for imperfect substitutes is a curving line, we face a *diminishing* marginal rate of substitution.

Viewed on a graph (Figure 5-3), a change in either input can be shown as a straight line in the direction of that change, such as the line segment

[4] Some prefer to refer to this as the *marginal rate of technical substitution* (MRTS), to indicate the technical nature of resource use in production and to distinguish this from the same phenomenon in consumption where one good is substituted for another, the measure there referred to as *MRS*. We prefer using *MRS* with the subscripts identifying whether the substitution is in production or consumption, because they are one and the same, having exactly the same meaning in their respective usages, thus maintaining their close identification.

[5] The *MRS* for perfect substitutes would be a constant, for perfect complements it is zero, and for imperfect substitutes a number varying in magnitude.

from point *a* to point *b* showing a reduction in the amount of $X_2$ employed. Reducing $X_2$ without a compensating increase in $X_1$ will drop output to a lower isoquant (that runs through point *b*), the amount of that drop being the marginal product of that change in $X_2$, or $\Delta X_2$ times $MPP_{x2}$. An increase in the use of input $X_1$ equal to the distance *b*-to-*c* (measured along the $X_1$ axis) will increase output by $\Delta X_1 \cdot MPP_{x1}$. Notice that as we have started in the upper left portion of the isoquant we have a high ratio of $X_2$ to $X_1$. There $MPP_{x2}$ is relatively low and $MPP_{x1}$ is relatively high, meaning that a given increment of $X_1$ replaces a relatively large amount of $X_2$. As we continue to add more and more $X_1$, its *MPP* falls (just as shown earlier in Figure 4-3) while the *MPP* of $X_2$ increases. Thus the vertical length of any line *a* – *b* gets shorter and shorter (compare with *a'* – *b'*, and *a"* – *b"*), the reason for calling it *diminishing MRS*.

Because these moves are being made *along* the isoquant, output is constant and any such move can be stated as $\Delta X_1 \cdot MPP_{x1} = \Delta X_2 \cdot MPP_{x2}$. As we change input $X_1$ (in the amount $\Delta X_1$), output quantity changes by the *MPP* of that change ($MPP_{x1}$), and similarly for compensating changes in the input $X_2$ with its resulting output changes. With the slope of the isoquant equal to $\Delta X_2/\Delta X_1$, these terms can be arranged to $\Delta X_2/\Delta X_1 = MPP_{x1}/MPP_{x2}$, showing that the slope of the isoquant ($MRS_{x1x2}$) is also equal to $MPP_{x1}/MPP_{x2}$, the ratio of their marginal products.

## Isocost Line

If a given amount of output is to be produced, we obviously can select any input ratio along the output contour so long as that contour line has a slope between zero and infinity, or between the horizontal and vertical. But of all those possibilities only *one* ratio can be best, or optimal, in terms of costs and returns. Any ratio other than that will cause us to spend too much on our resources to produce that output, therefore profits will be reduced. To find the proper ratio of resources, we must first relate their prices to their *MRS*.

Earlier (Figure 4-5) we looked at the firm's demand for a resource, so the problem hinges on viewing the whole set of resources this firm is using with a price–quantity demand curve for each resource. In this case our view is limited to just our two variable inputs [from $Y = f(X_1, X_2)$].

A clearer illustration of the problem makes use of an *isocost line*, which shows what amounts of these two resources can be bought for a given amount of money. Let's use $100 as the amount to be spent, given $P_{x1} = $5.00 per unit and $P_{x2} = $2.50 per unit. As Figure 5-4 shows, we can spend all this money for $X_1$, all of it for $X_2$, or some combination of the two as indicated by the diagonal straight-line isocost. The line is a straight line because of an implicit assumption about the resource market: No matter

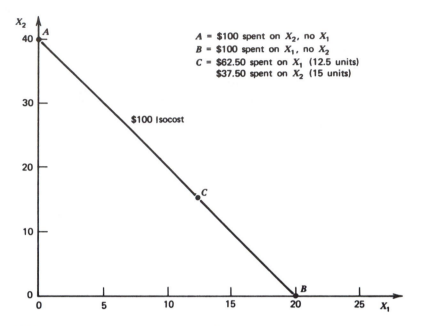

FIGURE 5-4   Cost allocation possibilities.

what this producer does in allocating expenditures, by buying nothing of either input, or buying all of either or both, input prices will not be affected at all.[6] (This is only one of a set of assumptions about pure competition—the topic of a later chapter—but not at all atypical for most producing firms in American agriculture, as they purchase resource inputs.)

## The Least-Cost Combination

Individually, Figures 5-3 and 5-4 show us, respectively, what happens to output as we change resource proportions and what amounts of the two resources can be purchased with a given dollar outlay. To minimize the

---

[6] Given $P_{x1}$ = $5.00 and $P_{x2}$ = $2.50, this $100 will buy 40 units of $X_2$, or 20 units of $X_1$, or any combination along the isocost line. If we use $C$ to indicate the $100 cost level, then $C = P_{x1} \cdot X_1 + P_{x2} \cdot X_2$, where ($P_{x1} \cdot X_1$) is the amount spent on $X_1$ and ($P_{x2} \cdot X_2$) the amount spent on $X_2$, with total spending being their sum. Rearranging our terms, we get $P_{x2} \cdot X_2 = C - P_{x1} \cdot X_1$, and $X_2 = (C/P_{x2}) - (P_{x1}/P_{x2}) \cdot X_1$. Thus, $X_2$ is the unit quantity along the vertical axis, $C/P_{x2}$ is the vertical axis intercept, $P_{x1}/P_{x2}$ the (negative) slope of the isocost line, and $X_1$ the unit measure along the horizontal axis. Our cost-line formula is identical to the straight-line formula $Y = a - bX$. We thus have an incremental measure of slope ($P_{x1}/P_{x2}$) showing how these resources substitute for one another in the market.

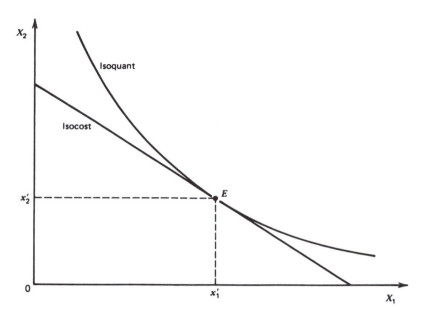

**FIGURE 5-5** Optimizing resource combinations to minimize costs.

cost of producing a given output (or what is the same thing stated in a different way, to maximize output from a given cost outlay), we must combine the type of information each provides us with, as shown graphically in Figure 5-5.

Given output and resource prices, we optimize the ratio of the inputs $X_1$ and $X_2$ by finding that one point along the entire isoquant where the slopes of the two curves (isoquant and isocost) are equal. In other words, for any given output the producer will find an optimum ratio in the use of these resources by equating their *MRS in production* with their price ratio, or the *rate at which they substitute in the market*. Resource prices tell what they are worth to other users in the competitive market, so the producer must use them in such a way that they each are worth *just that much* at the margin in order to make optimal use of those resources.

With the slope of the isocost line determined by the ratio of market prices $(P_{x1}/P_{x2})$ we equate this ratio with the slope of the isoquant $(MPP_{x1}/MPP_{x2})$.[7] When $P_{x1}/P_{x2} = MPP_{x1}/MPP_{x2}$, the slopes of the two curves are equal (at tangency point $E$). We have minimized the resource cost of producing that output, called the **least-cost combination** and have found that input *ratio* ($0–x_1'$ of $X_1$ and $0–x_2'$ of $X_2$) at which an additional

---

[7] $MRS_{x1x2} = \Delta X_2/\Delta X_1 = MPP_{x1}/MPP_{x2}$ and, when equated with their price ratio, can be rewritten as $MPP_{x1}/P_{x1} = MPP_{x2}/P_{x2}$.

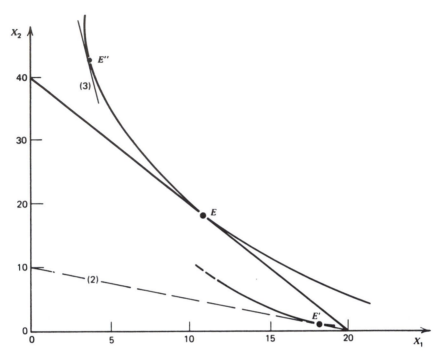

**FIGURE 5-6**  Illustrating the effect of resource price changes on the least-cost combination.

dollar spent on either variable input will yield the same output or value of product.[8]

Rewriting the least cost combination ratios as

$$\frac{MPP_{x1}}{P_{x1}} = \frac{MPP_{x2}}{P_{x2}}$$

demonstrates that whatever the relationship between $X_1$ and $X_2$, the $MPP$s per dollar's worth of these two resources must be in the same ratio in order that the cost of producing any quantity of output be minimized.

The effect of prices in prescribing resource use (resource allocation) is made more evident by noting how the least-cost combination changes as either or both resource prices change. Suppose, for instance, that the price of input $X_2$ increases to $10 per unit. The dashed line labeled (2) in Figure 5-6 is the new $100 isocost line. That sum can still buy 20 units of $X_1$

---

[8] You might verify this by choosing any point along the isocost curve toward either axis from point E. Another isoquant of lower output runs through that point. So a given dollar outlay produces less and less output as different proportions of the inputs are selected farther and farther away from the equilibrium point E.

but only 10 units of $X_2$, so it is a flatter line than our original isocost and will therefore shift the point of tangency to a different ratio of inputs. With relative prices now changed, the producer would be prompted to cut back on the use of the resource $X_2$ whose price has risen. The $100 isocost line would be tangent at a lower level of production and different input ratio (point $E'$). On the other hand, a price fall for $X_2$ means that more of this resource can now be bought with $100, the isocost line would be steeper and tangent to a higher isoquant than the one shown.

Another isocost line, labeled (3) shown as the steeper solid line, demonstrates the combined effect of a doubling in the price of input $X_1$ to $10 per unit and a fall in the price of $X_2$ to $1 per unit. A variety of other prices, isocost lines, and isoquants can be postulated with the resulting changes similarly identified.

Essentially, this sort of resource substitution in response to relative price changes has taken place over many decades in American agriculture. View, for example, input $X_1$ as capital inputs and $X_2$ as labor. As higher wages attracted agricultural labor to industries, the opportunity cost of using labor in agriculture increased relative to the cost of capital inputs. Being imperfect substitutes for one another (see Figure 5-2C), we should expect labor use to be reduced relative to the capital inputs employed. And this is what happened, especially during the 1940s through the 1960s. (See Table 13-6 for relative changes in the use of selected important inputs since 1960.)

# PRODUCT–PRODUCT RELATIONSHIPS IN COMBINING ENTERPRISES _____

## Combining Enterprises

A third choice—what to produce—is often a very complex decision. Grain farmers seldom are so specialized that they can ignore the profit potential offered by other types of crops, or in some cases, adding a livestock enterprise. Stock ranchers most frequently find that adding forage or grain crops will enhance the firm's earnings.

Land and other resources are usually sufficiently versatile that they can be put to a number of different uses. So the decision boils down to one of combining two or more enterprises to maximize profits from the resource set available to the firm.

Just as the choice in the rate and proportion of resource use is optimal only so long as resource and product prices remain unchanged, so also is the enterprise combination proper only until product prices change. Shifts in demand for the firm's products will change the relative profitability of enterprises and will require adjustment in emphasis.

Except for the occasional complete reorganization of a farm or ranch, year-to-year enterprise changes may not appear especially severe. But widespread enterprise adjustments in response to changing economic conditions have taken place over longer periods of time. Changes in technology and product prices have forced many farmers to make sharp adjustments from earlier production efforts.

Ohio has slipped far from its once prominent position in the sheep industry. Cotton, once produced exclusively in the South, has shifted so that much of its production is on irrigated land in the arid Southwest. Few farmers in Aroostook County, Maine, trouble themselves with raising wheat, since they have become especially well known for potatoes. Yet an 1838 report of a study of the Aroostook territory concluded that "the staple crop is, and must ever be, wheat."[9]

## The Firm's Production Possibilities

To illustrate the principle of enterprise choice we must again back away from the complexity of real-life problems. Begin with a farm that has a given amount of each of the resources—land, labor, capital, and management. The operator must choose what product(s) to produce from a number of possible enterprises, adopting only those that will contribute toward maximizing the firm's profits. It is not unlikely that an operator may have a dislike for some particular enterprise. A dislike for handling hogs, milking cows, or caring for poultry, and so on, may be so strong that the individual knowingly sacrifices higher earnings in favor of a more pleasing enterprise.

In spite of the fact that the farmer will face choices involving a large number of potential crop and livestock enterprises that *could* be produced, let's simplify the problem to one of deciding between just two enterprises. Suppose this is a 320-acre farm with the operator deciding whether to produce only grain sorghum or soybeans, or some combination of the two. The product output information for these two enterprises is shown in Table 5-1. To maintain consistency with our earlier use of the label $Y$ to indicate a product, we will call grain sorghum $Y_1$ and soybeans $Y_2$.

The table demonstrates the principles of *sacrifice* and *cost*. In order to produce more grain sorghum, the acreage devoted to producing soybeans must be reduced. Soybeans are sacrificed and therefore are a cost of any additional grain sorghum produced.

Given full utilization of this firm's resources, our farmer could devote all of the farm's resources to producing soybeans, in which case output

---

[9] By 1930, 85 percent of all farms in Aroostook County were classified as potato farms. Cited in John D. Black, et al., *Farm Management*, New York: Macmillan, 1947, p. 136.

**TABLE 5-1** Enterprise Combination Possibilities

| Choices | $Y_1$<br>Grain Sorghum<br>(*bu*) | $Y_2$<br>Soybeans<br>(*bu*) | $MRPS_{y1y2}$<br>$(= \Delta Y_2/\Delta Y_1)$ | Total Revenue When |  |
|---|---|---|---|---|---|
| | | | | $P_{y1} = \$1.00$<br>$P_{y2} = \$5.00$ | $P_{y1} = \$2.50$<br>$P_{y2} = \$3.75$ |
| A | 0 | 7500 | | \$37,500 | \$28,125 |
| | | | $\dfrac{500}{2000} = 0.25$ | | |
| B | 2000 | 7000 | | \$37,000 | \$31,250 |
| | | | $\dfrac{650}{2000} = 0.33$ | | |
| C | 4000 | 6350 | | \$35,750 | \$33,812.50 |
| | | | $\dfrac{775}{2000} = 0.39$ | | |
| D | 6000 | 5575 | | \$33,875 | \$35,906.50 |
| | | | $\dfrac{1025}{2000} = 0.51$ | | |
| E | 8000 | 4550 | | \$30,750 | \$37,062.50 |
| | | | $\dfrac{1450}{2000} = 0.73$ | | |
| F | 10000 | 3100 | | \$25,500 | \$36,625 |
| | | | $\dfrac{3100}{2000} = 1.55$ | | |
| G | 12000 | 0 | | \$12,000 | \$30,000 |

of that crop would be 7500 bushels. But with all the land and other resources used to produce soybeans, there is none available for grain sorghum, so the output of that crop is zero. The other extreme allocation devotes all the resources to producing grain sorghum (12,000 bushels), with zero soybean output.

For ease in visualizing the decision problem and its consequences, we have set up only seven different output proportions, Choices *A* through *G*. In addition to the one-product-only resource allocations (*A* and *G*) with their resulting values of production, five other possibilities are shown. The operator could produce some of both enterprises (any one of *B* through *F*), with the total value of the products produced changing as a result.

The full range of these possible allocations is called the firm's *production possibilities*.[10] The production possibilities show all the possible combinations of two products, $Y_1$ and $Y_2$, that can be produced, given the set of resources in the firm's control. As this is a case of, first, assuming full utilization of the firm's resources, then using those resources in one enterprise or the other, costs are constant; they do not change with different allocations of resources to the two enterprises.

---

[10] Sometimes also called the "product-transformation curve," that demonstrates the rate at which one product can be transformed into the other.

Note carefully the effect of changing relative product prices on the amounts of each one produced, and the logical basis of the statement that prices allocate resources. When relative product prices change, we *must* respond by adjusting resource quantities devoted to the different products or needlessly suffer a loss in the firm's profits.

When the price of grain sorghum ($Y_1$) is $1.00 per bushel, and soybeans ($Y_2$) are $5.00 per bushel, the farm's net income is maximized by producing only soybeans (choice A). No other allocation will yield as much income. Attempting to produce any amount of grain sorghum, given these prices, will serve only to reduce total revenue and profits by the same amount because total costs, whatever their level may be, are unchanged. Therefore, any increase or decrease in total revenue will increase or decrease profits by the same amount.

If, on the other hand, prices were $2.50 per bushel for grain sorghum and $3.75 per bushel for soybeans, a different allocation of resources would be required. Profits would be maximized with resources allocated to produce 8000 bushels of grain sorghum and 4550 bushels of soybeans. Total revenue would then be a maximum, $37,062.50, and any other allocation would simply throw away potential earnings. And if the price of grain sorghum were to rise to, say, $5.00 per bushel, with soybeans still $3.00, one would not produce any soybeans at all. Producing only grain sorghum (12,000 bushels) would maximize the firm's earnings.

In the table we have shown output $Y_1$ in increments of 2000 bushels to aid in understanding the computations we must make and their economic meaning. This information is also graphed in Figure 5-7 to aid in visualizing the relationships between enterprises.

Start at point A with zero $Y_1$ and 7500 bushel of $Y_2$, then change to point B with 2000 bushels of $Y_1$; product $Y_2$ drops from 7500 to 7000 bushels. An increase of 2000 bushels of $Y_1$ causes a reduction of 500 bushels of $Y_2$ because resources have been shifted from producing $Y_2$ to $Y_1$. We have substituted $Y_1$ for $Y_2$.

The *marginal rate of product substitution* (MRPS) measures the differing rates at which either of these products will replace (substitute for) the other along the production possibilities curve. That calculation tells us the amount by which one product output is decreased in obtaining a unit increase in the alternative product. If we divide the reduced $Y_2$ by the 2000 bushel increase in $Y_1$, we have the marginal rate of product substitution of $Y_1$ for $Y_2$. The formula for this determination is

$$MRPS_{y1y2} = \frac{\Delta Y_2}{\Delta Y_1}$$

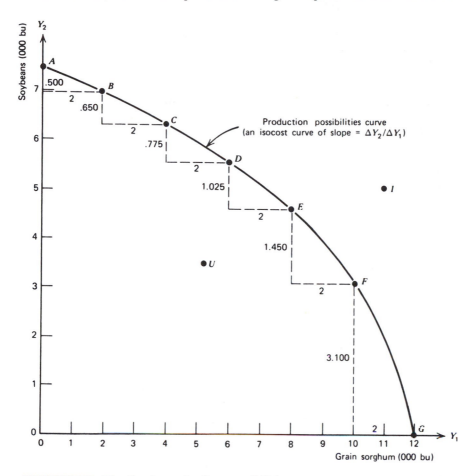

**FIGURE 5-7** The firm's production possibilities.

which reads "the marginal rate of product substitution of $Y_1$ for $Y_2$."[11] This calculation is a measure of the *slope* of the production possibilities curve between two points on the graph, or between the same two sets of data in Table 5-1.

In going from $A$ to $B$, $MRPS_{y1y2} = 500/2000 = 0.25$, which means that (within the $A$–$B$ range) one bushel of $Y_1$ will replace 0.25 bushels of $Y_2$. Since costs are unchanged, one bushel of $Y_1$ added has the same production cost as 0.25 bushels of $Y_2$ given up.

In moving from $B$ to $C$, the $MRPS_{y1y2} = 650/2000 = 0.33$, a greater sacrifice of $Y_2$ per bushel of $Y_1$ added than in going from $A$ to $B$. As we con-

---

[11] The marginal rate of product substitution of $Y_2$ for $Y_1$ simply measures these changes in a counterclockwise direction: $MRPS_{y2y1} = \Delta Y_1/\Delta Y_2$, giving a set of numbers that are the reciprocals of those measuring the $MRPS_{y1y2}$, the sign being ignored in both cases.

tinue to move clockwise along the production possibilities curve, more and more of $Y_2$ must be given up for each additional unit of $Y_1$. Diminishing returns causes proportionately more resources to be used for each increment of $Y_1$ added.[12] What this also says is that the *opportunity cost* of producing additional amounts of $Y_1$ increases. The more resources that are committed to producing $Y_1$, the greater are the sacrifices of $Y_2$ given up to get those greater amounts of $Y_1$. This characteristic causes the production possibilities curve to bulge outward from the origin.

The production possibilities curve is sometimes also called a "frontier" because it identifies the maximum output for each and every combination of the two products that the firm's resources can produce. It is a frontier because it identifies the boundary between what is and what is not possible. The firm's resources are incapable of producing more goods than those indicated by the production possibilities curve. Any combination outside the curve, such as the one labeled *I*, is impossible. Only an increase in the overall productivity of the firm—from a new technology or a greater quantity of resources, or both—will push the curve further out into the quadrant space.

A point such as *U* is possible, however. At any point *U*, some resources are idle, or unemployed, so production of both products is less than what is possible. It is an inefficient organization.

As we consider the *MRPS* derivations, we are unable to decide which quantities of those that could be produced would provide the greatest net return to the firm. This return can be determined only after taking into account the effect of product prices.

At $P_{y1} = \$1.00$ and $P_{y2} = \$5.00$, the market says that it takes 5 bushels of $Y_1$ to equal the value of 1 bushel of $Y_2$; 1 bushel of $Y_1$ is worth 0.20 bushels of $Y_2$. Given these prices, producing at B instead of A means that \$2500 worth of $Y_2$ has been sacrificed to gain \$2000 of $Y_1$: An unnecessary loss of \$500 has been incurred. Whatever the profit level that would have obtained at A, the firm is \$500 worse off at B, and worse off still at other points farther down the curve.

The seven output choices (A through G) with only two pairs of product prices was a necessary abstraction from all the real-life production and price possibilities. We would be hard pressed to compute all the revenue capabilities were we to expand the number of different price combinations to the number that might occur. A full accounting of all output and

---

[12] The previously mentioned cause of diminishing returns—changing resource proportions—may not be so obvious here. But other factors take their toll. It is unlikely that resources would be of equal productivity in both enterprises, so yields will decline as more and more (increasingly less productive) resources are shifted to either enterprise.

price possibilities would cause the choices to balloon to an impossibly large number. We need a simplifying rule capable of incorporating both ratios of production and prices into the decision mechanism.

## The Firm's Revenue Possibilities

Just as in consumer and producer choices among things used to produce utility or goods, we need to compare the rates at which goods substitute for one another in consumption or production with their rate of substitution in the market. And just as an objective-maximizing rule could be given in each of those cases, a similar rule should determine the profit-maximizing allocation of the output from two different enterprises. Figure 5-8 adds the price dimension to our choice problem.

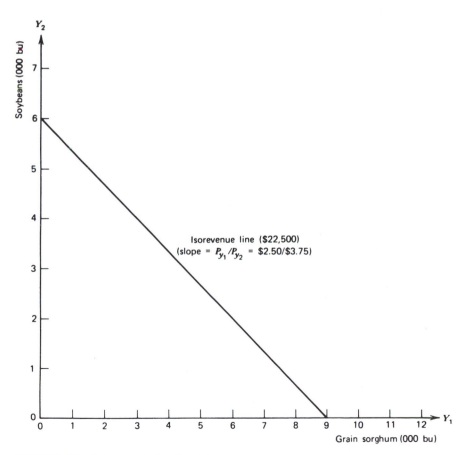

**FIGURE 5-8**   Prices and the firm's isorevenue.

An *isorevenue line* shows all the possible combinations of two products sold that will bring the same total revenue. The isorevenue line is a straight line because of an assumption regarding this firm's size in relation to the market for its products; it is too small to have any influence on price regardless of the firm's production and marketing decisions. If this firm produced and sold all it could produce of either $Y_1$ or $Y_2$, and none of the other product, neither product's price would be affected. Thus, if this firm produces and sells 9000 bushels of $Y_1$ at $2.50 per bushel, revenue is $22,500. That same revenue could be obtained from 6000 bushels of $Y_2$ at $3.75 per bushel. A straight line drawn between 6000 $Y_2$ and 9000 $Y_1$ describes all the possible combinations of these two products that will bring a total revenue of $22,500. Such quantity pairs at 5000 $Y_2$ and 1500 $Y_1$, 3000 $Y_2$ and 4500 $Y_1$, and 1500 $Y_2$ and 6750 $Y_1$ are readily identifiable from among the many others along the isorevenue line that would total $22,500 in revenue.

Note that the slope of the isorevenue line is determined by the ratio $P_{y1}/P_{y2}$.[13] Since it would take 9000 bushels of grain sorghum to equal the value of 6000 bushels of soybeans, the isorevenue line has a slope of 1:1.5. Each unit measured along the vertical axis equals the value of 1.5 units measured along the horizontal axis, a direct result of these product prices.

For any higher or lower product prices, with their ratio unchanged from the $2.50 and $3.75, we could draw a great many other isorevenue lines of identical slope anywhere in the quadrant space. For each even slightly different ratio of prices another similar number of isorevenue lines, all of different slope, could be drawn.

If $P_{y1}$ were to rise to $3.00 per bushel, given $P_{y2} = $3.75, only 7500 bushels of $Y_1$ would have to be sold to return $22,500. And with $Y_2$ still at 6000 units, the isorevenue line is now steeper than the one drawn in Figure 5-8. Given any increase in $P_{y1}$ relative to the price of $Y_2$, the slope of the isorevenue line is increased; a reduction in $P_{y1}$ relative to $P_{y2}$ will reduce (or flatten) the slope of that line.

## Optimizing Output

We can now put these two sets of information together and determine that *one* allocation of resources to these products that will maximize the firm's profits, as in Figure 5-9. With axis labels the same as in Figure 5-7, but scales expanded to aid in reading quantities, we plot the data for the production possibilities curve. The ratio of the first pair of prices used in Table 5-1 ($P_{y1} = $1.00/bu and $P_{y2} = $5.00/bu) results in Isorevenue line I. With

---

[13] That $P_{y1}/P_{y2}$ is the determinant of the (negative) slope of the isorevenue line is proved in the same way as was shown earlier in this chapter (in footnote 11) for the slope of an isocost line.

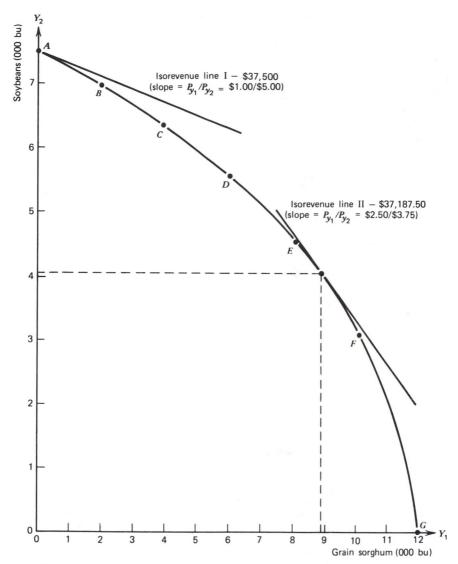

**FIGURE 5-9**  Optimizing the firm's output.

this specific isorevenue line there really is no clearly seen point of tangency, but being the only output proportion available, the firm's optimum is at point A. As can be confirmed by the data in Table 5-1, no greater revenue can be obtained than by devoting all the firm's resources to producing soybeans only, worth $37,500. Any other proportion simply reduces total revenue without a reduction of costs, so profits suffer as a consequence.

As product prices change, so that the price of $Y_1$ rises relative to the price of $Y_2$, the point of tangency between the isorevenue line and the production possibilities curve rotates clockwise along the production possibilities curve. When the isorevenue line slope increases, less $Y_2$ is produced while the output of $Y_1$ is being increased.

Such a change in price ratios is shown by Isorevenue line II. Assume the price of $Y_2$ has fallen to $3.75 per bushel while the price of $Y_1$ has risen to $2.50 per bushel. As the slope of the isorevenue line is determined by $P_{y1}/P_{y2}$, this change in product prices causes Isorevenue line II to be steeper than Isorevenue line I. With these price changes, the revenue-producing capabilities of the two products also changes, de-pending on the magnitude of price changes. The point of tangency now falls between points $E$ and $F$ on the production possibilities curve. At this point, sufficient resources have been shifted from producing $Y_2$ to producing $Y_1$ that the output proportions are now about 4050 bushels of $Y_2$ and 8800 bushels of $Y_1$. These quantities at the given prices produce a total revenue of $37,187.50, which is greater than the revenue at either point $E$ or point $F$.[14]

We stated earlier that equating the $MRPS_{y1y2}$ ($\Delta Y_2/\Delta Y_1$) with the price ratio ($P_{y1}/P_{y2}$) would be an optimum in that no greater revenue could be earned from this set of resources. When we have located the point at which $\Delta Y_2/\Delta Y_1 = P_{y1}/P_{y2}$ the rates of substitution at the margin for both production and the market have been equated. Their opportunity costs in production are now identical to the values the market puts on these products. Costs at the margin equal revenues at the margin. And because of the curvature of the production-possibilities curve, any movement from the point of tangency would cause more to be sacrificed in value of the product given up than is obtained from the value of the other product's increased output.

Thus far, we have considered only one specific set of resources. What happens when the resource base itself is changed? For instance, an increase or decrease in available labor and management may come about through the occasion of marriage, divorce, or death. Some of the owned land may be sold to acquire other resources. The operator may rent more land or lose the lease on some of the presently used acres. Some labor may be lost as offspring grow and leave the farm to other pursuits.

Each of these changes, among a great many other possibilities, will cause the firm's total output capabilities to change. And they may or may not result in proportionate increases or decreases in total resources. When resource proportions change, there may also be changes caused in the

---

[14] Again, as in Chapter 4, optimizing from the discrete numbers in Table 5-1 will only by chance give us the same answer as will the smooth production possibilities curve in Figure 5-9 with its assumed divisibility.

shape of the production possibilities curve as well, each different set of resources having its own specific production possibilities curve.

## The Expansion Path

How the firm optimizes output as the production possibilities curve expands may be viewed as a question of firm growth. The path along which a firm expands (or contracts) its output optima, given product prices, is shown in Figure 5-10.

Suppose that instead of the resource set we have been using, as shown by the more heavily drawn production possibilities curve, we had started with a firm having fewer of some or all the resources land, labor, capital, and management. The output of soybeans and grain sorghum would be reduced if this farm had only 160 acres of land, say, and would be able to produce even less with 80 acres of land. The production possibilities curves for these smaller sizes of the firm would lie within the more heavily drawn curve representing the 320-acre farm. For each different level of available resources there will be a different production possibilities curve, only three of which we have sketched in the diagram. Given just one pair of product prices, there will be only one most profitable combination of outputs for each different production possibilities curve. The line drawn through these points is called the *expansion path*, showing the revenue (and profit) maximizing proportions of $Y_1$ and $Y_2$ as the firm expands or contracts output when the firm's size is changed.

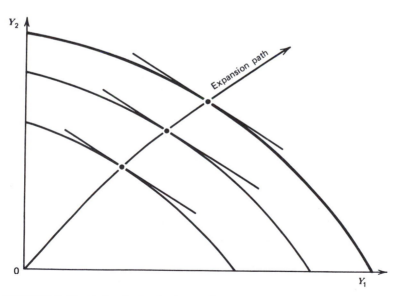

**FIGURE 5-10** A firm's expansion path.

To summarize, let's use the symbolism $MVP_{x1}^{y1}$ to mean the marginal value product to resource $X_1$ in the production of product $Y_1$, and $MVP_{x1}^{y2}$ to mean exactly the same for that resource in another product $Y_2$. The resource-allocating problem has been solved for resource $X_1$ when, for two products,

$$\frac{MVP_{x1}^{y1}}{P_{x1}} = \frac{MVP_{x1}^{y2}}{P_{x1}} = 1$$

Because $P_{x1}$ is unchanged, this relationship can be stated simply as $MVP_{x1}^{y1} = MVP_{x1}^{y2}$. The resource $X_1$ has been allocated to its most profitable use because no reallocation of that resource among products $Y_1$ and $Y_2$ will increase its value productivity.

With this set of relationships true for each and every variable resource in all possible enterprises, the firm is earning its highest possible profit. Any reallocation of resources between the different enterprises within that firm will reduce its profits.

Some, impatient to get on with cost analysis, may feel that we have been tediously long in identifying the physical and economic relationships in production. Too frequently, however, the student misses the firm ground of cost analysis in the production function. In dealing with the question of what happens to cost when output is changed, we forget too easily that the only way the operator can change output quantity is to reallocate variable resources between two or more enterprises. Thus the need exists to lay the foundation carefully for the next chapter's analysis of costs.

## SUMMARY

A two-variable input function, with other resources fixed, is used as the simplest device with which to describe the solution to the problem of how to produce. The physical response to changing proportions of the variable inputs results in the isoquants. The slope of an isoquant shows the marginal rate at which these resources substitute for one another— $MRS_{x1x2} = \Delta X_2/\Delta X_1$.

The slope of an isocost line $(P_{x1}/P_{x2})$ is determined by the ratio of market prices for $X_1$ and $X_2$. Equating their marginal rate of substitution with their price ratio $(\Delta X_2/\Delta X_1 = P_{x1}/P_{x2})$ derives the cost-minimizing ratio for these two inputs. Combining variable resources in this ratio results in an equal value of output per added dollar of cost for each resource.

A production possibilities curve for two products (enterprises) is used to describe the solution to the question of what to produce. That line of relationship is drawn convex from above to reflect diminishing returns as more and more of the firm's resources are devoted to one or the other enterprises.

The slope of the production possibilities curve is indicated by $\Delta Y_2/\Delta Y_1$. When that ratio is equated with the ratio of the product prices ($\Delta Y_2/\Delta Y_1$ = $P_{y1}/P_{y2}$), the firm's profits are a maximum. The firm's resources can produce no more than the quantities indicated by that point of equality. Increasing the output of either of these products would cause a reduction in the output of the other, also with a reduction in total value output and profit.

An optimum (maximum profit point) has been reached along the expansion path when the difference between total value output and total cost of the variable resources is maximized.

## CHAPTER HIGHLIGHTS

1. A producer faces three production choices: (1) factor–product—how much to produce; (2) factor–factor—what resource combination to use; and (3) product–product—what product(s) to produce.
2. Constant returns to scale means that output increases in a direct (constant) proportion to the increase in inputs.
3. Diminishing returns are the result of changing resource proportions.
4. An isoproduct contour (or isoquant) shows the different combinations of inputs that will produce the same quantity of output.
5. The marginal rate of substitution shows the rate at which one resource can be substituted for another without changing the level of output. *MRS* is thus a measure of the slope of the isoquant.
6. An isocost line shows the different combinations of resources that can be bought with a given cost outlay.
7. The least-cost combination of inputs is determined at that point where the isocost line is tangent to an isoquant. At that point,

$$MRS_{x1\,x2} = \frac{\Delta X_2}{\Delta X_1} = \frac{P_{x1}}{P_{x2}}$$

8. A production possibilities curve shows all the combinations of products that a firm can produce given its resources and technology.
9. The marginal rate of product substitution (*MRPS*) is shown by the slope of the production possibilities curve, which also shows the opportunity cost of producing more of either product.
10. The isorevenue line shows all the combinations of products sold that will bring the same total revenue.
11. A firm will maximize profits by producing that combination of products where the isorevenue line is tangent to its production possibilities curve.

12. The proper (profit maximizing) scale of the firm is determined along the expansion path.
13. The production function provides the basis for the firm's costs.

# APPENDIX _____

The need to classify resources into groups was discussed in Chapter 4. We must do this so that conclusions can be drawn regarding the contributions each resource makes in the production process. In the actual operation of a farm you may have to recognize that there are many different kinds of labor (each with different productivities) used in a number of different enterprises, and each to be assigned a different $X$; various types and qualities of land, each with a different $X$; and many capital items, each also requiring a separately identified $X$. A function such as $Y = f(X_1 \ldots X_{62})$ might then be appropriate for your operation. For our purpose, however, it is much more convenient to classify all resources into the four basic categories and avoid the confusion of trying to keep track of many dozens of different inputs $X$.

## The Constant Returns Function

Begin first with a long-run function using all four general resources (land, labor, capital, and management) and call a given amount of each of *labor* and *management* a unit of $X_1$, and a given amount of *land* and *capital* a unit of $X_2$. If we vary both of our composite inputs, the function is then $Y = f(X_1 X_2)$. An infinite number of alternatives exist for the ratios in which these two variables can be combined, with only a few of those alternatives sketched in Figure 5-11, plotting units of $X_1$ on the horizontal axis and units of $X_2$ on the vertical axis.[15]

Recall from our earlier discussion of graphics that no matter how one unit of $X_1$ is composed, two units of $X_1$ must be exactly twice as much of the identical resources making up a unit of that input. The same strict unit requirements hold true for input $X_2$.

If, for instance, we move out from the origin along the heavily drawn ray (the 1:1 ratio between inputs) to the point where we have used one unit each of $X_1$ and $X_2$, a specific amount of output $Y$ will be produced; two units of each input will be twice as far along the ray, with output $Y$ being doubled; three units of each input will be three times as far out

---

[15] As was stated in the previous chapter, all four general resources must be used to be able to produce anything, therefore, both $X_1$ and $X_2$ must be used to produce the output $Y$. Thus, if we move out along the $X_1$ axis we are not using any $X_2$, so no output can result. The same holds true along the $X_2$ axis—no $X_1$, no product.

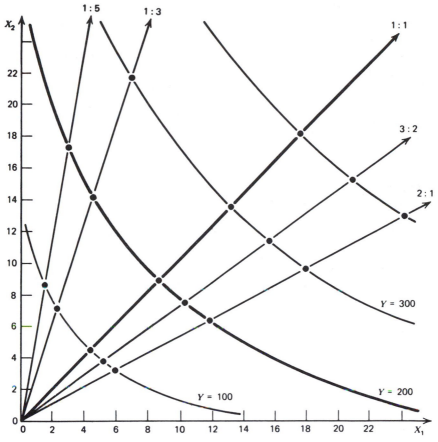

**FIGURE 5-11**   Production surface from a two-variable input function, $Y = f(X_1 X_2)$.

along the ray, with $Y$ being tripled, and so on.[16] We can get no other result. All other rays will exhibit the same input–output characteristic, but
with different output rates. And any of these rays will generate input–
output data[17] that plot a two-dimensional graph of the function $Y = f(X_1$
$X_2)$ with the inputs $X_1$ and $X_2$ combined in a ratio that differs from their
ratio along any other ray.

---

[16] Holding constant the proportions of $X_1$ to $X_2$ as we move outward along any ray, our inputs $X_1$ and $X_2$ could just as well be reidentified as an input $X$, with a general functional
notation of $Y = f(X)$, and with results exactly as shown in Figure 4-1(*A*).

[17] A graph of the function for any of these rays would be similar to the one in Figure 4-1(*A*),
all having an identical *shape* but with the *slope* of the function being determined by the ray
selected for plotting.

As we select a number of different output quantities along each ray, we are able to identify points of equal output on different rays. Connecting all points of equal output identifies the different isoquants in the graph space. With each of these isoquants showing different levels of output, our diagram now measures output vertically above the plane of the graph. This third-dimensional measure depicts and quantifies (in units of $Y$) height above the plane in the same way, and with exactly the same intent, that a cartographer uses contours to show the surface elevation of a hill.

It should be clear that if nothing directed us to select any one ray to describe the functional relationship between the inputs $X_1$ and $X_2$, nor were we forced to select any one isoquant, an infinite number of choices exist in looking at either (or both) rays and isoquants. Put the point of your pencil anywhere in the quadrant space and a ray of specific ratio between the inputs passes through that point. The ***production surface*** can be visualized in the same way and is diagrammed as a family of isoquants. In Figure 5-11, we could draw isoquants between those shown, and between those again, continuing to draw isoquants between others until we have a solid surface so dense with isoquants that we would be unable to distinguish one isoquant from another. Isoquants are everywhere in our product space, each one showing a different quantity of output.[18]

Note that a movement *along an isoquant* changes the ratio between $X_1$ and $X_2$ with the quantity of output constant, while a movement *along a ray* changes output but holds the input *proportion* constant. Almost hidden in a movement along an isoquant is the old problem of diminishing returns, caused by *changing input proportions*. As we select rays closer to the $X_1$ axis, we are increasing the ratio of $X_1$ to $X_2$, with the result that $X_1$ becomes less productive. There simply is too much of that resource being used with any given amount of $X_2$. The same holds true for input $X_2$ and its productivity as we select rays tilting closer and closer to the $X_2$ axis.[19]

---

[18] A special characteristic of isoquants is that they do not intersect. Were an intersection of isoquants possible, it would mean that two or more different amounts of output could be produced by one level of input use. This impossible contradiction of the law of diminishing returns, further, would destroy the meaning and predictability of any functional relationship.

[19] Even though these isoquants might be viewed as most typical of the real world when all inputs are varied, there are important (and realistic) exceptions based entirely on how the inputs are put together. Visualize the type of production surface that would result if a unit of $X_1$ were composed of ten days of labor, one day of management, $100 of capital, and one acre of land; and a unit of $X_2$ included five days of labor, two days of management, $500 of capital, and three acres of land. Each of these inputs would be capable of producing output on its own (the other need not be present at all because all four of the basic resources are included in a unit of either input); therefore, $Y$ would rise continually along either axis rather

Because there are relatively unproductive resource combinations toward the vertical and horizontal axes, we should expect to have to move farther out from the origin along those outer rays to locate some specific quantity of output. Nearer the outer extremes of the production surface, inputs have been combined in a (physically) inefficient manner, while a better physical balance of inputs is established between those outer rays. All points along any isoquant are technically possible, but only one point along that isoquant can be an economic best, giving rise to the problem of having to select that one combination of inputs that will minimize the cost of producing that output.

The optimal proportions of inputs $X_1$ and $X_2$ are determined when each variable resource is used to the point where its addition to value of output (*MVP*) just equals its addition to cost (*MFC*). If the *MVP* of a resource exceeds its *MFC*, the amount of that resource used should be expanded. Its use should be reduced if its *MVP* is less than its *MFC*. When these criteria are met for all variable resources, the ratio of inputs is at an optimum, and

$$\frac{MVP_{x1}}{P_{x1}} = \frac{MVP_{x2}}{P_{x2}} = \cdots = \frac{MVP_{xn}}{P_{xn}}$$

## From Constant Returns to Diminishing Returns

Note that to here our consideration of resource use optimizing has been limited to the two-variable input function $Y = f(X_1\ X_2)$ with the four basic resources included in these two variables. The production surface has been that as described by Figure 5-11. We earlier portrayed a single variable input function with fixed factors, and developed appropriate data (Table 4-1), with no logical contradiction between constant returns and diminishing returns.

Consider Figure 5-12, which more fully projects the production surface from $Y = f(X_1\ X_2)$, the two-variable, constant returns function that was first shown in Figure 5-11. Since constant returns are unique to varying all resources in a constant proportion, and diminishing returns come into play when one or more resources in a production activity are held constant, both these phenomena should be demonstrable from this constant

---

than be zero along the axes. Note carefully, however, that such exceptions have nothing to do with attempting to prove invalid the law of diminishing returns, or proving that these are only hypothetical examples lacking real-life observation or meaning. They result purely from a different arrangement of the inputs and the manner in which they are combined with another, problems for which every producer must find answers in the everyday use of resources.

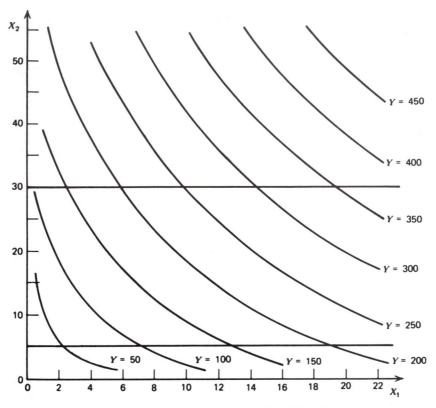

**FIGURE 5-12** A two-variable input function, $Y = f(X_1 X_2)$.

returns production surface if, as we claim, there is no logical inconsistency between the two concepts.

Fix $X_2$ at the 5-unit level and vary $X_1$; the function now is $Y = f(X_1 \mid X_2)$. Diminishing returns are evident in the curvature of the function in Figure 5-13, as the data plot the *TPP* curve of $X_1$ variable with $X_2$ fixed at five units.[20] Multiplying output by the price of the product (as was done in Table 4-2) makes it possible to determine an optimal amount of $X_1$ to use by locating the input level at which $MVP = MFC$. Fixing $X_2$ at a different level causes a different production function, as is shown for $X_2$ fixed at 30 units. Input $X_2$ may be fixed variously at 10, 15, 20, units and so on, with as many different functions as there are possible numbers of units

----

[20] Visualize slicing down through the production surface at the fifth unit of $X_2$ and parallel to the $X_1$ axis, then turning the exposed side of that cut so that you are looking at the third dimension of the model—a single-variable input subfunction of the two-variable function surface.

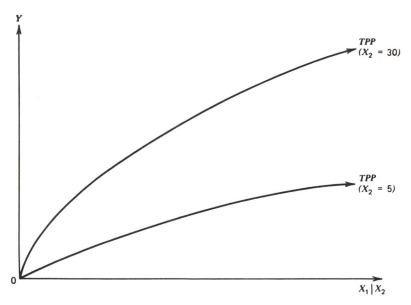

**FIGURE 5-13** Diminishing returns subfunctions of the two-variable input function, $Y = f(X_1 X_2)$.

of $X_2$.[21] It should be clear that individual units of the variable resource are made more (or less) productive as the quantity of the other resource is changed, demonstrating the substantial dependence of resources on one another for their productivity. The different production functions that may be identified also illustrate the importance to the producer of productivity limitations imposed when an inadequate scale of operations is established.

## The Diminishing Returns Surface

Earlier we held that a more realistic view of producer options would be one in which some resources would be fixed and a few others could be varied. This, in general, is the sort of thing the foregoing section dealt with but where all resources were packaged into a two-variable input function.

Suppose we take a somewhat different approach with the four basic resources, calling *labor* the variable input $X_1$, *capital* in all its different forms the variable input $X_2$, and *land* ($X_3$) and *management* ($X_4$) as the resources

---

[21] The results would be similar if, on the other hand, $X_1$ were fixed at any level and $X_2$ were varied. We would just be cutting slices through the production surface in a direction parallel to the $X_2$ axis.

held constant.[22] We have exactly the same basic resources (with different unit packages, however) as in the constant returns case—exactly the same firm, if you wish—but now the function has something fixed before we even begin considering production alternatives.

With $Y = f(X_1, X_2 \mid X_3, X_4)$, a set of output quantities must result such as those shown in Figure 5-14. These numbers then permit a plotting of isoquants that gives the production surface an appearance such as that in Figure 5-15 (rather than as Figure 5-12, the constant returns case). The reason for this is that no matter in what ratios we may expand the use of the two variable resources, their proportion to the fixed inputs $X_3$ and $X_4$ is being changed and we can't escape diminishing returns. The diagram must reflect this in all directions in the quadrant space, rather than just when moving parallel to either axis as in the constant returns case. Furthermore, since there are fixed inputs ($X_3$ and $X_4$), those fixed resources become limiting. As more of $X_1$ and $X_2$ are used there is a maximum output (916 $Y$) that can be produced by these resources at 110 $X_1$ and 90 $X_2$. Total physical product falls away from the maximum in all directions as the quantities of $X_1$ and $X_2$ are increased individually or together because of negative *MPP*s for both $X_1$ and $X_2$.

Diminishing returns are indicated by the spacing of the isoquants. Their increased spacing farther out in the quadrant results from diminishing *MPP*. Progressively greater amounts of the variable resources are required to produce the increment to product represented by the more distant isoquants.

To confirm the general shape of the production surface, fix $X_2$ at different levels. We demonstrate this with $X_2$ fixed at two levels only—10 units and 40 units as in Figures 5- 14 and 5-15. Varying $X_1$ with $X_2 = 10$, for instance, will cause output changes from 0 to 20 $Y$, then 100, 200, and so on, rising to a maximum of 340 units of output. From that point (where $MPP_{x1} = 0$), total product actually declines with additional amounts of $X_1$ used because its *MPP* is less than zero. The 340 $Y$ isoquant must therefore bend away from the horizontal axis at amounts of $X_1$ used beyond 100 units.

The same holds true for all other isoquants with $X_2$ held fixed at other levels. The $X_2 = 40$ subfunction (our source of the data for the production function used as the primary example of diminishing returns in Chapter 4) performs in a manner similar to the one just discussed. With $X_1 = 0$, output is zero, rises at a rate determined by the marginal productivity of

---

[22] Land, a fixed resource, for an ordinary crop/livestock ranch may itself be a composite of given acreages of many different types of land—dry rangeland; grazed forestland; unimproved meadowland for grazing; improved meadow hayland; irrigated hayland; tilled dry cropland; irrigated cropland; and so on—each separately identifiable with an X of its own, each with different productivities and resource combinations that enable it to produce a valued product.

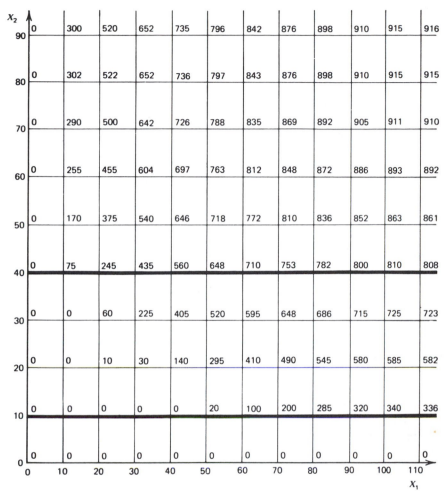

**FIGURE 5-14** Production data derived from the function, $Y = f(X_1X_2 \mid X_3X_4)$.

$X_1$ up to a maximum $Y$ of 810 units, and declines from that point as greater amounts of $X_1$ are applied.

The heavily drawn line rising vertically from near 100 units on the $X_1$ axis is called a *ridge line*. This line connects all points of zero $MPP$ of the input $X_1$, thus separating Stage II from Stage III for all subfunctions with only $X_1$ varied.

Similar results are obtained when $X_2$ is varied, while $X_1$ is fixed at different levels. All isoquants eventually bend away from the vertical axis. Stages II and III for $X_2$ are separated by the horizontal ridge line that connects all the points where isoquants reach and pass the vertical (i.e., points of $MPP_{x2} = 0$).

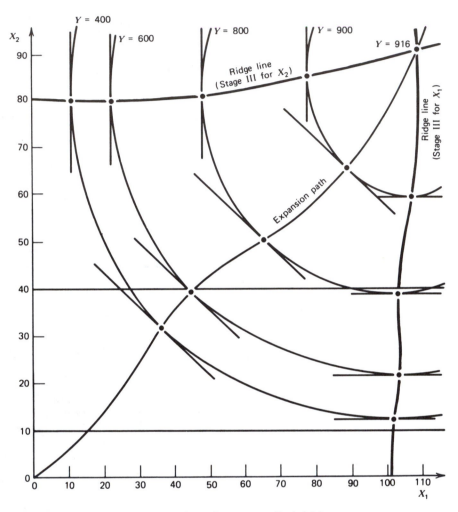

**FIGURE 5-15** Production surface illustrating diminishing returns.

The *expansion path* is determined by connecting the least-cost combination points along the production surface. For any given pair of prices
for $X_1$ and $X_2$, a least-cost combination exists for each isoquant. Numerous expansion paths that extend outward from the origin are possible,

bounded at the extremes by the axes and the ridge lines, as determined by the ratio of the prices for $X_1$ and $X_2$. If the price of $X_1$ were zero (a free good) and that of $X_2$ some positive number, the expansion line would follow the vertically rising ridge line; and if $X_2$ were free with a positive price for $X_1$, the expansion path would follow the horizontal ridge line. With any positive price for both resources the expansion path would lie somewhere between the ridge lines. The specific location of the expansion path within the area is determined, as in Figure 5-15, by the points at which the slope of the isocost and isoquant curve are equal $(P_{x1}/P_{x2} = \Delta X_2/\Delta X_1)$.

Recognize that as only one of these inputs is varied, we are dealing with subfunctions of that resource set which generated the full production surface, with the same solution to the problem of optimality as discussed in Chapter 4. So there really is no need to maintain fixed resource identification—we could just as well call all the nonvaried resources an input $X_2$, with the relevant function then being $Y = f(X_1 \mid X_2)$. The reason for ignoring fixed resources in this way will become clearer in Chapter 6. We pay no attention to fixed resource costs because they have nothing to do with deciding how much of a variable resource to use. The value productivity of a variable resource and its price are the only determinants of that.[23]

When all resources are variable as in $Y = f(X_1 \ldots X_n)$, where $X_n$ may be the tenth different resource, the 100th, or an even greater number, we used $Y = f(X_1 X_2)$ as sufficiently descriptive of the particular input–output relationship. The larger function would be too cumbersome to describe and comprehend, and it would be no more able than the simpler form to answer meaningful economic questions. Where one or more resources might be variable with the remaining resources fixed, such as $Y = f(X_1 \ldots X_g \mid X_h \ldots X_n)$, $X_g$ might be a large number of resources and $X_n$ much larger still. And the simplifying fact is that resource proportions are being changed, so diminishing returns cannot be avoided. Thus, we can use a much simpler appearing function, $Y = f(X_1 \mid X_2)$, for the single-variable input type; and with two-variable inputs, $Y = f(X_1 X_2 \mid X_3 X_4)$. These two general functions are correctly and fully descriptive of the economic meaning of resource–product relationships.

---

[23] Restudy Table 4-2 and Figure 4-4 to make this point clear.

# KEY TERMS AND CONCEPTS TO REMEMBER

Expansion path

Factor–factor relationship

Factor–product relationship

Imperfect substitutes

Isocost line

Isorevenue line

Isoquant

Least-cost combination

Marginal rate of product
    substitution

Marginal rate of substitution

Perfect complements

Perfect substitutes

Product–product relationship

Production possibilities

Production surface

Resource substitution

Ridge line

# REVIEW QUESTIONS

1. Draw an isoquant to show how a farmer in your area can choose between commercial fertilizer and a legume crop to provide nitrogen for a growing crop. By adding an isocost line to your diagram, show how that person can determine the proper amount of each to use.

2. Show how the decision in review question 1 would be changed if the price of nitrogen fertilizer or the cost of raising alfalfa changed.

3. Imagine two farms in a particular type-of-farming area of your state with equal land acres and identical soils in their farms. Should these farms have the same crop and livestock enterprises to maximize their profits? Why, or why not? What affects the decision of what enterprises to include in the firm, and the relative sizes of each?

4. Given a farm with its two major enterprises being cotton and corn/livestock. Suppose that by shifting more of its resources to cotton, a reduction of $4000 worth of corn and $5000 worth of beef permitted an increase of 12,000 pounds of cotton. What was the cost of that additional cotton? If the farm price of cotton were $.85 per pound, would this have been a profitable change?

5. In what way is an isoquant (an isoproduct contour) similar to an indifference curve?

# SUGGESTED READINGS

Bishop, C. E., and W. D. Toussaint. *Agricultural Economic Analysis*, New York: John Wiley & Sons, 1958, Chapters 9 and 10.

Bradford, Lawrence A., and Glenn L. Johnson. *Farm Management Analysis*, New York: John Wiley & Sons, 1953, Chapters 9, 10, and 11.

Brehm, Carl. *Introduction to Economics*, New York: Random House, 1970, Chapter 7.

Castle, Emery N., Manning H. Becker, and A. Gene Nelson. *Farm Business Management: The Decision-Making Process*, 3rd ed. New York: Macmillan, 1987, Chapter 14.

Ferguson, C. E. *Microeconomic Theory*, rev. ed. Homewood, Ill.: Richard D. Irwin, 1969, Chapter 6.

Leftwich, Richard H. *Introduction to Microeconomics*, New York: Holt, Rinehart, & Winston, 1970, Chapter 7.

Peterson, Willis L. *Principles of Economics: Micro*, 8th ed. Homewood, Ill.: Richard D. Irwin, 1991, Chapter 5.

# AN OUTSTANDING CONTRIBUTOR

**Vernon R. Eidman** Vernon Eidman is professor of Agricultural and Applied Economics at the University of Minnesota. He was born and raised on a 220-acre grain/livestock farm in southwestern Illinois. He earned his B.S. degree in agricultural economics at the University of Illinois in 1958. After completing six months active military service, he operated the home farm in partnership with his brother. Eidman returned to the University of Illinois in 1960, completing his M.S. in Agricultural Economics in 1961. He received his Ph.D. degree in 1964 at the University of California, Berkeley.

Eidman's Ph.D. dissertation applied decision theory to the California turkey industry. The American Agricultural Economics Association recognized a journal article he based on that dissertation as the outstanding article in the *Journal of Farm Economics* in 1968. A Giannini Monograph based on this work also was selected by the Western Agricultural Economics Association in 1969 for its Outstanding Published Research Award.

Eidman joined the faculty at Oklahoma State University in 1964 as assistant professor of Agricultural Economics. He advanced to associate professor in 1968 and professor in 1971. He joined the faculty at the University of Minnesota in 1975. He was visiting professor at the Swedish Agricultural University in 1972 and at the University of Maryland in 1989–90.

His research has focused on decision making at the firm level. Major efforts were devoted to the economics of poultry and livestock production, irrigation economics, technology assessment, decision making under uncertainty, and the integration of production, marketing, and finance decisions in farm planning and control. Additional research has dealt with water resources and the structure of agriculture. He served as a consultant to the Caribbean Agricultural Extension Project in 1987 and to the South African Agricultural Economics Association in 1989. He has authored and coauthored numerous refereed journal articles, book chapters, and experiment station and extension service publications.

Eidman teaches undergraduate courses in management and graduate-level courses in mathematical programming and production economics. He received awards for outstanding undergraduate teaching in the College of Agriculture at Oklahoma State University in 1970 and at the University of Minnesota in 1988. He coauthored the textbook *Farm Management*, published in 1984.

Eidman served as Chairman of the Graduate Committee at Oklahoma State University in 1970–72 and as Director of Graduate Studies at the University of Minnesota from 1986 to 1989. He served on the Editorial Council of the *Southern Journal of Agricultural Economics* from 1969 to 1972 and as vice president of the SAEA, 1975–76. He served on the Editorial Council of the *American Journal of Agricultural Economics* in 1972–74 and 1984–87. He was elected to a three year term on the AAEA Executive Board in 1989.

He is married to the former Bonnie Jean Klingelhoefer, and they have three children.

# CHAPTER 6

# Production Costs, Supply, and Price Determination

$\mathbf{A}$ firm's costs are incurred by using valuable resources to produce its product. Individual firms can change their total output only within the restraints imposed on them by the resources under their control. Hence the concentration on resource relationships in Chapters 4 and 5, are one element affecting a firm's costs. The other factor affecting production costs is the values of the resources used, an important part of this chapter's discussions.

We identify costs, as viewed in economics, by studying the methods by which costs are measured, in order to make intelligent production decisions. It is necessary to combine consumer valuations of a good expressed by the market demand curve with producer costs demonstrated by the market supply curve. These two curves together—demand and supply—are the determinants of the equilibrium market price for all goods bought and sold in the open market. Changes in these prices are signals to consumers and producers to adjust their consumption and production, reestablishing their optimal positions.

In a market economy, prices are the signals guiding decision makers to do those things that will be to their best advantage. As was shown in Chapter 4, when the price of a product increases or decreases relative to other prices, operators are being told by the market to increase or decrease their output of that good. Likewise, when resource prices change, resource cost changes force adjustments in the rates of resource use.

The operator of a firm develops plans and carries them out on the basis of expected future product prices, resource costs, and technical production relationships. A resource mix should be used to produce such output quantities that the difference between revenues and costs is the greatest possible. But what cost items do we tally up to derive the true costs of any operation in order to examine a firm's economic profitability? Because the economic meaning of a number of cost-identifying terms differs from common usage, their special meanings must be made clear.

## IDENTIFICATION OF COSTS _____

All production costs can be divided into two general groups, *explicit costs* and *implicit costs*.[1] An *explicit cost* has been incurred when money is spent to hire labor, repair machinery, buy seed, fuel, or other things for which cash expenditures are made. These expenditures have been made

---

[1] These costs are sometimes referred to as "direct" and "indirect," "cash" and "noncash," or "operating" and "overhead," with the same basic meaning intended.

to enhance product output, but a simple totaling of all such money spent is inadequate when trying to determine the costs of production, since explicit costs account for only about one-third of all costs on a typical American family-operated farm.

For all variable resource services purchased,[2] their costs are explicit, incurred directly to buy or hire those resources, with each being paid its market price. In Figure 4-4 we considered use of a variable resource $X_1$ whose market price was $5, meaning that others stood ready to buy that resource for $5 per unit. The market price reflects a resource's alternative employment opportunities, which must at least be matched by the buying firm to be able to obtain any of that resource.

An *implicit cost* has been incurred in using any resource for which there was no cash outlay during the period that resource was being used, and it is necessary to determine that type of resource cost. One such type of cost is for a resource lasting for two or more years. For instance, a $40,000 tractor may be purchased with the full purchase price paid on delivery, but since it will last for several years the $40,000 cost cannot be charged against one year's operations. The cost of the tractor must be allocated in such a way that the flow of its costs matches the flow of its services over the tractor's productive life. If this isn't done properly, costs will be over- or understated and apparent losses or gains also erroneously computed.

Another type of implicit cost is that incurred for the operator's own labor and management, which, because of circumstances, must accept whatever is left after other resources have been paid. Some even look on that leftover reward as something akin to profit, but this would be inconsistent with the last chapter's four-way classification of resources. Worse, this says nothing about the economic worth or value contribution of labor and management and provides no information for determining the true costs of production.

## Opportunity Cost

Common practice in appraising the cost of owned land gets closer to the economic meaning of implicit costs (not just for land, but for all owned resources). Taxes paid on owned land are correctly viewed as an explicit cost. But what about the value of one's investment in that land? Surely a cost exists that somehow relates to the value of that resource. The land, whatever its market value might be, may be owned free and clear, but most operators correctly feel that an investment interest charge should be made as a cost in the use of that land. This reflects the basic foundation

---

[2] Since one's own unpaid labor and management may be used in varying amounts, and certain fixed resources may be rented for cash, we are not able to state that explicit costs and the costs of all variable resources used are the same thing.

on which resource costing must rest by asking the question, "What would this investment earn if it were allocated to an alternative use?"[3] Thus, alternative earnings possibilities have very much to do with the cost of using that land.

Suppose your land could be sold for $500 per acre and that this money could be invested in a land mortgage (a loan even to the buyer of your land) at 8 percent per year. The annual interest yield would thus be $40 per acre, so if your present use of land returns only $25 per acre you simply are throwing away $15 per acre, a cost in the purest sense. This cost concept is a simple one that economists refer to as *opportunity cost* (or alternative cost). It considers the value of other opportunities foregone as a cost. So the cost of using any asset is the value of output that could have been obtained from a different, forfeited use.

Now, if your land is capable of producing either corn or wheat, and a commitment of $100 cash costs will yield $300 worth of corn or $200 worth of wheat per acre, what is the cost of producing wheat? Not the $100 spent to raise wheat! It is the $300 worth of corn sacrificed (the value of a foregone alternative) that is the true cost of producing wheat.

Maybe you are raising your own hay for livestock feed. Even though you may have made no direct payment to someone in producing that hay, you still wouldn't price the hay at zero cost to your livestock, because at least one alternative is to sell it. Therefore, the opportunity cost for the hay is its selling price when fed to your livestock.

The same is true for all the firm's owned resources. Returns for each resource, in any use, must equal the next best alternative for each of them or the resource is being used inefficiently and more profitable earnings are being bypassed. So we may go even further and state that *opportunity costs are the true costs of production*.

A summary view of owned-resource costs would then run something like this: If I could be paid $5000 for my labor and management doing this same work for someone else, and my land investment could earn $15,000 if the money value of the land were loaned to someone else, and if the value of machinery, buildings, brood stock, and other capital assets could earn $10,000 in another use, each of these amounts is the opportunity cost of that particular resource type, and their sum ($30,000) is the full opportunity cost of the present use of these resources.

## Profit

The reason for our hesitancy to mention profit may now be made more clear. From whatever viewpoint, the word profit carries the implication

---

[3] Whatever the alternative investment might be, it must be one of equal risk in order to arrive at a realistic estimate of alternative earnings.

of a surplus of receipts over expenses. Thus, an accounting profit has been made when all operating costs (direct expenditures for labor hired, rentals, and other purchased inputs) and overhead costs (taxes, insurance, depreciation, and so on) are exceeded by revenues so that a net balance remains.

A bookkeeping profit, however, is not necessarily the same as an economic profit. A net surplus of revenue over expenses in the corn-wheat example just mentioned, is possible even if only wheat is produced, yet an economic loss is incurred because a higher return (with the same costs) from corn production was forfeited. An *economic profit* (or *pure profit*)[4] has been made when a firm's revenues exceed the total of its explicit and implicit (opportunity) costs. From the standpoint of any resource, an economic profit is the amount by which its net earnings exceed the payment required to attract it to (or keep it in) its present use. This concept of profit thus includes what might be called "normal profits" as a cost to the firm because each resource is priced to the firm at its opportunity cost.

## Fixed and Variable Costs

After looking at the true meaning of costs, it now becomes necessary to make a different classification of costs—one that coincides with the fixed and variable resources as categorized earlier in the production function.

We will call *variable* those costs that increase or decrease as output changes, and *fixed* those costs incurred for the resources that do not change as output is changed. Our costs now include both explicit and implicit costs, and the sum of variable and fixed costs is the total cost of the firm's operations.

This arrangement of resources and their costs permits an emphasis clarifying just how the costs of production affect the firm's output. We now are able to demonstrate the operator's response to market price changes.

## Length-of-Run

In order to emphasize the source of production costs, look back to the discussion (in Chapter 4) of how some resources come to be fixed and others variable. Two possibilities were discussed in relation to the law of diminishing returns.

One possibility was that we simply could choose any desired functional relationship by classifying some resources as fixed and others as variable, then handling them accordingly, and noting the outcome. There's

---

[4] Resource economists have traditionally used yet another term, *economic rent*, as we will do in Chapter 11, with exactly the same meaning as economic (or pure) profit.

nothing wrong with this, it is scientifically correct. But you might consider this an unexciting exercise with little real-life application.

The distinction between fixed and variable resources need not be an arbitrary one, however. Whether resources are fixed or variable is an everday circumstance of the planning process *all* producers face, one that is related to the passage of time, yet not dependent on days, months, or years as basic to the distinction.

Many farmers plan their cropping season operations during the preceding months. And each has a set of expenses that would continue at the same (fixed) level whether the farm produces to its maximum output or shuts down completely. These expenses are rightly regarded by the operator as fixed.

Other expenses would be variable because of choices available in the rates of use for such resources as labor hired, machine use, seed, fertilizers, pesticides, and herbicides. These are all variable expenses because the amounts of each of them can be varied. And whether their number is large or small they will influence the *length-of-run* of the production function and the consequent cost function.

Length-of-run is a planning concept reflecting that there exist differences in one's ability to change input use. Some resources cannot be changed during one span of time; but those same resources may not be fixed given a longer period of time. We may then look at two extremes of length-of-run with innumerable possibilities between.

## The Short Run

One length-of-run we might call the *immediate short run,* meaning right now, a time span so short that no resource changes can be made. Everything, therefore, is fixed. A truck gardener, for example, may have completed harvesting and have the produce ready for sale. The operator faces the problem of disposing of that produce in as short a time as possible to prevent spoilage losses. Nothing can be done to change the quantities of resources used or the amount of produce to be marketed.

A somewhat longer time period is involved when a producer considers what to do next year. Certain changes might be made in the use of land without changing total land acres. The acreage of a particular crop might be adjusted, resulting in some other resource cost changes but with the large majority of costs still remaining unchanged.

An even longer period might be one covering the next two, five, or even ten years. A change in total land acres may or may not be planned, with no change planned for certain capital improvements, and with some machinery and equipment being replaced. The production function still has some resources fixed and some variable, but with more resources varied than in the examples just given. What we are considering then, is just different short-run periods, some being shorter or longer than others, and still within the framework of the law of diminishing returns, a short-run concept itself.

Figure 5-14 can be used as an explanation of two different short-run functions. The two-variable function that yielded the production surface would itself be a short-run function; the subfunction slice of that surface, with a single-variable input, would be an even shorter length-of-run with costs of production affected accordingly.

## The Long Run

The other extreme, the *ultimate long run,* is then a period long enough that all resources may be varied. The quantities of all resources may be changed, including management itself, either by special training to make it more capable of directing the planned operation, or by hiring the service. In its strictest application, this length-of-run gives rise to the special constant returns function described in Figures 4-1A and 5-11, where all resources are varied. Thus, the distinction between the long run and any shorter run hinges on whether or not there are fixed factors in the function. Since there may be an infinite number of possibilities between now and some period far into the future, we may summarize by simply saying that the *long run* is a period of time so long that everything is variable, whereas in the *short run* one or more of the factors cannot be varied, giving rise to diminishing returns.

## Short-Run Costs of Production

To maintain sight of the specific relationships between the production function and the costs of production, we will retain an example from the previous chapter, using the (short-run) functional information from Table 4-1. Those data came from the production surface with two variable inputs (Figures 5-14 and 5-15) in which the input $X_2$ was fixed at 40 units, yielding a subfunction of the general form $Y = f(X_1 \mid X_2, X_3 \ldots X_n)$. As the production relationships were analyzed all measures were *input oriented*, with averages and marginals derived on a per-unit-of-input basis. Because the input $X_1$ was the *independent variable*, units of $X_1$ were measured along the horizontal axis, and output, the *dependent variable*, along the vertical axis.

We switch now to an *output oriented* set of cost measures, all of which are derived from the production function, generating cost of production information. Output is now the *independent variable* plotted along the horizontal axis, and input $X_1$ (now) the *dependent variable* plotted along the vertical axis.

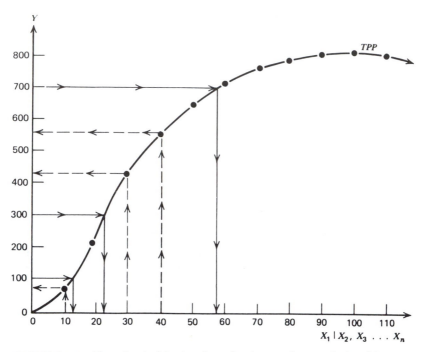

**FIGURE 6-1**   The physical basis of production and cost relationships.

Figure 6-1 is a graphic demonstration of the reciprocal nature of production and cost functions (data from Table 4-1). Earlier, output quantities were determined in a manner equivalent to saying, "If 10 units of $X_1$ are used, how much output will be produced?" The same question is apparent for all other input levels. Arrows pointing upward from the X-axis to the production function, then to the Y-axis, demonstrate this for selected input levels. Cost analysis requires only that the question be restated as, "If a certain amount of product is to be produced, how many units of the variable input $X_1$ must be used?" Arrows pointing from the Y-axis to the production function, then to the X-axis, demonstrate the cause–effect nature of this relationship.

With output now used as the independent variable measured along the horizontal axis (using 100-unit increments for convenience only), the physical basis of the cost of producing each output quantity becomes clear. The physical basis is the amounts of $X_1$ used in producing those output quantities, as shown in the first two columns of Table 6-1.

When dollar amounts are computed for each observed quantity of output and input, given the same prices as were used in the earlier production example ($P_y$ = $1/unit, and $P_{x1}$ = $5/unit), we can determine total revenue and total costs for each level of output. Production costs (the three

**TABLE 6-1**  Short-Run Costs (With $X_2$ Fixed at 40 Units)

| Output[a] (Y) | Input ($X_1$) | Total Variable Cost (TVC) | Total Fixed Cost (TFC) | Total Cost (TC) |
|---|---|---|---|---|
| 0 | 0.0 | $ 0.00 | $150.00 | $150.00 |
| 100 | 11.6 | 58.00 | 150.00 | 208.00 |
| 200 | 17.6 | 88.00 | 150.00 | 238.00 |
| 300 | 22.8 | 114.00 | 150.00 | 264.00 |
| 400 | 28.0 | 140.00 | 150.00 | 290.00 |
| 500 | 34.5 | 172.50 | 150.00 | 322.50 |
| 600 | 43.9 | 219.50 | 150.00 | 369.50 |
| 700 | 57.8 | 289.00 | 150.00 | 439.00 |
| 800 | 90.0 | 450.00 | 150.00 | 600.00 |
| 810 | 100.0 | 500.00 | 150.00 | 650.00 |

[a] Be aware that our last increment to output is only 10 units of Y, rather than 100 units as in all other increments.

remaining columns in Table 6-1) are expenditures for the variable input, the costs of the fixed resources, and the total of these costs. Multiplying each quantity of $X_1$ used, at the different levels of output, by the market price of $X_1$ gives **total variable cost** (TVC), the total spending for the variable input. Note that the first 100 units of output required 11.6 units of $X_1$ ($58.00 worth) with variable resource requirements per unit of output declining through 400 units of output, then rising to 32.2 units of $X_1$ ($161.00 worth) required for the last full 100-unit increment of output (700Y to 800Y), a consequence of diminishing returns.

Since we have used the data from Table 4-1 (which was a subfunction slice of the production surface in Figure 5-15, with $X_2$ fixed at 40 units), the cost of $X_2$ is a component of fixed costs. At $2.50 per unit of $X_2$, this cost item amounts to $100. Assuming that the implicit opportunity costs of all *other* fixed resources amount to $50, **total fixed costs** (TFC) now are $150. This $150 cost is a constant for all levels of output.

**Total cost** (TC) for this firm is the sum of its total variable cost and total fixed costs, which, because of the increasing variable cost, also rises as output is increased. These cost measures are graphed in Figure 6-2.

The firm's **total revenue** (TR) is derived by multiplying the price of the product ($1 per unit) by the units of product, at each output level. TR plots as a straight line because of the market assumption that prices will not change as a result of this one firm's production decisions.

With all costs included, subtracting TC from TR shows the firm's net revenue at each level of output, as in Table 6-2. These net values can correctly be called *pure profits* because all production costs have been in-

**TABLE 6-2**    Revenues, Costs, and Profits

| Output (Y) | Total Revenue (TR) | Total Cost (TC) | Profit | |
|---|---|---|---|---|
| 0 | $ 0 | $150.00 | $–150.00 | |
| 100 | 100 | 208.00 | –108.00 | |
| 200 | 200 | 238.00 | –38.00 | |
| 300 | 300 | 264.00 | 36.00 | |
| 400 | 400 | 290.00 | 110.00 | |
| 500 | 500 | 322.50 | 177.50 | |
| 600 | 600 | 369.50 | 230.50 | |
| 700 | 700 | 439.00 | 261.00 | Maximum |
| 800 | 800 | 600.00 | 200.00 | |
| 810 | 810 | 650.00 | 160.00 | |

cluded. Because *TC* is a curving line (resulting from changes in productivity of the variable input as increasing amounts of that resource are combined with the fixed resource), there can be only one optimal output, 700 units of output with a profit of $261. Profits will be reduced by producing any output other than 700 units. Graphically (Figure 6-2), this is where the vertical distance between the *TR* and *TC* curves is a maximum. Any movement away from 700 *Y* will reduce that vertical difference and, therefore, reduce profit.

It was stated earlier that fixed costs have no influence in determining the optimal level of output. The correctness of this can be illustrated here. On the graph, imagine that fixed costs are zero: The total cost line will simply shift downward by $150 without changing its shape or slope, and the profit-maximizing output remains unchanged at 700. Increase fixed costs to $500: Now there is no net profit, only net losses. What can be done?

The first temptation might be to say "Quit!" maybe hoping that market conditions will improve later. But shutting down production would result in forfeiting any possible return on fixed costs, with a loss totaling $500. As levels of output greater than zero are considered, losses are reduced until we reach 700 units of output with losses minimized at $89.00, so an **optimum** output may be said to be *that output quantity at which profits are maximized or losses minimized*. And this optimum output has been determined without regard to fixed costs.

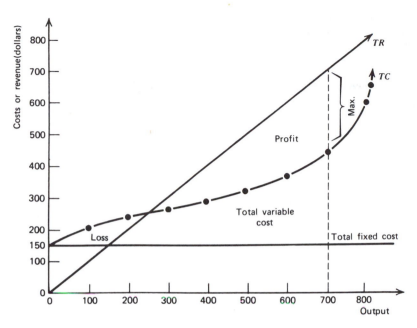

**FIGURE 6-2**  Short-run costs and revenue.

## Measuring Per-Unit Costs and Returns

An analysis such as the foregoing reveals aggregates of returns, costs, and profits. But market signals, being given only in dollars per unit of products and resources, require a further breakdown so that our information is in the same form. When put in the same unit measures as provided by the market, the operator can then determine whether the market price of the product is high enough to make it worth it to use valuable resources to produce that commodity, and what quantity produced will be an optimum.

Four additional cost measures are required to complete analysis on a per-unit-of-output basis—*average variable cost* (*AVC*), *average fixed cost* (*AFC*), *average total cost* (*ATC*), and **marginal cost** (*MC*). Each of these unit cost measures is derived from its total counterpart, as shown in Table 6-3 and graphed in Figure 6-3. Each computation furnishes the producer with useful information on the types and amounts of costs, all on a unit-of-output basis.

**TABLE 6-3**  Short-run Costs and Returns per Unit of Output

| Output (Y) | AVC $(TVC \div Y)$ | AFC $(TFC \div Y)$ | ATC[a] $(TC \div Y)$ | MC $(\Delta TC \div \Delta Y)$ | MR $(\Delta TR \div \Delta Y = P_Y)$ |
|---|---|---|---|---|---|
| 0 | $0.00 | $0.00 | $0.00 | | |
| | | | | $0.58 | $1.00 |
| 100 | 0.58 | 1.50 | 2.08 | | |
| | | | | 0.30 | 1.00 |
| 200 | 0.44 | 0.75 | 1.19 | | |
| | | | | 0.26 | 1.00 |
| 300 | 0.38 | 0.50 | 0.88 | | |
| | | | | 0.26 | 1.00 |
| 400 | 0.35 | 0.38 | 0.73 | | |
| | | | | 0.33 | 1.00 |
| 500 | 0.35 | 0.30 | 0.65 | | |
| | | | | 0.47 | 1.00 |
| 600 | 0.37 | 0.25 | 0.62 | | |
| | | | | 0.70 | 1.00 |
| 700 | 0.41 | 0.21 | 0.63 | MC = MR | |
| | | | | 1.61 | 1.00 |
| 800 | 0.56 | 0.19 | 0.75 | | |
| | | | | 5.00 | 1.00 |
| 810 | 0.62 | 0.18 | 0.80 | | |

[a] Numbers do not add exactly, because of rounding error.

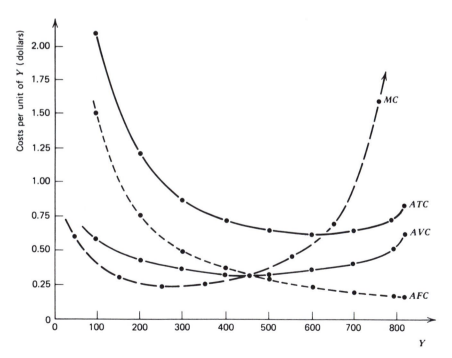

**FIGURE 6-3**  The firm's short-run cost curves.

*Average variable cost* (*AVC*), the amount spent on the variable input per unit of output, is derived as

$$AVC = \frac{\text{Total variable cost}}{\text{Output}} = \frac{TVC}{Y}$$

Because the variable input is relatively unproductive when only a small amount of that resource is being used, the cost of the variable input per unit of output produced will be relatively high. As the productivity of that resource increases with greater amounts used, *AVC* must fall, describing a U-shaped curve, as graphed, that is the reciprocal of the $APP_{x1}$ curve.[5] *AVC* falls to its minimum at the output quantity where $APP_{x1}$ is a maximum and rising thereafter as greater amounts of output are produced, becoming vertical at the firm's maximum output (810 $Y$ when $X_2 = 40$).

*Average fixed cost* (*AFC*), the cost of the fixed resources per unit of output, is derived as

$$AFC = \frac{\text{Total fixed cost}}{\text{Output}} = \frac{TFC}{Y}$$

*AFC* is a declining curve with increased output because a constant (*TFC*) is being divided by an increasingly larger number (*Y*), up to the maximum output of 810 units. For this firm, given its specific resource organization, *AFC* can fall no lower than \$0.18 because an output greater than 810 units is impossible without first increasing the fixed resources in the firm.

*Average total cost* (*ATC*), the total cost of all the resources used per unit of output produced, is the sum of *AVC* and *AFC* and is computed as

$$ATC = \frac{\text{Total cost}}{\text{Output}} = \frac{TC}{Y} \text{ or } = \frac{TVC + TFC}{Y}$$

Graphically, *ATC* is equal to the combined heights of *AVC* and *AFC*. Since both these curves are high and falling, beginning with the lowest output levels, *ATC* must also be high and falling through that same general range of output. Its minimum point occurs at a greater amount of output than for minimum *AVC* because *AFC* continues to decline as output is increased. A point is reached, past the point of minimum *AVC*, where the increase in *AVC* is just equal to the decrease in *AFC*. At that

---

[5] That changes in the productivity of the variable input give *AVC* its specific shape can be verified by looking at the source of each measure of *TVC* and output. *TVC* equals the units of $X_1$ used, times the price of $X_1$ (i.e., $TVC = X_1 \cdot P_{x1}$). Units of output produced equals the units of $X_1$ used times its *APP* (i.e., $Y = X_1 \cdot APP_{x1}$). So by substitution, $AVC = TVC/Y = (X_1 \cdot P_{x1})/(X_1 \cdot APP_{x1}) = P_{x1}/APP_{x1}$. With $APP_{x1}$ divided into a constant ($P_{x1}$), *AVC* must trace a U-shaped path, clearly demonstrating the reciprocal relationships between *AVC* and *APP*.

point (600 $Y$) $ATC$ must be at its minimum. $ATC$ will rise from that level of output because $AVC$ is rising, becoming vertical at the maximum output of 810 units. Like $AVC$, $ATC$ also traces a U-shaped curve between zero and maximum output.

The seventh and final cost concept, *marginal cost*, looks at production costs in the same incremental manner as was done in the last chapter to determine the marginal productivity of a variable input. Marginal cost ($MC$) is defined as the change in total cost when output is changed by one unit and is determined by

$$MC = \frac{\text{Change in total cost}}{\text{Change in output}} = \frac{\Delta TC}{\Delta Y} \text{ or } = \frac{\Delta TVC}{\Delta Y}$$

Remember that the output observations in Table 6-1 are in 100-unit groupings; therefore $\Delta Y$ is 100 for each observed change (except for the last increment of only 10 units of $Y$). With the *change in output* divided into the *change in total costs* accompanying that output change, the computed cost value is then on a per-unit-of-output basis. Note that in computing $MC$ we can use either $\Delta TC$ or $\Delta TVC$ because they are one and the same. As fixed costs cannot be changed, the only element of change in $TC$ is the change in $TVC$.

The reason for the specific shape of the $MC$ curve can be made more clear by looking at the physical basis of production as indicated by the marginal product of the variable resource. Because the productivity of the variable input increases when more of that resource is used, from zero $X_1$ up to its maximum $MPP$, $MC$ must be falling at the same time. No matter what the output produced by the first unit of $X_1$, given $P_{x1}$ unchanged, if a second unit yields more product than the first unit, the cost per unit of producing that additional output must fall—$MC$ must decline as long as $MPP$ increases. The same reverse relationship must hold when $MPP$ is declining: Smaller and smaller increments to output with each (equal) increment to input must cause $MC$ to increase. Where $MPP$ is a maximum, $MC$ is a minimum; and where $MPP$ is zero, $MC$ becomes vertical.[6]

---

[6] This relationship between marginal cost and marginal product may be restated in a similar manner as was done for $AVC$ and $APP$. Since $\Delta TC$ (= $\Delta TVC$) equals $\Delta X_1$ times the price of $X_1$, and $\Delta Y$ equals $\Delta X_1$ times $MPP_{x1}$, by substitution

$$MC = \frac{\Delta TC}{\Delta Y} = \frac{\Delta X_1 \cdot P_{x1}}{\Delta X_1 \cdot MPP_{x1}} = \frac{P_{x1}}{MPP_{x1}}$$

With $MPP_{x1}$ divided into a constant ($P_{x1}$), $MC$ must trace a U-shaped path that is a reciprocal of the $MPP$ curve. (Some prefer to call this a J-shaped curve, rather than U-shaped.)

## The Search for an Optimum

In the previous chapter, the question of how much of the variable input to use was answered after developing incremental measures of value productivity per unit of input (*MVP*) and cost per unit of input (*MFC*). Once having determined the *MVP* schedule, and the level of input use at which increments to revenue and cost were equal (*MVP* = *MFC*), the optimal input use is determined (Figure 4-4).

Our problem now is to determine that one *output level* at which increments to costs and revenue are equal. The one additional concept needed to make this determination is **marginal revenue** (*MR*). Marginal revenue can be defined as the amount added to total revenue when an additional unit of output is produced and sold.

An assumption about the market—that this producer's decisions will have no effect on the price of the product[7]—simplifies the problem of optimizing output. Because $P_y$ remains constant whether this operator decides not to produce at all, or expands output to the maximum, the market price can be plotted as a horizontal line in Figure 6-4. In algebraic symbols,

$$MR = \frac{\Delta TR}{\Delta Y} = P_y^8$$

Given $P_y$ = $1.00, each additional unit sold will add $1.00 to total revenue (the definition of *MR*). Because costs at the margin change (according to the *MC* curve) as output is changed, the operator must find that one output level at which *MC* = *MR*.[9] At that output (700 *Y*), profits are a maximum (or losses a minimum).[10]

---

[7] The market assumption we make here is not at all unrealistic; it simply says that this producer's firm is so small a part of the total market that decisions to either increase or decrease output will not be noticed by the market, that is, market price will not change as a result of this firm's decisions. As you might observe, very much of this nation's agricultural output is produced in this type of market. Whether it be grain crops, livestock, poultry, or numerous other commodities produced by large numbers of farms, the decision of producers, acting individually, cannot affect the market price of their product.

[8] Because $\Delta TR = \Delta Y \cdot P_y$, by substitution $MR = (\Delta Y \cdot P_y) / \Delta Y$. Since the $\Delta Y$'s cancel, we are left with the identity $MR = P_y$.

[9] For any output level to be an optimum, *MC* must be increasing and equal to or greater than *AVC*.

[10] This can be checked against the data in Tables 6-2 and 6-3. Profits are a maximum of $261 at 700 *Y*, where *MC* (= $1.00) = *MR* (= $1.00), and would fall to $230.50 if output were reduced to 600 *Y*—a needless forfeiture of $30.50 profit. Further reductions of output make that difference even greater, with further unnecessary sacrifices of profitability. Increase output to 800 Y and *MC* (= $1.61) exceeds *MR* by $0.61, an unnecessary loss of $61.00 on that 100-unit increment of output.

So the *MC* = *MR* optimizing rule forces adjustments in output because of inequalities in costs and returns at the margin. If *MR* at any level of output exceeds *MC*, that inequality

## The Firm's Short-Run Supply Curve

With a product price of $1.00 per unit, we discovered that the profit-maximizing output would be 700 $Y$. But what of the many other possible prices for this product? In the absence of strict price controls, there is no reason to expect price to be locked in at $1.00 only.

What would you as a producer do if the market price were to rise, say to $1.50 per unit? By applying the rule, produce to where $MC = MR$, you would find a new optimum output between 700 and 800 $Y$. You would employ a greater amount of the variable input to produce more $Y$, to where $MC = \$1.50$, because that would be more profitable than just maintaining the previous level of output. Instead of $P_y = \$1.50$, make it $2.00, $4.00, or $10.00, and the optimum output will be greater still (but don't forget that the fixed resource structure of this firm prevents any output greater than 810 $Y$).

Let the price fall to $0.50 per unit and we must find that output level where $MC$ also is $0.50 per unit ($MC = MR$ at 600 $Y$).[11] At that output, $ATC$ is $0.62, meaning that costs per unit total $0.12 more than the item brings in the market. Now you're losing $72.00. Should you just quit? No, at that output $TVC$ *is* $219.50 while $TR$ is $300. There is a net return of $80.50 over variable costs to apply toward fixed costs, and losses are minimized at 600 $Y$.

How about a market price for $Y$ of $0.30? At that price, $MC = MR$ at 400 $Y$. But at this output $AVC$ is $0.35, so you would be spending more for the variable resource ($TVC = \$140.00$) than you get back from its product ($TR = \$120.00$), for a net loss of $20.00 on variable costs alone. This loss, in addition to the $150.00 fixed costs (a total of $170.00), is worse than your losses would be if you stopped producing entirely.

Note that at 500 $Y$, $AVC$ is a minimum of $0.35. At any product price less than that, variable costs cannot even be covered because more would be spent on the variable resource than the product can be sold for. So we have found the minimum price below which this firm can't afford to produce—it pays more not to produce at all. At all prices greater than this, output will be determined at the level where $MC = MR$. We now have a *price–quantity schedule* that is the firm's short-run *supply curve*: the $MC$ curve above minimum $AVC$ is the firm's supply curve showing how much $Y$ will be produced at all possible prices for this product, as indicated by the heavily drawn, solid-line segment of the $MC$ curve in Figure 6-4.

As market price may fluctuate from a price that is equal to this firm's minimum $AVC$ to any level greater than that, the firm finds its profitabil-

---

simply tells the operator that an additional surplus can be captured and added to profits by simply increasing output. The opposite signal is equally forceful when $MC$ exceeds $MR$.
[11] It's really more like 575, but we'll use 600, because numbers are handy for an output of 600.

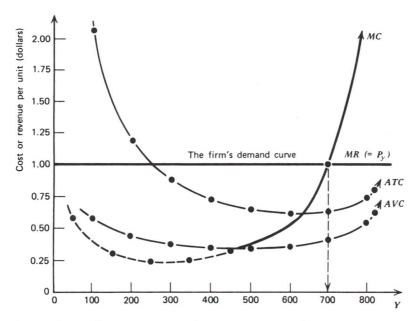

**FIGURE 6-4** Short-run costs and returns per unit of output.

ity affected accordingly. Note that earlier in this chapter we priced all the fixed resources to this firm at their opportunity costs, the only meaningful measure of the *true* costs of using resources to produce something. When we tally up the *explicit* costs for the variable resources used with the *implicit* costs of the fixed resources, then subtract this from total revenue, a balance exists that can be called *economic profit*. This profit can be negative, zero, or positive. A zero profit then is simply a situation where the present use of resources is neither more nor less profitable than the next-best-paying alternatives for those resources. Economic profit is thus a surplus, being either greater or less than the return necessary to attract or keep these resources in their present use. If a zero profit can be expected to keep these resources in their present use, either a negative or positive profit, on the other hand, should trigger adjustments within this firm in the quantities of resources it uses.

If the market price is greater than *AVC* but less than *ATC*, the firm will be incurring a loss. How long can it continue to do so? Until fixed asset depreciation has continued to the point where those assets must be replaced, or when the opportunity costs (alternative earnings) of the fixed assets cause these resources to be shifted to alternative, higher paying uses. At that time some or all of the fixed resources are no longer available to produce the product—either worn out and unproductive or shifted elsewhere—reducing output below that possible from the original function. This situation will reduce productivity of the variable resources and increase unit costs of production, leaving production even more unprofitable and squeezing this firm (with its present resource organization) out of the market.

The existence of a net surplus over full production costs and the changes this will cause are more easily visualized than the consequences of a deficit. Figure 6-5 shows a situation where price is greater than *ATC*. The firm's optimum output is $0$-$q_1$ (where $MC = MR$). Total revenue for the firm is the rectangle $0$-$b$-$c$-$q_1$ (price of $0$-$b$ times the quantity $0$-$q_1$) while total costs are only $0$-$a$-$d$-$q_1$. The shaded rectangular area $a$-$b$-$c$-$d$ is therefore a surplus (economic) profit. The operator would be encouraged to increase the scale of the operation by adding to the fixed resources and using a greater amount of the variable resource to take even better advantage of this profitable situation.

Such a situation may be unique to this one firm only, but it also could be somewhat general in a competitive industry in the short run. If so, the market for this product will cause other producing firms to adjust similarly, and attract others into the market, having an effect on both resource costs and product price, eliminating the economic profit. The final effects of these changes are shown in Figure 6-6 demonstrating market changes (*B*) as well as those for the individual firm (*A*).

As potentially competing firms see the profitability of this and similar firms, the prices of the variable resources will likely increase (more firms wanting and bidding for the existing supply of these resources, and others being attracted to this use only if their price and expected returns will increase and exceed their present returns). This circumstance will cause

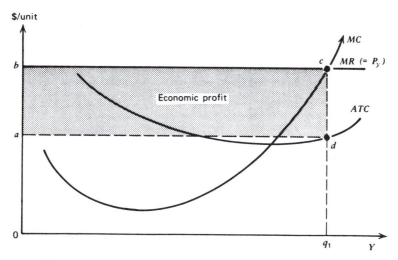

**FIGURE 6-5**    Production costs, revenue, and economic profit.

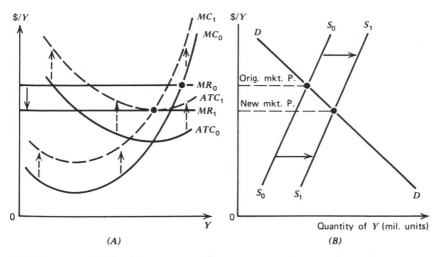

**FIGURE 6-6** Final adjustments in firm costs, market supply, and price in responses to economic profit. (*A*) The firm. (*B*) The market.

an upward shift of the firm's *AVC* and *MC* curves, resulting also in an upward shift of the *ATC* curve.

At the same time, the supply of fixed resources will find their prices bid up by the activities of the new firms coming into the market as well as by the existing firms wishing to expand the scale of their operations. This will cause an upward movement in *AFC*, and another source of upward pressure on *ATC*.[12] These two sets of forces will result in this firm's *ATC* moving upward to the new position shown in Figure 6-6A.

As more resources are committed to producing this product (new firms coming in, plus the expansion of existing firms), the price of the firm's product will be forced downward. Expanding resource use (both fixed and variable) means that the market output will expand as shown by the shift in supply from $S_0$ (original supply) to $S_1$ (new supply) in Figure 6-6B. Given demand unchanged, the new market price for this product will be lower than the original price, as determined by the intersection of the market supply and demand curves.

The consequences of all this is that, with costs being pushed upward and market price moving downward, our firm's original economic profit has been squeezed out, and it is now just normally profitable.

---

[12] Here's how: Suppose an operator had priced his own management at $5000 per year (the next-best-paying alternative use of this resource). The $5000 cost is built into the firm's fixed cost structure. Suppose now that someone else wishing to get into this type of business also can see the way clear to hiring this operator to do the same management job, but at a payment of $10,000 per year. Our operator's opportunity cost for his management, in the existing firm, has now gone up, with corresponding upward changes in *TFC*, *AFC*, and *ATC*.

Changes such as these presuppose a number of qualifications, which we will treat in Chapter 7. We have dealt here only with a market situation in which the producer and all others in this market are too small to have any influence whatever on market price. Along with this has been an implicit assumption that the individual is free to choose from among alternatives as market prices change. Our intent is not to describe a particularly desirable market, but simply to hold all other factors constant so that the effects of certain decisions can be isolated and attributed to that decision alone. The effects of market restrictions, by whatever source or cause, will be treated in Chapters 7 and 8 where specific market conditions can be handled separately.

# MARKET SUPPLY ————————————————

Market supply curves differ from demand curves in that they are determined by producers' costs and they normally slope upward and to the right rather than downward to the right. A supply curve is defined as the amount of a good or service producers are willing to offer for sale at different prices, *ceteris paribus*. The **market supply curve** is determined in the same manner as is done for a market demand curve. An individual firm's output response to price changes was described (Figure 6-4, and related text) as being determined by the intersection of the firm's *MR* (the product's market price) and *MC* curves. Thus, the *MC* curve above minimum *AVC* is the firm's supply curve.

All the firms producing a good for the market are the source of the market supply of that good. Individual *MC* curves are summed horizontally to obtain the total amount of that good those firms are willing to produce for the market at all possible prices. A graphic description of this procedure is shown in Figures 6-7 and 6-8, assuming input prices remain constant.

In Figure 6-7, three separate firms 1, 2, and *n* and their respective *MC* curves are used to represent the number of firms in this market. With one price shown ($1.00), each firm decides its output quantity by equating price (*MR*) and *MC*. The quantity for each firm (plus other firms not shown) becomes the total market supply at that price (Figure 6-8). The shape and slope of the market supply curve is determined by the quantities each firm would produce at all other prices.

Using wheat as an example, assume a situation as indicated in Figure 6-9. The market supply curve shows that wheat producers are willing to produce 500 million bushels of wheat at a price of $1.00 per bushel and two billion bushels at a price of $4.00 per bushel. This figure illustrates the concept that as wheat prices increase, producers are willing to commit more resources to wheat production and increase the output of wheat.

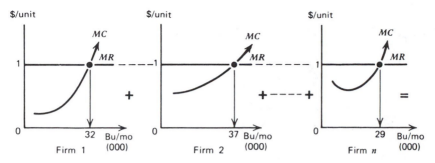

**FIGURE 6-7**  Supply curves for all firms producing Good A: Firm 1, Firm 2, Firm *n*.

This direct relationship between price and quantity produced exists due to the increased resource costs of increasing output. Given the resources and technology available to them, producers seek to maximize their returns within the legal framework in which they operate. This practice does not mean that everything a person does as a supplier of goods and services is related to profit, but only that it is an important influence. Other motivating factors are prestige, tradition, religion, and so on.

## Changes in Market Supply

Movement along a supply curve is called a *change in quantity supplied*, while a *change in supply* is a shift in the entire supply schedule. An increase in supply is shown in Figure 6-10 by a shift from *SS* to *S'S'*. This movement could be precipitated, for instance, by good grow-

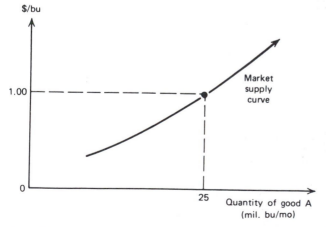

**FIGURE 6-8**  Market summation of firms' supply curves for Good A.

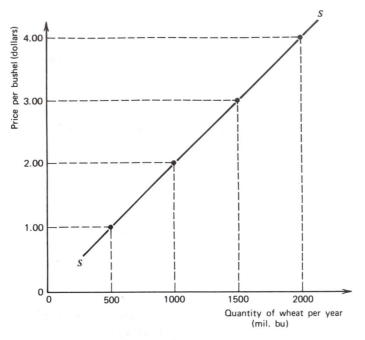

**FIGURE 6-9** A market supply curve.

ing conditions or by an improvement in technology (new crop variety that increases yield per acre). Other factors that could cause this shift include a reduction in resource prices (making it profitable to use more of them), a reduction in relative prices of other products (causing producers to increase their output of this commodity), or changes in institutional constraints such as increased acreage allotments under a farm program. On the other hand, a decrease in supply could be caused by such things as drought or crop diseases or opposite changes in the supply shifters mentioned above. The impact of these factors would be to shift the supply curve to the left as in Figure 6-11 so that fewer units of product are supplied at each price.

## Elasticity of Supply

Price elasticity of supply and demand are calculated with the use of the same algebraic expression:

$$E_S = \frac{(Q_1 - Q_2)/(Q_1 + Q_2)}{(P_1 - P_2)/(P_1 + P_2)}$$

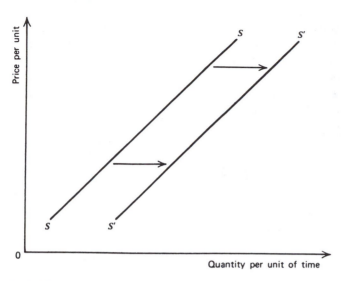

**FIGURE 6-10**  An increase in supply.

Price elasticity of supply is defined as a measure of the percentage change in quantity supplied in response to a percent change in price, *ceteris paribus*. A supply elasticity of 0.4 for cotton in the short run means that the quantity supplied increases 0.4 percent with a 1 percent increase in the price of cotton. The sign on the price elasticity coefficient is usually positive, as the supply curve is normally positively sloped.

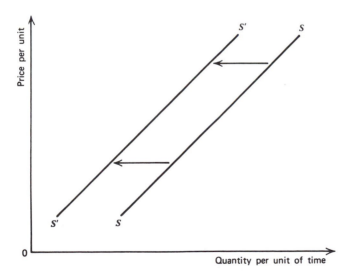

**FIGURE 6-11**  A decrease in supply

A perfectly vertical supply curve has a zero elasticity coefficient. A zero supply elasticity coefficient means the quantity supplied is not responsive to price changes. If the supply elasticity is between zero and one, the supply elasticity is referred to as being *inelastic*. The percentage change in quantity supplied is less than the corresponding percentage change in price. A supply elasticity coefficient greater than one defines an *elastic* supply. When the percentage increases in quantity supplied and price are the same, the coefficient is 1.0. That supply elasticity is called *unitary*.

Supply elasticities are highest for those crops and livestock where production adjustments are relatively easy to make, such as in potatoes, eggs, and poultry. Low supply elasticities are encountered for fruit, wheat, tobacco, cotton, feed grains, and milk.

## PRICE DETERMINATION _____

Individual demand curves for a product are added up to derive the market demand curve; individual supply curves for a product also sum horizontally to derive the market supply curve. The demand curve reflects the desires of the consumers whereas the supply curve indicates the motivations of producers. These two curves interact as in Figure 6-12 to determine market price.

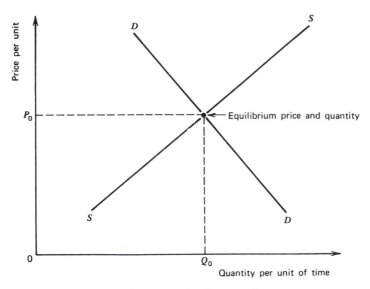

**FIGURE 6-12**   Price determination in a market.

## Equilibrium in the Market

At the point where the market demand curve (*DD*) intersects the market supply curve (*SS*), the quantity demanded by consumers ($Q_0$) equals the quantity supplied by producers ($Q_0$). This balance occurs at the equilibrium price $P_0$. At equilibrium, all buyers of this product who are willing to pay price $P_0$ for the commodity could buy the amount they wanted, and all producers who supplied quantity $Q_0$ could sell their product at $P_0$, the price they needed to receive in order to produce that quantity. There are no shortages or supluses in the market; the market is in *equilibrium*.

## Market Disequilibrium

There is a tendency for equilibrium to exist unless demand shifters or supply shifters cause price to change from the equilibrium position. If such changes occur, a new equilibrium will be found.

A price such as $P_1$ in Figure 6-13 is not an equilibrium and will cause a surplus to exist. At $P_1$ producers will wish to sell $Q_2$, but consumers are willing to buy only $Q_1$, leaving a surplus ($Q_1$ to $Q_2$) in the market at this price. Producers who want to sell this surplus must yield to the downward pressure on their asking price. Only when the price falls to the equilibrium price $P_0$ will consumers purchase all that suppliers want to sell.

On the other hand, if a price is initially established below equilibrium, at price $P_2$ (Figure 6-14), suppliers will supply only quantity $Q_1$, but consumers want quantity $Q_2$, as shown by the demand curve. Therefore, there

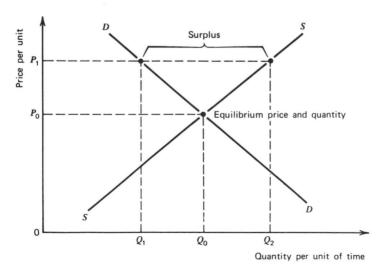

**FIGURE 6-13**   Disequilibrium in the market: A surplus.

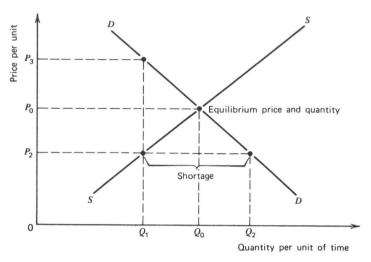

**FIGURE 6-14**    Disequilibrium in the market: A shortage.

is a market shortage equal to $Q_1$ to $Q_2$. In order for consumers to purchase the short quantity supplied, they must bid the price up to $P_0$. Only at the equilibrium price, $P_0$, is the amount producers supply equal to the amount consumers demand.

An example of this situation existed during World War II, when rationing was put into effect. Many items such as sugar, shoes, and tires were rationed in order to meet the needs of the public, as well as the war effort. Ceiling prices were established by the government at prices below the equilibrium level, as at $P_2$ in Figure 6-14. This decision resulted in some "black market" operations because at price $P_2$ producers would supply quantity $Q_1$, but for quantity $Q_1$ consumers were willing to pay price $P_3$. Thus, some rationed items were sold on the black market at prices higher than the price ceilings set by the government.

Using the preceding analysis, the effect of changes in supply or demand may be determined. The impact on the United States of the 1974 oil embargo can be shown in general terms (Figure 6-15). Approximately 20 percent of the United States' crude oil supply then came from the Arab States. Figure 6-15 shows the American supply (*SS*) and demand (*DD*) for crude oil before the embargo. The equilibrium price was $P_0$, and the quantity consumed was $Q_0$. The oil embargo cut off Arab oil shipments to the United States and caused the supply curve to shift to $S'S'$. The result was a new equilibrium at $P_1$ and $Q_1$. Consumers now had to pay price $P_1$ for their oil and received only quantity $Q_1$ rather than $Q_0$.

A change in the demand schedule occurred with the 1972 sale of wheat to the Soviet Union. The USSR purchased about 400 million bushels of

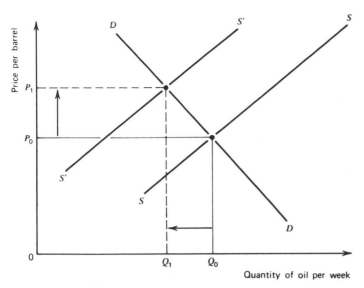

**FIGURE 6-15**   Impact of a supply shift on equilibrium price.

wheat, which caused an increase in the world wheat price from $2.00 per bushel to about $4.00 per bushel. The United States was unaware of the total volume of wheat the Soviet Union intended to buy. It was assumed that the purchases were for feed grains in order to increase livestock production. Instead, because of poor weather conditions in the Soviet Union, wheat was needed and purchased for human consumption. The poorer quality Soviet wheat was then used for feed grain purposes in the Soviet Union.

Before the wheat deal, the equilibrium world price was at price $P_0$, as shown in Figure 6-16, and the quantity consumed was $Q_0$. The Soviet purchase increased the demand for wheat from $DD$ to $D'D'$. This new demand increased the price of wheat to $P_1$ and increased the quantity consumed to $Q_1$.

# SUMMARY _____

The true costs of production are opportunity costs. Opportunity costs arise because using resources to produce any output causes a sacrifice of other goods that could have been produced with those resources.

Production costs can be quantified only after recognizing the specific resource relationships on which any product output is based. We began

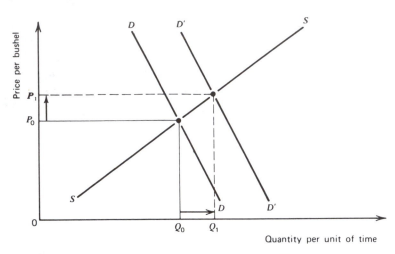

**FIGURE 6-16**    Impact of a demand shift on equilibrium price.

the analysis of a firm's costs by again using a single-variable production function as a point of departure. Given the rigid tie between production functions and cost functions, we made use of the production function data in Table 4-1 (and graphed in Figure 4-2) as the starting point.

Short-run cost curves are defined within a given period of time. Each of these curves (*AVC*, *AFC*, *ATC*, and *MC*) shows which cost items change and their directions (increasing or decreasing) as the firm changes its rates of resource use and output.

The firm optimizes its rate of output by producing a quantity of output that is determined by the point at which *MC* and *MR* are equal. The point of minimum *AVC* determines the lowest price at which the firm will produce for the market. From that point upward the *MC* curve is the firm's supply curve. As price may rise above minimum *AVC*, the firm is experiencing a loss until price equals *ATC*. Higher prices will result in economic profit for the firm (*TR* in excess of all opportunity costs or, what is the same thing, in excess of all explicit and implicit costs).

As firms in an industry earn economic profits, other firms will enter the industry, bidding up the prices of both variable and fixed resources in those firms. With more firms producing, market output will increase (a market supply shift). Given demand unchanged, price will fall and cause economic profits to decline.

If, on the other hand, firms are earning negative economic profits, some firms will leave the industry. As firms leave there will be an output-reducing shift in market supply causing price and economic profit to rise. Thus long-run market equilibrium can be described as one of zero economic profits.

# CHAPTER HIGHLIGHTS

1. Product supply results from the use of resources, giving rise to two general types of costs: (1) explicit costs—direct or cash costs paid for resources bought or hired; and (2) implicit costs—indirect or noncash costs of the owned resources.

2. Opportunity costs are the true costs of production. In a free market, the payments to variable resources must at least equal their best-paying alternative in order to obtain the services of these resources. Fixed resources are being used inefficiently if they are not earning at least as much as they could in the next-best alternative.

3. What we ordinarily like to call "profit," economics includes as a cost of production. So if receipts exceed the full costs of production, there is a surplus, called "economic (or pure) profit."

4. The shapes of the cost curves are determined by the firm's production function.

5. The short run stems from a production function with one or more fixed resources.

6. The longer run is derived from a production function that has a larger proportion of resources that are variable. The ultimate long run is when all resources are variable.

7. Economics utilizes seven cost concepts: $TVC$, $TFC$, $TC$, $AVC$, $AFC$, $ATC$, and $MC$.

8. Two concepts are especially useful in determining an optimal output: $MC$, which is defined as the change in total costs when output is changed by one unit; and $MR$, which is defined as the amount added to total revenue when an additional unit of output is produced and sold.

9. Since the price that the competitive firm gets for its product does not change as it adjusts output, the firm's demand curve is its $MR$ curve.

10. The firm's profit-maximizing output is determined by the point at which $MC = MR$, an output that can also be determined by using the $TR$ and $TC$ curves.

11. In the short run, a firm will shut down operations if it cannot at least cover its variable costs.

12. The short-run supply curve of the competitive firm is its $MC$ curve above minimum $AVC$.

13. The existence of economic profit will cause present firms to expand output (by increasing firm size) and will also attract new firms into the industry, both of which cause the market supply curve to shift to the right and market price to fall.

14. Factors that shift a supply curve are changes in growing conditions (weather), technology, resource prices, product prices and profitability of substitute products, and institutional constraints.

15. Individual demand curves for a commodity are added horizontally to derive the market demand curve. Individual supply curves are summed in the same manner to derive the market supply curve.

16. Equilibrium market price and quantity are determined by the interaction of market demand and supply curves.

17. At market equilibrium all buyers of a commodity who are willing and able to pay the equilibrium market price will obtain the amount of product they desire. Also, all sellers will supply the amount purchasers want at that price.

18. A new market equilibrium will be established if any of the demand or supply shifters change.

## KEY TERMS AND CONCEPTS TO REMEMBER

| | |
|---|---|
| Average fixed cost | Market supply curve |
| Average total cost | Optimum |
| Average variable cost | Pure profit |
| Economic profit | Short run |
| Explicit costs | Supply curve |
| Implicit costs | Total cost |
| Length-of-run | Total fixed cost |
| Long run | Total variable cost |
| Marginal cost | Total revenue |
| Marginal revenue | |

## REVIEW QUESTIONS

1. What is the economic meaning of costs?

2. Suppose that you maintain a full set of books for your firm, which meet minimum requirements for recording income and costs. In what way are your book's costs different from economic costs?

3. What is the special meaning of each type of cost (*TVC, TFC, TC, AVC, AFC, ATC,* and *MC*)?

4. Suppose you are the operator of a firm with the following short-run schedule of output and total cost:

| Output | Total Cost |
| --- | --- |
| 0 | $10,000 |
| 1000 | 15,000 |
| 2000 | 25,000 |
| 3000 | 40,000 |
| 4000 | 60,000 |
| 5000 | 90,000 |
| 6000 | 130,000 |

a. What is your firm's total fixed cost? total variable costs? average variable costs? average total costs? marginal cost?

b. How much of this product would you produce at a market price of $2.50 per unit?

c. How much (if any) economic profit would you earn if the market price were $10 per unit?

5. Many people think that profits are wrongly taken from consumers. Justify an economic profit as a reward for producing something that society wants so strongly that you were able to claim that amount as yours. Did you cheat your hired labor to get that economic profit? Did you cheat your consumers?

6. Are economic losses (negative economic profits) wasteful? Is this loss in any way a use of the wrong resources to produce the wrong product, thus "wasting" society's resources?

7. Obviously, shifts in supply and demand can counteract one another when both curves shift or accentuate the impact of the other's change. Draw several graphs with increasing and decreasing shifts of both supply and demand and analyze the effects on equilibrium price and quantity.

# SUGGESTED READINGS

Bishop, C. E., and W. D. Toussaint. *Introduction to Agricultural Economic Analysis*, New York: John Wiley & Sons, 1958, Chapters 7 and 8.

Bradford, Lawrence A., and Glenn L. Johnson. *Farm Management Analysis*, New York: John Wiley & Sons, 1953, Chapter 12.

Brehm, Carl. *Introduction to Economics*, New York: Random House, 1970, Chapter 6.

Gwartney, James D., and Richard Stroup. *Economics: Private and Public Choice*, 6th ed. Fort Worth: Dryden Press, 1992, Chapter 8.

Leftwich, Richard H. *Introduction to Microeconomics*, New York: Holt, Rinehart & Winston, 1970, Chapters 8 and 9.

Peterson, Willis L. *Principles of Economics: Micro*, 8th ed. Homewood, Ill.: Richard D. Irwin, 1991, Chapters 5 and 6.

# AN OUTSTANDING CONTRIBUTOR

**C. Robert Taylor**   Robert Taylor is the Alfa / Alabama Farmers' Federation Eminent Scholar in Agricultural and Public Policy at Auburn University, a position he has held since 1988.

Taylor was reared on a farm in southeastern Oklahoma. He earned the B.S. degree in agricultural economics from Oklahoma State University in 1968. He served as an assistant county agent in Kansas in 1969, then obtained a Master's degree in applied economics from Kansas State University in 1970. His Ph.D. degree was received in agricultural economics from the University of Missouri at Columbia in 1972.

He held a postdoctoral research associateship from 1972 to 1974, then was assistant professor of agricultural economics, 1974–76, at the University of Illinois. Dr. Taylor served at Texas A&M University from 1976 to 1978, first as assistant, then as associate professor. From 1980 to 1985, he was professor of agricultural economics at Montana State University. In 1985 he returned to the University of Illinois as professor of agricultural economics, with a joint appointment in the National Center for Supercomputing Applications.

Dr. Taylor's research has focused on developing and using regionalized econometric and programming models to estimate the aggregate economic impacts of production and resource policies as well as conventional farm programs. Aggregate analyses that he has conducted include proposed pesticide withdrawals, integrated pest management, restrictions on fertilizer use, a conservation reserve program, hail suppression technology, livestock growth hormones, expanded ethanol production, and changes in farm programs.

Another thrust of his research is on applying stochastic and dynamic optimization techniques to agricultural decision problems. Applications of these techniques at the firm level include pest management and eradication, firm growth, grain marketing and storage, and farm program participation.

Dr. Taylor was a visiting research scholar at the International Institute for Applied System Analysis in Austria. He received the Economic Research Service, USDA Administrator's Special Merit Award for excellence in implementation and evaluation of boll weevil insect management programs. He also received the American Agricultural Economics Association award for Quality of Research Communication in 1978 and the Western Agricultural Economics Association Published Research Award in 1985. In addition, he was coadvisor to a recipient of the 1983 AAEA Ph.D. Dissertation Award.

Most of Taylor's teaching has been in the areas of graduate production economics, microeconomic theory, mathematical economics, and applied dynamic programming. He coauthored the graduate level text *The Economics of Production* in 1985.

He is married to the former Claireda Weaver, and they have one daughter.

# CHAPTER 7

# Competition and the Market

$T$his chapter deals with the competitive conditions in a market in order to explain a firm's pricing and output decisions. The competitive model is presented so that other noncompetitive models can be examined. The role of scarcity and rationing and the manner in which they influence competition is also discussed.

The problem of scarcity is basic to much of the topical material of this book for a reason: Because of scarcity some means of allocating (rationing) limited resources among their alternative uses, and of distributing limited goods and services among those desiring them, must be devised. There simply are too few of the desired things to fully satisfy all of our desires, thus we have scarcity and the need for "rationing." The criteria by which the allocation problem is solved only reflects the fact of scarcity, which, in itself, is the cause of competition.

Rationing can be accomplished in a number of ways, any of which require discriminatory criteria, by discriminating against those who, no matter how strong their desire for the particular good, are unable to meet requirements to get it. If rationing is done administratively, the basic criteria may simply be sex, age, height, weight, family size, agility, willingness to wait in line, level of education, or any of a great many other possibilities and combinations. The point is that specific criteria are established (somehow, or by someone); then those eligible will compete in whatever manner is required to gain their share of that desired item, whatever it might be.[1]

But scarcity and rationing are hardly the only causes of competition. A Robinson Crusoe, alone on an island, will expend effort to improve fishhooks, for example, if the expected benefits of that effort exceed the sacrifices required to gain benefits. If improved fishhooks ("capital," with cost measured as the value of other desired things foregone while producing fishhooks) will reduce the time spent in fishing, and permit more time to be spent in other activities (an overall increase in desired goods), the benefits are determinable. The worth of these benefits (and sacrifices) will be influenced by the values of what Crusoe considers to be good or bad, desirable or undesirable. Thus, the system has not been the cause of competition. Given the freedom to choose among alternatives, the driving force of competition is the individual's own set of preferences and de-

---

[1] Should a large, heavy person get more food than a small person? Should an adult get less milk than a child, with the needs of all those under five years of age, say, being met first? If individuals performing the most difficult physical labor get more of certain foods than others not so employed, there will be competition for jobs, with their performance in those jobs designed to continue employment, depending on the degree of scarcity and the amount of discrimination in their favor.

sires, and the relative scarcities of the desired things, not someone else trying to outdo him.

## THE FUNCTION OF PRICE _____

An isolated Robinson Crusoe causes no concern for others. Crusoe alone gains or suffers from his decisions. Only his well-being is improved or worsened by what is done. By economizing on scarce resources, including his time and abilities, an optimal balance is achieved between the sacrifices made to achieve desired benefits.

But this is too simple and private a problem. Proper or improper rates of resource use and product output impact only on Crusoe's own well-being, with no one else affected in the process. When a number of people are involved—when the decisions made, and the efficiency with which any one person operates, have an effect on the well-being of others—the problem becomes one of organizing the economic system and ordering decisions in such a way that undesirable effects on others are minimized or eliminated. One answer to this need for controlling individual actions is a "market," as was discussed earlier.

In a free enterprise economy the market is decentralized, with decisions being made and carried out by individuals responding to their preferences and market prices, rather than by conscious direction from elsewhere in the system. In such a market-oriented economy, prices play a key role in directing the allocation of resources among alternative uses and in causing the produced goods to be divided among consumers according to their individual preferences.

When a good becomes more scarce relative to the demand for it, the price of that good will increase. A relative price increase is the market's signal to both producers and consumers that changes are required. Penalties are inflicted on those who make the wrong decisions or who simply refuse to change: Producers needlessly forfeit potential profits, and consumers find increased sacrifices of other desired goods, as their penalty for not making the proper adjustments in their purchases.

An increase in the relative price of a good conveys to both buyers and sellers the information that this good is now more scarce than it was formerly. No individual needs to know why the good is now more scarce, nor does the individual necessarily need to contemplate how that increased scarcity can best be alleviated. The message of increased scarcity, as transmitted by price alone, suggests the solution. Individual buyers who now value the good lower than does the market are induced to reduce purchases by shifting more spending to acceptable substitutes. Thus, demand for the good falls because marginal utility per dollar spent has fallen and causes the spending shift. Resource owners are encouraged by the prospect of increased profits to divert resources from other goods because the marginal revenue per dollar of resource commitment is now enhanced. The supply of the good is increased. The market is thus an efficient clearinghouse bringing order out of conflicting desires, by giving the appropriate price signals to all who by their actions are able to help correct the problem. As the market functions efficiently it is then possible for prices to be efficient indicators of the relative values of all things traded in the market. But if the market itself is not efficient (if it is unable to reflect changes in market forces because of restrictions stemming from the ability of individuals, groups, or government actions to regulate prices or production), the ability of prices to indicate values correctly is diminished.

## MARKET CLASSIFICATION ————————————

As we consider how prices of goods and services are determined, and how earnings of all types of resources are established, it is necessary to look more carefully at different types of markets within which all economic activity takes place, where price is used as the rationing device. It is in this area that economic efficiency has its roots.

Some firms appear to be at the mercy of the market, with fluctuating prices seemingly unrelated to the production activities of those firms; many agricultural producers fall within such a market situation. It is a market characterized by a high degree of competition between producing firms.

Numerous other industries appear to be made up of firms that seem able to manipulate price to their advantage. A market such as this is loosely referred to as being "monopolistic" to some degree. Competition among these firms appears to be minimal, or even absent, because there are so few firms in that industry.[2] A pure monopoly, in fact, is defined as a single seller.

Because economic efficiency involves production of goods and services in the quantities and proportions that people want, and that with a minimum of resource expenditure, we may classify markets in a way having some usefulness in making judgments about the efficiency with which a system operates. Extreme opposites (models), based on the degree of competition between firms in a market, are generalized in Figure 7-1. The mar-

---

[2] Don't let the discussion mislead you into thinking that the labels and distinctions used here apply only to producing firms. Identical conditions may also be applied to the buyer's side of the market. If, for instance, there is only one buyer for an industry's output, we would call that firm a monopsony (rather than monopoly). Purity on both sides of the competitive market occurs when there are many buyers as well as sellers.

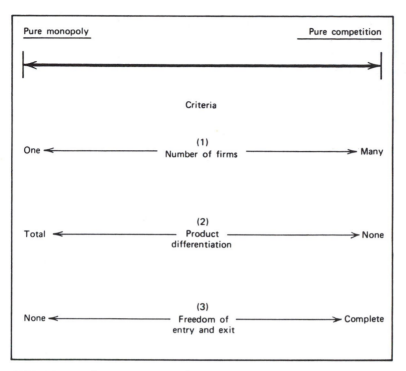

**FIGURE 7-1** Criteria extremes for classifying markets.

ket situation within which the firm operates and the firm's reactions to those conditions form the basis for this classification.

Given identifying characteristics, firms may be found anywhere along a continuum ranging from the purely competitive firm at one extreme to the pure monopolist firm at the other, with its particular location along the continuum depending upon the degree to which the firm fits some or all of the criteria.[3]

## Pure Competition

The necessary conditions for the existence of *pure competition* are (1) many firms in the industry, (2) a homogeneous product, and (3) individual freedom to enter or leave the industry.

---

[3] The words *pure* and *perfect* are sometimes loosely used synonymously in these market descriptions. But for pure to become perfect requires two additional criteria: (1) perfect mobility (of all goods and services) and (2) perfect knowledge and foresight (for all decision makers).

The actual number of firms that it takes to constitute *many* is relative. We simply need to realize that it means so many of them that no individual firm can have any influence whatsoever on the market price of its product as a result of its own decisions and actions. If a firm is only a minute part of the total market, then whether it produces the maximum amount that it is capable of producing, or shuts down completely, the market will not be affected. The firm's output is such an insignificant bit of the total market supply of that product that it cannot affect the market price.

Many farms and ranches in the United States fit this requirement. Producers of grain, livestock, and many other food and feed products are especially incapable of influencing price by their individual actions. And they frequently are used as examples of a highly competitive industry, unable on their own, to do anything about the prices they get for their products in the market place.

A *homogeneous product* may be achieved either by uniformity in all physical characteristics of the product as it leaves producing firms or by the market classifying and separating the product into distinctive groups according to specific grades and standards. In either case there will be no favoring or discriminating against any firm's product in the market. Product homogeneity eliminates the possibility of buyers preferring one firm's product over that of another.

Any attempt by the individual firm to obtain a premium for its product would be futile because the market can obtain all of that good it wants from other producing firms at the market price. Any attempted discounting of the price of one firm's product would be rejected because the firm could sell all the product at the going market price. The reduced price would unnecessarily reduce that firm's revenue. Pure competition thus results in the products of competing firms being perfect substitutes for one another.

Homogeneity by grade-standardization is evident in many agricultural product markets. Specific grade standards are sufficiently precise that the buyer need not even see and inspect the commodity being purchased. All firms able to deliver the particular grade of product will receive the market-determined price for their product at the time of the transaction, because the market is indifferent as to who produced it.

The third condition, *freedom of entry and exit*, permits the individual firm to enter or leave a market as its own costs and returns might dictate, without other restrictions or encouragement of any kind. Such freedom permits the firm to produce or not, or to produce more or less, as its own decision criteria and objectives might direct, without restrictions beyond the pure force of market price. Thus, if the firm sees an opportunity to profit by switching from corn to wheat, for instance, it may do so without restraint or direction either from other firms in the market or by government institutions.

Given the three criteria for pure competition, a firm operating in such a market may be described as being a ***price taker***: price is taken as given, with no opportunity for influence by the individual firm. The firm competes by organizing its mix of land, labor, capital, and management to establish the rate of resource use and product output that will maximize its net revenue.

This characteristic of the firm in pure competition is significant to the larger (social) objective of an efficient economic system. Even though the *market* demand curve slopes downward to the right, the *firm's* demand curve is horizontal (i.e., perfectly elastic). Because of this characteristic, the product price will not change as a consequence of the individual firm's decision making; price is beyond the firm's influence. It is a boundary on the firm's options: The firm recognizes its inability to affect or manipulate the price of its product. It concentrates instead on the area in which individual control may be exercised with beneficial results—achieving an economic optimum in the firm's productive activities. This management environment is one reason why you see farm or ranch operators spending so much of their time and effort in direct management activities and so little (individually) in economic activities "beyond the farm gate."

## Pure Monopoly

At the other extreme of our continuum is the ***pure monopolist***, which may be typecast as a ***price searcher***. This firm, unlike those in pure competition, need not accept price as given. It "searches" for that price for its product that will balance its rate of output (and sales) with its cost structure so that its profitability is maximized.

The classification as a pure monopoly hinges on the three basic criteria in Figure 7-1. Where there must be many firms in pure competition, monopoly, in its purest extreme, means there is a single firm selling the product. The *market* demand curve and the *firm's* demand curve are one and the same, which adds another dimension to the monopolist's decision making and market powers.

Being the only firm producing and selling its product, the monopolist has no competitor producing a similar good—that firm's product is differentiated from all others—so competition in that product market is absent. This lack of competition carries with it some incorrect implications about a monopolist's pricing policies.

A monopolist appears to have a strangling control over the price that its product may command in the market place. But that strength cannot be so absolutely exercised. The power to set the price at any level *does* exist; however, an unconcerned use of that ability would require that the demand for its good also be perfectly inelastic.[4] Granted that an uncontrolled monopoly is free to set its product price at any level it might wish. But if the monopolist is profit oriented (and not just flexing muscle in a demonstration of its market power), it will recognize that a price too high for the product will impose its own penalty by causing so large a reduction in sales that profits are being forfeited.

Without the ability to protect and maintain its unique position, a monopoly could not be continued for long. Given the market's price signals, a highly profitable business could expect competitors to appear from elsewhere in the system in the expectation of higher profits for them. Those competitors must be kept out, and they may be by a variety of devices (legal and otherwise) that prevent others from becoming established in the monopolist's market.[5]

Conceivably, a monopolist could prevent competitive entry by other firms where the firm owns or controls the only available source of a necessary resource used in the manufacture or production of the good that

---

[4] The elasticity of demand for any good reflects, among other things, the availability of acceptable substitutes for that good, ranging from the perfect substitutes (perfectly elastic demand curve for a homogeneous product) of pure competition to the complete opposite of no substitutes whatever. The monopolist's perfectly inelastic demand curve would graph as a vertical line, which says that a given quantity of that good will be demanded no matter what the price might be. But this situation is a real-life impossibility because consumers would have to possess infinite incomes to pay the infinitely high price that such a demand curve shows a pure monopolist could charge for this product.

We are misled if we attempt to base the meaning of substitutes on the similarity of physical characteristics, or on the desires satisfied, by two different goods. Rather, it must be remembered that sacrifices of other goods means that the higher the price of *any* good (monopolist's product, or otherwise) the more of other goods that must be sacrificed if one is to continue buying it.

No matter how strong our desire for a good may be, the economics of its consumption prevents us from consuming a specific quantity of that good on the basis of some physical criterion. For example, we may have an intense desire for baked goods made from wheat, and might even convince ourselves that we absolutely must have some of that product, no matter what (a good without substitutes, having an apparently perfectly inelastic demand curve). Then imagine if you will, the price of wheat going higher and higher. Eventually, the physical fact of life (that other goods can, maybe in lesser degree, satisfy that need) and the economic fact of life (that its cost, in terms of other goods sacrificed has gotten too high to be worth buying any more of it) will force us to accept a substitute made from oats, barley, corn, rye, or some other cereal grain, or even a synthetic substitute. Our preferences may lean strongly toward consuming a wheat food, but our bodies need only food that provides the proper nutrition, not necessarily wheat. Therein lies the basis for a demand curve that is anything but vertical. We are forced to the realization that there are substitutes for everything and that the monopolist's perfectly inelastic demand curve is an economically impossible fiction.

[5] In terms of the monopoly model itself, the method by which entry is prevented is immaterial. The theoretical elimination of competition is essential, however, to permit an analysis of the specific operation of a pure monopoly and its effects in the economy.

it sells in the market.[6] Unless prohibited by law, the firm could protect its monopoly position simply by refusing to share its supply of this ingredient with any other firm.

Many other means of preventing entry also exist, some with powers developed by the firm and others actually granted by the government. Economies of scale may prevent entry of new firms in the monopolist's market. Large initial investment outlays for fixed capital items may rule out potential competitors because they simply can't amass sufficient start–up capital. And the technology of production may be such that a very large-scale operation is the only possibility, preventing another from beginning as a small firm, then growing larger as it succeeds in its efforts. For whatever internal reason, the *ATC* curve may fall through such a wide range of output that smaller firms cannot compete. The declining *ATC* curve may also be over such a wide range of output that the single firm is capable of supplying the entire market demand without experiencing rising average costs. A firm that produces under these cost and demand conditions has sufficient market power to keep potential competitors out of that market because of its ability to price its product below any other firm's costs of production.[7]

The firm may hold a patent on the production process with exclusive use of that process being granted and protected for a period of years by the government, under the conditions of its patent laws. Franchise or license fees to operate may also be set so high, or the number so restricted, as to prevent other firms from competing in that market.

To this point in the discussion, we have quite rigidly defined and described two opposite market types that we would not even expect to find (in their pure forms) in the real world. Then why study them? Must we now admit to a kind of academic pretension that has simply been toying with a set of artificial concepts that are without real-life meaning? Hardly. Their usefulness becomes apparent when they stand as benchmarks, or yardsticks, against which we may measure the efficiency of markets and firms as they use valuable resources to produce want-satisfying goods and services.[8] These pure models make clear the manner by which firms

---

[6] An example of this was the market power that the Aluminum Company of America had in the aluminum industry, through its control of 90 percent of the nation's bauxite supply. See John Ise, *Economics*, New York: Harper and Brothers, 1946, p. 135.

[7] In such a situation, fostering (or requiring) competition from other firms would be an economic waste in that the product cost to consumers would be greater with two or more competing firms than with the single monopoly producing the commodity. Public utility companies are frequently used as examples of this type of situation. Granting them exclusive market rights prevents the economic waste involved in competitively duplicating facilities.

[8] Although we limit ourselves here to the price effects of these two market types, other, broader, social evaluations may also be made in which the point of concern is directed to-

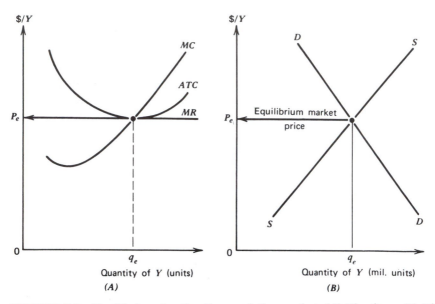

**FIGURE 7-2**   Equilibrium for the firm and the market. (A) The firm. (B) The market.

search for an optimum in their operations and the social effects in terms of resource allocation, production, and prices.

## THE EFFICIENCY OF PURE COMPETITION _____

The efficiency with which the competitive market determines production decisions in the short run was presented in the chapters on production and costs. When the firm has established the rate of output at which its profits are maximized, as in Figure 7-2A, given the assumed market conditions of pure competition, no adjustment to a higher profit level is possible without first changing the number and mix of resources within the firm's control, or improving the technology of using those resources. With all firms in the market at their optima the resulting price is an equilibrium price because quantities demanded and supplied are equal (Figure 7-2B). At that price, firms are willing to supply just the amount that consumers are willing and able to buy. This situation is called an equilibrium because the firm is not encouraged, by expectations of greater profits, to

---

ward the efficiency of existing or proposed economic policies, programs, institutions, or even the organization of an economic system itself.

make any changes in its present operations. An equilibrium situation is sometimes referred to as being one of *normal profits*. Profits are *normal* because each resource within the firm is earning a return that is neither greater nor less than its next-best employment possibility (its opportunity cost). No firm is encouraged to enter or leave the industry because earnings are neither better nor worse than they would be elsewhere in the system. Resources are thus efficiently allocated.

We may summarize the pure market's adjustment solution to the questions of resource allocation, production rates, and prices by studying the individual firm's solution to its own problems under conditions of disequilibrium. Either of two types of disequilibrium is possible: (1) Firms are experiencing surplus earnings—*economic rent*; or (2) they are incurring losses—*negative economic rent*. These situations are shown in Figure 7-3A and B.

## Longer-Run Changes Caused by Economic Rent

Since the firm's *ATC* curve includes a normal rate of return (profit) as a cost of production, any return per unit of output that is greater than *ATC* at that output is therefore an excess or surplus over costs. It is a surplus in that it exceeds the expected return that had previously attracted each of the firm's resources into its present use and is greater than the return needed to keep the firm and its resources in their present employment.

Suppose the firm shown in Figure 7-3A is a corn farm that, under its current costs and product price conditions, is producing 5000 bushels of corn per year. Because the firm is a price taker, its demand curve is its *MR* curve and the optimal output for the firm is 5000 bushels, as that is the output at which this firm's *MR* and *MC* are equal. At this level of output, however, the firm's total costs per unit of output are only $P_2$ while its revenue per unit is $P_0$; it is earning a surplus equal to $P_2$ to $P_0$ (measured along the vertical axis). If this firm is typical of others in the industry, the pure market will cause longer-run changes to be made by individual firms, all acting on their own initiative, that will eliminate this surplus.[9] This rate of return exceeds the earnings of other firms elsewhere in the system, which will encourage them to shift from their present employment to corn production in the expectations that they, too, might

---

[9] To present a more easily recognized picture of longer-run market adjustments, another set of market forces has been disregarded here: When new firms enter the industry, their effect is not solely on the market price of the product (the result of a shift in the market supply), but also affects the prices of resources within the industry as well. New firms will be bidding for resources already in the industry and, excepting the possibility of a perfectly elastic input supply curve, will increase the existing firms' resource costs because of their now increased opportunity costs (as discussed in Chapter 6). Thus there are two simultaneous changes occurring: product price falls and resource costs increase.

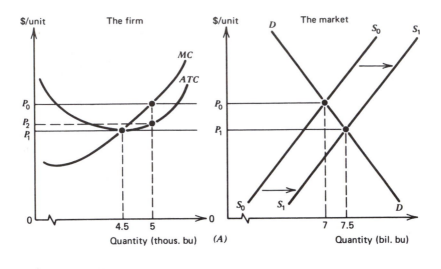

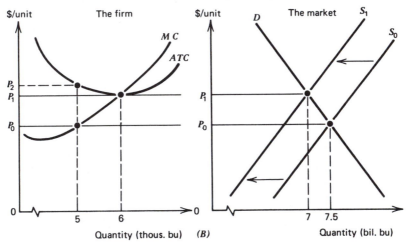

**FIGURE 7-3**   From short-run disequilibrium to longer-run adjustments. (*A*) Economic rent and shift in supply as firms enter the industry. (*B*) Economic losses and shift in supply as firms leave the industry.

reap such earnings. As other firms enter the market, the supply curve shifts to the right from the original supply ($S_0$) to the new supply ($S_1$). Given the demand for corn (*D*), the increased supply causes the new equilibrium price ($P_1$) to be lower than the original price ($P_0$). Firms will continue entering the market until the price of corn has fallen to the point where price (*MR*) equals the firm's *MC* at minimum *ATC*, a longer-run equilibrium position. Economic rent has now been eliminated, and consumers are supplied with the amount of corn they are willing and able to buy at a minimum opportunity cost to them. No other good or service more valuable than the last unit of corn produced has been sacrificed to produce that last unit of corn.

This equilibrium is the *economic efficiency* that is so much to be desired. With all industries so structured (i.e., $P = MC$), and each at an equilibrium, society's costs for all its goods and services are minimized; the system is efficient in that no greater output of any good or service can be obtained from the available resources without having to sacrifice something else more highly valued by consumers.

## Longer-Run Adjustments to Losses

When firms in an industry are earning less than normal returns (i.e., not covering their full opportunity costs) in the short run, as in Figure 7-3B, changes will be made that are opposite in directions from those just discussed. In the market, supply and demand are equal at $P_0$. But for the price-taking firm, a produce price of $P_0$ is not sufficient to cover all costs. The short-run loss-minimizing output for the firm is 5000 bushels when price is $P_0$. At 5000 bushels of output, the firm is losing an amount per bushel that is equal to the vertical distance (measured along the vertical axis) from the point where $MR$ and $MC$ are equal up to the $ATC$ curve ($P_0$ to $P_2$). Since the firm's $ATC$ curve is the sum of all per-unit opportunity costs of the resources used to produce corn, a longer-run adjustment for this firm is to shift some or all of its resources to other uses—such as a shift of land from producing corn to alternative crops or other uses that will yield greater returns than corn.[10] As firms carry out these changes, the market supply curve shifts to the left, causing the market price of corn to rise until a new equilibrium price ($P_1$) is reached. At this price level all opportunity costs are once again just fully covered, and firms are experiencing zero economic rent.

## Longer-Run Market Supply Curves

To focus our attention more clearly on the direction of changes in market supply, straight-line market supply curves were used in Figure 7-3A and B, with no change in the slopes of those shifting curves. Let's now exam-

---

[10] The shifting of rangeland from livestock to wheat production, by some ranch operators in areas of the West, is a vivid demonstration of how producers react to changes in relative prices. As the price of beef fell between 1973 and 1977, the relative profitability of wheat increased to the point where it paid to plow up some of that rangeland and use it to produce wheat instead. No matter how optimistic beef producers' longer-run expectations might have been, short-run losses were forcing some operators to make such an adjustment.

ine more carefully the market supply changes that would result as firms make their longer-run adjustments to economic profits or losses.

The firms' short-run *MC* curves sum horizontally to form the market supply curve. But when firms react to economic rent by changing the fixed resources used, those short-run cost curves are no longer the relevant length-of-run for either those firms or the market. As the length-of-run is increased, the firm's *MC* curve becomes more elastic because a greater proportion of the firm's resources is variable than in the shorter run.

Figure 7-4 shows the different market supply curves that result from different length-of-run production and cost functions for the firms that make up the industry. The basis for defining length-of-run is the proportion of resources that can be varied by the producing firm. So the longer the length-of-run for the firms, the greater will be the elasticity of the market supply curve.

A perfectly inelastic market supply ($S_1$) occurs when no changes in resources can be made by the producing firms because the time period is too short to make resource and product changes—the immediate short run. Its opposite is the ultimate long run, which is so long a time period that everything is variable, resulting in the perfectly elastic market supply curve ($S_4$).

Many intermediate length-of-run possibilities and their consequent supply elasticities are possible, depending on the proportions of the resources that are variable for the firm. If firms are able to vary only a few of their resources a relatively inelastic market supply curve (such as $S_2$) results. However, if more resources are variable, the market supply curve is more elastic ($S_3$).

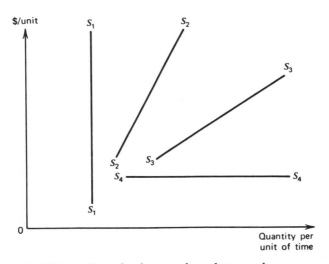

**FIGURE 7-4**   Length-of-run and market supply.

# SUMMARY

Markets are classified by the number of firms in an industry— whether or not products are differentiated—and the entry and exit conditions of the firms. A purely competitive market is one in which there are many firms in the industry, a homogeneous product is produced, and each firm has the freedom to enter or leave the industry. A firm operating in a purely competitive market is described as being a price taker. The firm's demand curve is perfectly elastic or horizontal. The firm cannot influence price by the amount of product it sells. Remember, however, that the market demand curve in a competitive industry slopes downward and to the right.

A monopolist, on the other hand, is a single firm selling the product. The market demand curve and the firm's demand curve are identical. Entry is blocked or restricted in some way.

Firms in pure competition are more efficient than are monopolies. In pure competition, firms can only make a normal profit in the long run because firms can enter or leave the industry. Entry conditions influence the supply of the product to ensure that only a normal profit is made. However, in monopoly this may not occur.

# CHAPTER HIGHLIGHTS

1. Scarcity causes rationing by discriminating: (1) Through the market, where those able to pay for the goods are the ones who get them; or (2) by using nonmarket criteria that direct goods and services to those who get them.
2. Competition is caused by scarcity and not by the economic system or by someone else.
3. Market prices are signals that direct production and consumption decisions.
4. The conditions necessary for "pure competition" are (1) many firms, (2) a homogeneous product, and (3) complete freedom of entry and exit.
5. A purely competitive firm is called a *price taker*.
6. The market demand curve slopes downward to the right. The firm in pure competition faces a horizontal (or perfectly elastic) demand curve; the pure monopolist's demand curve is the market demand curve.
7. The conditions necessary for a "pure monopoly" are (1) a single seller, (2) a differentiated product, and (3) no freedom of entry.
8. A pure monopolist is called a *price searcher*.

9. Market models such as pure competition and pure monopoly are useful in evaluating the economic efficiency of an industry.
10. Purely competitive firms in equilibrium, with zero economic rent, are economically efficient because the costs of society's goods and services are minimized.
11. The concept of length-of-run derives from the resource-use options available to producing firms in a market affecting market supply elasticity.

## KEY TERMS AND CONCEPTS TO REMEMBER

Economic efficiency
Economic rent
Normal profits
Price searcher

Price taker
Pure competition
Pure monopoly

## REVIEW QUESTIONS

1. What is the role of prices in a free enterprise economy? What happens if prices are changed artificially?
2. Define pure competition. Is agriculture a purely competitive industry? Discuss.
3. Do firms such as General Motors and Ford operate in a competitive market? Why or why not?
4. Why do firms like to become price searchers? Explain how a firm might become a price searcher.
5. What is a "normal profit"? Are normal profits made in agriculture? Explain.
6. Does the length-of-run of firms influence the elasticity of the market supply curve? Why?
7. Draw the firm and industry demand curves for a firm producing a product under purely competitive conditions.
8. Draw the industry demand and supply curves for a product. Now draw another diagram showing the competitive firm's demand and cost curves for the same product.
9. How does competition reduce the cost of goods to customers?
10. How does competition increase the quality of consumer goods?

## SUGGESTED READINGS

Hirshleifer, Jack. *Price Theory and Applications*, 4th ed. Englewood Cliffs, N.J.: Prentice Hall, 1988, Chapters 9-11.

Leftwich, Richard H., and Ross D. Eckert. *The Price System and Resource Allocation*, 8th ed. Hinsdale, Ill.: The Dryden Press, 1982, Chapter 7.

Quirk, James P. *Intermediate Microeconomics*, 2nd ed. Chicago: Science Research Associates, 1982, Chapter 8.

Samuelson, Paul A., William D. Nordhaus and Michael J. Mandelo. *Economics*, 15th ed. New York: McGraw-Hill, 1995, Chapter 25.

Watson, Donald S. and Malcolm Getz. *Price Theory and Its Use*, 5th ed. Lanham, Maryland: University Press of America, 1991, Chapters 13 and 14.

## AN OUTSTANDING CONTRIBUTOR

**Emery N. Castle**   First appointed in 1976 to a position as senior fellow and vice president of Resources for the Future, Dr. Castle served as president of that organization from 1979 until his retirement in 1985. Before joining the organization, he had spent most of his professional career at Oregon State University as professor, head of the Department of Agricultural Economics, director of the Water Resources Institute, dean of Faculty, and dean of the Graduate School.

Emery Castle was born in Kansas and completed the work for his B.S. and M.S. degrees in agricultural economics at Kansas State University in 1948 and 1950. He earned his Ph.D. degree in 1952 at Iowa State University. Castle served two years as an agricultural economist with the Federal Reserve Bank of Kansas City, then joined the Department of Agricultural Economics at Oregon State University in 1954 as an agricultural and natural resource economist. While earning distinctions as a teacher, researcher, and administrator at Oregon State University, he also worked with the Water Resources Board of Oregon, and later with the Water Policy Review Board of Oregon.

Castle's current research includes an analysis of the behavior of farm real estate prices and a survey of U.S./Japanese agricultural trade relations. When he was at Oregon State he taught courses in farm management, production economics, resource economics, and research methodology. He has co-authored a widely used textbook (*Farm Business Management*) and many journal articles and bulletins in those interest areas.

A fellow of the American Association for the Advancement of Science, of the American Academy of Arts and Sciences, and of the American Agricultural Economics Association (1976), he also has served as president of the American

Agricultural Economics Association (1973) and of the Western Agricultural Economics Association (1962). He is a member of the American Economics Association, on the boards of the Agricultural Development Council and Duke University's School of Forestry and Environmental Studies, and has served on the editorial board of _Land Economics_.

The excellence of Dr. Castle's work has been given special recognition through such awards as the Distinguished Professor Award from Oregon State University and the Distinguished Service Award from Kansas State University. In 1979, he was invited to present the Kellogg Foundation lecture to the National Association of State Universities and Land Grant Colleges. Castle has twice been given the Award for Excellence in Published Research (1972 and 1973) by the American Agricultural Economics Association.

He is married to the former Merab Weber, and they have a daughter.

CHAPTER **8**

# Imperfect Competition and Market Regulation

$\mathbf{G}$iven the criteria for pure competition, we can draw important conclusions regarding the economic efficiency with which an economy so organized could operate. Such a system constitutes an ideal in that no greater efficiency in the organization of an economy can be visualized. Thus, it is useful as a standard of performance for appraising other market types.

## THE PURE MONOPOLY

We defined a monopoly in Chapter 7 not only on the basis of its failure to meet the requirements for pure competition but also because of its position at the opposite end of the continuum with special criteria of its own.

As we considered decision-making principles in pure competition, it was necessary to distinguish between the market's demand curve and the firm's demand curve. No such need exists in the case of pure monopoly because the monopoly firm and market are the same. The market demand curve *is* the pure monopolist's demand curve.

Because the market demand curve is always downward sloping, the simple criterion for profit maximization, of equating the firm's costs and returns at the margin ($MC = MR$), cannot result in an efficient allocation of resources for society. An important difference from pure competition is the problem a monopoly faces in selling its products. Although the market will take all of the price taker's output at the market price, the only way a monopolist can sell more output is to reduce the price of the product. The monopolist's marginal revenue and demand are thus two different curves and are the underlying cause of resource misallocation in that market.

Let's look at a hypothetical set of data appropriate to a monopolist's market that demonstrates the relationship between price and sales, Table 8-1. The downward sloping demand curve is evident in that more of this good will be purchased by consumers only if its price is reduced.

Since we define marginal revenue as the amount by which total revenue changes when another unit of product is sold, we derive *MR* by (first) determining the firm's total revenue schedule (*TR*). *TR* is simply derived by multiplying the price times the number of units sold at that price.[1] Marginal revenue is then computed by determining the change in

---

[1] Because elasticity of demand reflects the responsiveness of quantity demanded to a change in the price of the product, we note that *TR* is not a constant function of units sold, but increases throughout the range of the demand curve, where its elasticity coefficient is greater than one, to a maximum *TR* at the point where price elasticity of demand is unity, and ac-

**TABLE 8-1**  A Monopolist's Demand and Revenue

| Price per Unit = AR | Quantity Demanded | Total Revenue | Marginal Revenue |
|---|---|---|---|
| $6 | 2 | $12 | |
| | | | $4 |
| 5 | 4 | 20 | |
| | | | 2 |
| 4 | 6 | 24 | |
| | | | 0 |
| 3 | 8 | 24 | |
| | | | -2 |
| 2 | 10 | 20 | |
| | | | -4 |
| 1 | 12 | 12 | |
| | | | -6 |
| 0 | 14 | 0 | |

*TR* that accompanies each change in output and sales: $MR = \Delta TR/\Delta Q$. These resulting schedules, $D(= AR)$, and *MR* plot as shown in Figure 8-1. Since demand is a declining schedule, and the price at each quantity demanded applies to all units sold and not just to the last unit, *MR* lies below the demand curve and is more steeply sloped than *D*.

We now superimpose on the same graph a set of cost curves necessary for decision making. As in the data for Table 8-1, the actual numbers used are of no special importance, as long as they properly reflect the total market price–quantity relationships. The numbers used to demonstrate the firm's costs are also, in themselves, unimportant, as long as they are sufficiently realistic as to portray a firm subject to the law of diminishing returns and one capable of supplying the entire market demand for the commodity. These conditions are all met by the curves drawn in the graph.

# THE PROFIT-MAXIMIZING MONOPOLIST ——————————————————

Whether the firm be in pure competition or monopoly, its production decisions are based on its costs and returns *at the margin*. The price taker can influence only its costs by the decisions the firm makes—price is taken as a given and is beyond the firm's influence—but the monopolist's decisions affect both its costs and price. Both firms, however, find an optimum at that output where marginal revenue and marginal cost are equal.

Any unit which adds more to revenue than it adds to costs is a profitable unit to produce and sell, no matter what we might have classified that firm to be; profits are greater from the sale of that unit by the differ-

---

tually declines with further price reductions. This relationship between revenue and units sold causes an additional problem for the monopolist—the effect of price on sales prevents an equating of price and *MC* as in pure competition, as well as the effect of the level of output on the firm's production costs.

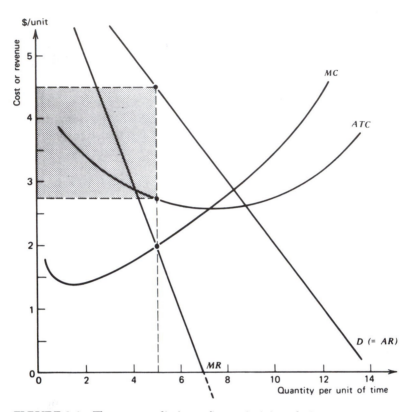

**FIGURE 8-1**  The monopolist's profit-maximizing choice.

ence between *MR* and *MC* than they would be without that unit. The monopolist maximizes profits in exactly the same way as the pure competitor—expand output and sales until that quantity is reached at which *MR* = *MC*. Any inequality between *MR* and *MC* dictates an adjustment in output, with equilibrium for the firm occurring only at the point where *MR* = *MC*.

The monopolist, depicted by the graph, discovers an optimum at five units of output. That many units will sell in the market for $4.50 per unit, the number that consumers are willing to buy for that price, as shown by the demand curve. But the cost of producing those five units is only $2.75 per unit, so this firm obtains a surplus over all costs (economic rent) of $1.75 per unit, the difference between *AR* and *ATC*. Total revenue for this firm is the rectangular area under the demand curve ($4.50 × 5 units), whereas total cost is the rectangular area under the *ATC* curve ($2.75 × 5 units), with the lightly shaded area being the economic rent at this output ($1.75 × 5 units).

Because market demand is given, being determined by the consumer's demand functions, its location, shape, and slope are independent of the firm or firms serving a market. Thus the location of the demand curve in the graph space, with respect to the location of the firm's cost curves, need only be such that it is apparent this firm can supply the market's demand for the good in question.

One might shift the *D* and *MR* curves right or left to reflect a larger or smaller market demand for the good, relative to the size of the firm, and we will find the same profit possibilities as for any other type of firm—economic rent may be positive, zero, or negative depending on demand and production costs.

If we increase the size of the market, by moving the demand curve farther and farther to the right, economic rent to the firm increases. Were a situation of this sort to occur in an actual market, either potentially competitive firms would be faced with such excessively high production costs that their entry into this market would be impossible, or, barring that, other restrictive devices would have to be employed to protect the market for this monopolist.

As shown in Figure 8-2*A* and *B*, we can identify two separately unique situations in the monopoly case that result in (1) zero economic rent for the firm and (2) minimized production costs, out of an almost infinite number of demand possibilities.

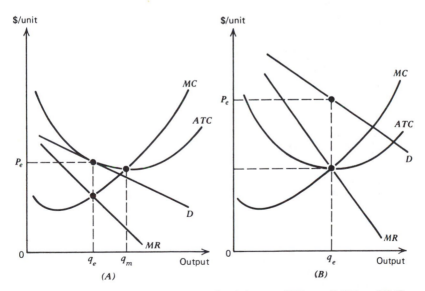

**FIGURE 8-2** Zero economic rent and minimum *ATC* possibilities. (*A*) Zero economic rent. (*B*) Production at minimum *ATC*.

In Figure 8-2*A*, the firm is sufficiently large relative to the total market, that the demand curve is tangent to the *ATC* curve somewhere along the downward sloping portion of that curve. In such a situation the price that consumers are willing to pay for this good, and the firm's cost of producing it, are both the same at the optimal output, $q_e$, and there is no economic rent for the firm.

A number of possibilities could cause this situation to exist. For one thing, the firm may have been serving a larger market at the time its facilities were constructed and now finds that the demand curve has shifted to this position. A further shift to the left would cause this firm to leave the market eventually, as some or all of its fixed resources would find greater earnings elsewhere (because of the failure to cover their opportunity costs). The word *eventually* requires emphasis because short-run resource fixity means that no changes can be made in these unprofitable resources during this length-of-run. As long as price is greater than *AVC*, the firm faces two choices: (1) Continue to produce until its fixed resources are no longer of use, which will force this firm to withdraw from the market, or (2) reorganize its fixed resources so as to reduce its unit costs of production, both of which require more time than the short run defines.

Another possibility might be that the firm, given this demand curve as its beginning market, built this particular scale of plant in the expectation of sufficient future growth of its market (a later shift in the demand curve to the right) that it would soon begin capturing economic rent while also protecting its monopoly position.[2]

Figure 8-2*B* describes another possible market situation in which both the demand curve and the firm's size are such that the *MR* curve passes through the *ATC* curve at its minimum point. Equating *MR* and *MC* to determine the firm's optimal output ($q_e$) means that the cost per unit of good produced, in terms of resources used, has been minimized. Even this circumstance does not maximize social preferences, however. Misallocation of resources and social waste are still present.

## EFFICIENCY COMPARISONS _____

We stated previously that setting price equal to *MC* leads to an efficient allocation of resources and their products by finding that one equilib-

---

[2] Examples of a shrinking market demand can be found in certain low demand–high cost facilities such as airports or rail lines that have been abandoned despite public service commission attempts to perpetuate those operations. The demand–growth expectation is more frequently observed in city water or sewer systems, electric power plants, etc., where, because of the long lead time for planning and construction, and the large capital outlays required, the plant is presently overbuilt in the expectation of a larger population to serve (sometime) in the future.

rium level of output at which the value of an additional unit produced for consumers is just equal to the value of other goods sacrificed to get that additional unit. Any divergence from this market type leads to misallocation and is therefore inefficient. To demonstrate the reason for this inefficiency we need to review the basic meanings of the demand and marginal cost curves, as in Figure 8-3*A* and *B*.

The demand curve shows the amounts of a good that consumers will buy at various prices; it also implicitly reflects the values of other goods that must be sacrificed in order to buy these quantities. Thus in Figure 8-3*A*, the area under the demand curve between a pair of points such as *a* and *b* (between 150 and 250 units) represents the value of satisfactions to be derived from those 100 units of the good. If, as we have stated, a monopolist always produces less than a firm setting $P = MC$, and that output also is at a higher price, the value of goods *not* received by consumers (the area under the demand curve between 150 and 250 units) amounts to $250, the shaded area in the figure.

The area under the *MC* curve, on the other hand, represents the value of other goods that would have to be given up to produce those 100 units, since the *MC* curve is the opportunity cost (the value of alternative outputs sacrificed) of producing each unit of this good. That value amounts to $175, the shaded area under the *MC* curve in Figure 8-3*B*. Thus, as Figure 8-4 shows, a monopolist setting $P = MC$ would find an equilibrium

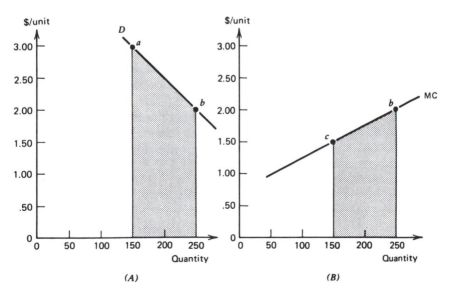

**FIGURE 8-3** Values in consumption and production. (*A*) Demand and consumer valuation. (*B*) MC and producer values.

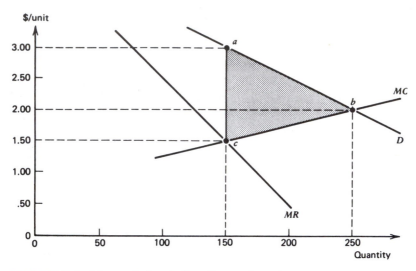

**FIGURE 8-4**   Monopoly's misallocation and waste.

(point *b*, where *P* = *MC*) at 250 units and a price of $2.00 per unit. But given freedom to determine its own production and pricing policies, the monopolist finds *MR* and *MC* are equal at 150 units at a market price of $3.00 per unit, the price consumers are willing to pay for 150 units of the good. Thus consumers are forced to give up 100 units of a good worth $250 to them while the value of other goods that would have been sacrificed to produce those 100 units amounts to only $175. Society has thus incurred a net loss of $75, the area of the shaded triangle (*abc*), which is the difference in the two areas described separately in the previous figure.[3]

The waste of misallocation in monopoly is that the resources that could have been used to produce an additional 100 units of this good, worth $250 to consumers, have been forced into lower-valued alternative production.

This conclusion is valid for any market deviation from the condition of *P* = *MC*. Because the demand curve is less than perfectly elastic, *MR* is less than market price, and the firm produces a lower optimal output with a higher price than would be the rule if *P* = *MC*, which forces consumers to sacrifice some of this preferred good for other, less desired goods.

---

[3] We could have drawn the *D* and *MC* curves as curving lines, to more correctly reflect diminishing marginal utility in consumption and diminishing marginal productivity in production, but this illustration would have added nothing to our ability to estimate the areas (and values) involved.

## Imperfect Competition in the Market

For purposes of theoretical clarity we have, to this point, ignored the large middle ground between pure competition and pure monopoly. The workaday world is not to be found at either of these two extremes, with firms, instead, falling somewhere along the continuum in Figure 7-1, between these two rigidly defined models, in situations exhibiting some (but not all) of the characteristics specific to one extreme or the other.

The label *imperfect competition* is used here to cover a wide variety of real-life possibilities. Although correctly including pure monopoly, we will limit ourselves in the discussion that follows to the general area covering a great many possibilities from a situation with two or more firms exhibiting some degree of competition, however limited that might be, toward the other end of the spectrum with a large number of competing firms that do not, however, meet all the requirements of pure competition.[4]

The word "imperfect" indicates a lack of conformity to the conditions for pure competition. And this departure causes definitional problems. To define an industry from the immense variety of goods available to consumers is arbitrary, at best. A distinction must be made for similar goods somewhere along a scale from *exactly the same* to *totally different* that says these two products are sufficiently alike to be viewed as competing goods in industry X, but a third product is so different that it cannot be included in that industry.

Because of wide differences in the number of sellers, which may range from a few to many, the degrees of product differentiation and control over price, and the manner in which firms respond to these situations, we are forced to search for common characteristics where differences are a matter of degree only.

Similarities in these firms' operations can be noted, particularly in that the competing firms' demand curves are *interdependent,* causing each firm to take other firms' possible actions into account when making their decisions.

Suppose Figure 8-5 represents the market position of Firm A in an imperfectly competitive market. Its demand curve will have a slope between the vertical and horizontal (the greater the number of good substitutes for its product, the less steeply sloped the demand curve will be). With interdependent demand curves, any change in the price of Firm B's

---

[4] Technically more correct distinctions (within this imperfect competition grouping) are made as between *monopolistic competition* and *oligopoly,* based on the number of firms in the market, ease of entry and exit, the degree of product differentiation, and each firm's expectations of other firm's retaliation to their product pricing decisions. Delving into the special characteristics and market behavior of monopolistic and monopsonistic competition, oligopoly and oligopsony, although present in the structure of the agricultural industry, is better left to more advanced courses.

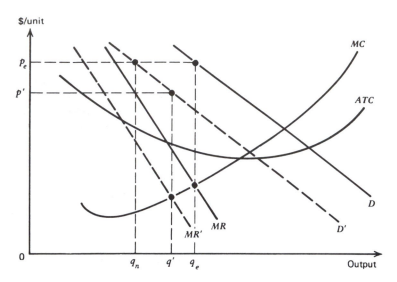

**FIGURE 8-5**   Price-competition with interdependent demand curves.

product will cause Firm A's demand curve to shift because B has now attracted some of A's customers and will call for retaliatory action by A to
recapture its share of the market.[5]

Given Firm A's beginning situation indicated by the solidly drawn lines
labeled $D$ and $MR$, and the later situation by the dashed lines $D'$ and $MR'$,
let's assume this shift of demand has been caused by Firm B cutting the
price of its product and that this has resulted in A losing some of its customers to B—its demand curve has shifted to the left. Firm A was originally at an equilibrium with its product selling for price $p_e$ and $q_e$ units
being sold. Firm A can no longer maintain its price at $p_e$ because that
would cause its sales to fall to $q_n$. Firm A will be encouraged to drop its
price to $p'$ because its $MR$ (from the new demand curve $D'$) at price $p_e$ is
greater than $MC$. We cannot conclude, however, that $q'$ will be optimal
(i.e., that $MR = MC$ at $q'$) because this price change will have brought back
some or all of its original customers (and maybe even some others be-
sides, depending on whether $p'$ is less than B's price). Our vagueness here
is caused by the addition of a second element to Firm A's optimizing decisions. When the firm's change in $TR$ results from a *shift* in its demand
curve, as well as a movement *along* the curve, we are unable to indicate
clearly what Firm A's $MR$ will be until both demand curves are clearly
specified (requiring additional, and more tenuous, assumptions about the
consumers in this market).

---

[5] This sort of competition by price leads to the price wars seen in a number of industries,
such as the gasoline price wars of local filling stations.

Such competition by price results in what is often called "cut-throat competition," which, if continued, can be so destructive as to drive some (or a large number) of the competing firms out of the market. When price competition is so severe that firms in the market will avoid using price as a means of competing with one another, a second characteristic of the imperfectly competitive market is evident—that of *nonprice competition*. Nonprice competition is engaged in when a firm advertises to create a real (or imagined) special image about its product in the minds of consumers. If buyers can be convinced that a firm's product has a special characteristic not to be found in competing products, the advertiser has effectively shifted the firm's demand curve to the right and made it more inelastic in the process, giving the firm a greater degree of control over its price.

Other types of nonprice competition take the form of product design, improving its serviceability, or making it more appealing or more easily recognized by the consumer. These changes are the outcome of research and market development expenditures that become important components of a firm's costs, in preference to the more immediate response from competing firms that can be expected when direct price competition is attempted.

Some improvements may only be illusory, at best, yet "Foaming Bubbles Soap" may find difficult an early response to the "New, New, Improved Suds" of one of its rivals if its own improved variety is not ready to be marketed. Until that response materializes, a market advantage has been obtained for the initiating firm's product. On the other hand, improvements may be significant additions in that they become more desirable products of greater value to the users. One person's frills may be another's necessities. But such innovations as the change from the steel wheels of earlier model tractors to the softer-riding rubber-tired equipment of today with its power steering and hydraulically controlled attachments are more than just conveniences. The transformation from the tractor operator's direct exposure to the elements, to the much more comfortable tractor of today with its sound (and air) conditioned cabs with radio and other conveniences, is hardly an undesirable change for many of today's operators. As one firm has added certain attachments and conveniences that have attracted a larger share of the market, other firms have been forced to follow suit with their own improvements.

Firms in an imperfectly competitive market may also offer different kinds and amounts of special services in addition to the good itself—"specials" such as free delivery, postage-paid shipping, or allowing charges without interest for a specified period of time, and so on.

Because potential *MR* per unit is so large, relative to *MC* for each firm in an imperfect market, there is strong incentive to act alone in determining price and output policies. Yet, the consequences of retaliatory pricing by competing firms are severe. Thus, these firms are encouraged to seek alternative solutions such as the nonprice competition just discussed, or to act in collusion so as to avoid competing by price. Aside from the illegality of acting in consort to set market prices and divide the market among the firms in the industry, widened differences between *MR* and *MC* carry the seeds of destruction for such arrangements because of the attractiveness of larger gains to be reaped by breaking away from the group. And the larger the group, the more difficult it would be to maintain discipline among members because the group is unable to reward individual firms adequately for acting in the interests of the whole rather than their own.

When the structure of markets in an economy is such that price competition is weakened, society has cause for concern because of the undesirable consequences of the noncompetitive behavior of producing or marketing firms. If a high degree of competition cannot be obtained from the functioning of the firms in the market, such markets will fail to allocate resources efficiently and their regulation by the government is sought, in the hope of eliminating those undesirable effects. How that regulation is best carried out and in what particular segments of these markets to place the greatest control so as to be of benefit to the public is the subject of much continuing debate, and the topic of the following section.

# PUBLIC REGULATION OF MARKETS _____

Through much of this nation's history many markets have fallen under the domination of a single large firm or a few relatively large firms. We characterize a market noted for the small number of firms in that market as *oligopolistic,* with each of those firms called an *oligopoly.*

The greater the size of such firms, the greater is their opportunity to exert a special influence in the market. As wealth becomes concentrated in the hands of a few people or firms in the marketplace, their ability to control or manipulate the market for selfish purposes is enhanced and the benefits to be obtained from competition are also threatened.

## The Growth of Firms

Some business firms have been able to grow with their market (both in absolute and relative terms) internally or externally, or by a combination of these methods. *Internal growth* is indicated by the increased value of a firm's capital that permits the construction of additional facilities and increasing business volume as the market for its products has grown. *External growth* has occurred when, through purchase or other means, a formerly independent firm becomes merged with another. As firms are so absorbed, growth is achieved through a change in asset ownership, with many different avenues by which such growth is accomplished.[6]

Three distinct periods of business mergers have been identified by students of industrial organization: 1898–1903, 1926–1929, and 1940–1947,[7] followed more recently by three other new waves starting during the 1960s, the 1980s and the 1990s. The merger movement in 1940–1947 was rather small compared to the much more massive volume of mergers occurring in the other five periods. Each period had its basic emphasis on horizontal, vertical, or conglomerate combinations, each its industries in which the activity was most intense, and each with its type of vehicle most often used to combine firms.

*Horizontal mergers* are combinations of firms in the same industry. An example is a dairy distributor taking over a competing distributor. A *vertical merger* is one that involves two or more firms in different production or marketing stages within the same industry. For instance, a tractor tire manufacturer combining with a firm producing tire cord. *Conglomerate mergers* are among firms in unrelated industries. An apple packing company merging with a cotton processor is one such example.

The first merger period has received particular attention because this cycle "gave to America its characteristic twentieth-century concentration of control."[8] Many of today's very large industrial corporations were formed at that time primarily by purchasing the stock instruments of competing firms. To name but a few, mergers in steel, copper, tobacco, meat packing, chemicals, and farm machinery saw the organization of U.S. Steel, Anaconda Copper, American Tobacco, Swift, duPont, and International Harvester. Few of the large horizontal combinations of the early merger period resulted in monopolies, but it is evident that it did increase concentration in their respective markets.

In the second merger period (1926–1929) more firms were involved than during the earlier period, yet the effect on industry concentration was less

---

[6] We loosely categorize this latter type of growth as a "merger," although technical distinctions are more correctly made between one firm's assets being acquired by another through lease or purchase, the consolidation of two or more firms into a new firm, or the holding company with its parent and subsidiary structure: See, J. Fred Weston, *The Role of Mergers in the Growth of Large Firms*, Berkeley, Calif.: University of California Press, 1953, p. 3. For a discussion of the growth of agricultural cooperatives by merger and other means see Leon Garoian and Gail L. Cramer, "Cooperative Mergers: Their Objectives, Success, and Impact on Growth," Corvallis, Oreg.: *Oregon Agricultural Experiment Station Bulletin* 605, February 1969.
[7] Weston, op. cit., p. 9. This classification of merger periods does not mean that no mergers were formed before or following the dates specified, only that the consolidations were much greater in number during those years.
[8] Paul T. Homan, "Trusts: Early Development," *Encyclopedia of the Social Sciences*, 15:114, 1935; cited in J. Fred Weston, ibid., p. 31.

than the former. The second period mergers occurred in less concentrated industries and involved many small firms.

One difference from the earlier period was that the merged firm's capital assets were acquired, rather than their stocks as in the first period. And whereas expansions during the first period were primarily horizontal (acquiring competing firms), mergers in the 1920s were both horizontal and vertical (expansion backward into materials supplies, and forward into consumer goods manufacturing and distribution outlets).[9] Most of the merger activity of the 1920s took place in the food industry, public utilities, banking, petroleum, and chemicals instead of the heavy industries of the first period.

The third merger period (1940–1947), although widespread, occurred primarily in the metals industries, textiles, food, liquor, and petroleum. Most of the firms absorbed were small, with little measurable impact on industrial concentration.[10]

Although a number of the mergers during the 1960s were horizontal combinations, the more spectacular mergers were of the conglomerate type. Conglomerates of the period—such as Ling–TV acquiring the Wilson Co., itself a conglomerate with branches in meat packing, sporting goods, and pharmaceuticals, among other products—combined firms with no product-line connections.

The 1980s merger growth period ended abruptly in 1989 when large defaults in junk bonds affected some corporations. The merger growth started in 1980–1981 had been triggered by widespread deregulation of industry such as airline companies, by increased use of debt to buy out companies and use of leveraged buyouts. The decline in mergers extended to 1990 and 1991 until the recession ended.

The fifth major merger growth wave began in 1992 with the resurgence of the stock market from the 1990–1991 recession. The key component of this most recent merger growth is the continuous movement toward industry consolidation in order to benefit from economies of scale and meet the challenges of technological change and regulatory action. Filings with the Department of Justice for mergers increased from 1589 in fiscal 1992, to 2816 in 1995. Filings in the first half of 1996 are already at the level of all of 1995. Total spending in 1995 for mergers, acquisitions, and restructuring activity was $271 billion. About $40 billion in 1995 transactons was the result of European companies buying into the United States. The media and banks are at the top of the recent merger list, partly as a result of the Capital Cities/Disney, Chase/Chemical, and First Fidelity/First Union

---

[9] John M. Blair, *Economic Concentration, Structure, Behavior and Public Policy,* New York: Harcourt Brace Jovanovich, 1972, p. 264.
[10] Weston, op. cit., p. 61.

deals. Mergers and acquisitions in the media business have accounted for 43 percent of total industry value.

## THE ANTITRUST LAWS ————————————————

The institution of government is an integral part of all human activity, devising the rules within which we act. Regulative institutional devices within a democratic system develop slowly, after the fact, and legislated reactions to undesirable business practices are no exception.

By the mid-1800s, state laws permitted the incorporation of business firms in limited form. With further legislative actions by a number of states to make incorporation easier (liability limited to the value of the shares of stock held by the owners, free transferability of those shares, one vote per share, and voting by proxy), the ability to manage the affairs of the corporation rested in the hands of directors elected by the stockholders. In addition to numerous other liberalizing laws passed by one or more of the states, New Jersey made an important change in its laws (in the 1880s) that gave corporations the right to own the stock instruments of other corporations. With reciprocity among states honoring the institutions created by any one of them, businesses incorporated under New Jersey's more liberal corporation law could operate nationwide.[11] The door was now opened for drastically reorganizing the structure and operational methods of business firms.

Public resentment and reaction to the development of large trusts and business combines following liberalization of the incorporation laws became so intense that both political parties, during the 1888 presidential campaign, expressed their intent to remedy the situation.[12] Two years later, the U.S. Congress produced the Sherman Antitrust Act, the foundation of the United States' business regulatory policy. The first two sections of this act contain the primary weapons against monopoly and other restrictive business practices.

Section 1 makes it illegal to act in restraint of trade (either interstate or internationally) by conspiring with other individuals or firms to do so, whether by contract, the formation of a trust, or other means. It forbids restraining trade through price-fixing arrangements or controlling and sharing industry output by collusive agreement.

Section 2 makes it illegal to monopolize interstate or international trade, or even to attempt to do so, by combining or conspiring with others to

---

[11] Harry L. Purdy, M. L. Lindahl, and W. A. Carter, *Corporate Concentration and Public Policy,* New York: Prentice Hall, 1942, p. 47.

[12] Ibid., p. 302.

monopolize the channels of trade. This section forbids the use of economic power to exclude competitors from the market.

Except for a few early landmark decisions where dissolution of the offending firms was ordered, deficiencies appeared in the act, and in its application. Enforcement was not pressed with equal vigor by succeeding administrations, and when proceedings were instituted, decisions of the court made differing interpretations of the law.

A Supreme Court decision in 1911 adopted the "rule of reason" that resulted in a great deal of criticism. That rule softened the interpretation of trade "restraint" by questioning whether agreements that restrained trade were "unreasonable." A firm might, by virtue of internal growth made possible by large-scale economies, become dominant in an industry, yet be immune to prosecution. But a strict interpretation of the law would cause the same size of operation to be subject to prosecution if it resulted from a combination of competing firms.[13] That contradiction was eliminated in the 1945 decision against the Aluminum Company of America. The court ruled that a high level of seller concentration in the market could, in itself, be a violation of Section 2 of the Sherman Act. With 90 percent of U.S. virgin aluminum production, Alcoa was ruled to be monopolizing the market, although a smaller market share might not be in violation of that section of the act.[14]

Even though court action under the Sherman Antitrust Act continued, a number of large firms also were formed during the same time period. Monopolistic business combines and trusts once again became a political issue, with both major parties proposing changes in federal antitrust policy during the 1912 presidential campaign. After much controversy over the content of the proposed legislation, two important measures were enacted in 1914—the Federal Trade Commission Act and the Clayton Act.[15] Given broad powers, the Commission was charged with the responsibility of investigating business organization and practices and with carrying out the provisions of the Clayton Act.

Whereas the Sherman Act was general in its identification of what actions were illegal, the Clayton Act was made specific. Section 2 prohibits discrimination between purchasers of a firm's products (except where differences in price are caused by differences in the grade, quantity, or quality of the commodity sold), tie-in sales (i.e., preventing later sellers from also dealing in a competitor's product), and interlocking directorates. Section 7 of the act also prevents a company from holding stock in compet-

---

[13] Ibid., pp. 332–334.

[14] Richard Caves, *American Industry: Structure, Conduct, Performance*, 4th ed., Englewood Cliffs, N.J.: Prentice Hall, 1964, pp. 90–95.

[15] Purdy, Lindahl, and Carter, op. cit., pp. 360–361.

ing companies, or the combining of two or more companies, where such ownership or combination might create a monopoly or reduce competition.

Although the Clayton Act prohibited corporate mergers by means of stock acquisitions that might reduce competition between the merging firms, no such ban was made clear regarding the acquisition of another firm's assets. That loophole was plugged in 1950 with passage of the Cellar-Kefauver Anti-Merger Act.[16]

As the number of merging firms increased through the years, the problem of enforcing the nation's antitrust laws also multiplied and after–the–fact prosecution in the courts often took many years to settle. This problem was sharply reduced for the majority of mergers following passage of the Hart-Scott-Rodino Act in 1976, amending Section 7 of the Clayton Act. The 1976 Act requires all the involved firms in a merger to give premerger notification of the intended merger to both the Federal Trade Commission and the Department of Justice. If either agency questions the proposed merger, the other agency is notified and only one agency handles the investigation. Such premerger notifications grew rapidly over the past decade, from 824 in 1980 to 2883 such notifications in 1989.[17]

Premerger notification carries with it a 30–day waiting period (15 days if the merger is by cash tender offer) before the merger may be completed. The waiting period is extended another 20 days (10 days for cash tender mergers) if the regulatory agency officially requests further information from the merging firms. Further, the enforcing agencies may obtain a preliminary injunction forbidding the merger until its legality is determined. Because of the economic and legal costs faced by the merging companies, the uncertainty of its outcome, and the years often required to reach a decision, few challenged merger proposals get as far as trial in court. They either drop their proposals or settle the contested issues by negotiating with the involved enforcement agency.[18]

# AGRICULTURAL BARGAINING _____

As any part of the market system for agricultural products becomes more concentrated, the power of large firms in the market to influence the prices of farm products also increases. Dominance of a market by one or a few

---

[16] Louis B. Schwartz, *Free Enterprise and Economic Organization*, 4th ed., Mineola, N.Y.: Foundation Press, 1972, pp. 136, 170–171.

[17] Antitrust Division, U.S. Department of Justice, "Antitrust Division 10–Year Workload Statistical Report," Washington, D.C., December 1989.

[18] William L. Baldwin, *Market Power, Competition and Antitrust Policy*, Homewood, Ill.: Richard D. Irwin, 1987, pp. 383–385.

large firms gives them a manipulative ability that may not only raise consumer prices, but widen the market spread by lowering farm prices as well.

Earlier (in Chapter 7) we used a set of criteria as a theoretical model to describe specific market structures, with one extreme of the continuum representing price takers and varying degrees of price-searching toward the opposite extreme of the pure monopolist.

The more typical agricultural situation has frequently been one of a large number of farmers facing a single buyer (a monopsonist) for their products. The theoretical and practical outcome of such a market situation is that producers are forced to accept the price of their product as "given," yet the monopsonist has sufficient market power to be able to hold down the offered price.

Historically, these "middlemen" have handled large volumes of agricultural products and, in attempting to maximize the profitability of their own operations, buy less of the product at a lower price to the producer than would be the case where both buyers and sellers were price takers. Agricultural producers, consequently, have been pressed to action in two directions. On the one hand, they have long felt the need to "fight fire with fire" by organizing themselves into *agricultural bargaining* groups to offset this disparity in market strength. On the other, many farm people have been strong supporters of antitrust and other legislation designed to prevent individual firms from attaining market power and to protect them from unfair practices in their markets.

Fears that the Sherman Antitrust Act might be broadly interpreted to include agricultural organizations as well were not unfounded. Words and phrases in that Act, such as "every" person, "every" contract, and a 1911 Supreme Court decision in the Danbury Hatter's case specifically including agricultural organizations under the Act,[19] did nothing to dispel farmers' fears of that possibility.

That Congress intended to give agricultural associations (limited) freedom to operate in a manner that would be illegal for industrial firms has been made clear in a later series of statutes beginning with the Clayton Act, one of the most important measures providing farm organizations a degree of immunity from the Sherman Act. Section 6 of the Clayton Act permits agricultural organizations to be formed for the purposes of mutually buying supplies or selling products for their members, but with the provision that they may not issue capital stock nor may they be operated for profit.

A number of cooperatives, however, had issued capital stock in organizing before passage of the Clayton Act. Congress then passed the Capper–Volstead Act (1922) to clarify that section of the Clayton Act as it

---

[19] Schwartz, op. cit., pp. 311–312, 320–321.

applied to agriculture. This act allows producers to establish "marketing agencies in common" with or without stock, provided:

1. That they are operated for the mutual benefit of their members,
2. That they do not deal in the products of nonmembers to an amount greater in value than such as are handled by it for its members,
3. That they conform to one or both of the following requirements:
   a. That no member of the association is allowed more than one vote because of the amount of stock or membership capital owned, or
   b. That the association does not pay dividends on stock or membership capital in excess of 8 percent per year.[20]

The Capper–Volstead Act does not provide cooperatives with blanket immunity from the antitrust laws. Yet an agricultural cooperative could gain control over a substantial share (even 100 percent) of the total supply of a commodity without being prosecuted for being a monopolist. How that cooperative gained its control and how it operates in the market are important determinants of whether or not it would be challenged in court.

Such market control would be illegal if the cooperative were to be developed as a combination or conspiracy with noncooperative firms. If the cooperative was properly organized under the Capper–Volstead Act, it would be illegal to use its market power to enhance the prices of its products. Further, the cooperative may not use its power to restrain trade by use of predatory practices designed to eliminate competitors. Any such actions would make the cooperative subject to legal action under the same antitrust laws as any other monopolizing firm.

A number of other acts favorable to agricultural organizations have been passed by Congress. These laws were enacted to protect agricultural producers and their associations from the superior powers of firms handling their products, or to give them a better balance of market power as they deal with large firms in their markets. Most notable of these laws are the Packers and Stockyards Act of 1921, the Cooperative Marketing Act of 1926, the Robinson–Patman Act of 1936, and the Agricultural Marketing Agreement Act of 1937.[21]

The Packers and Stockyards Act reinforced antitrust laws regarding livestock marketing, making certain stockyards public utilities, regulating the buying and selling of livestock, service charges and commission rates. The larger meatpacking companies had previously agreed to divest some of

---

[20] Ewell P. Roy, *Cooperatives: Today and Tomorrow,* Danville, Ill.: The Interstate Printers and Publishers, 1964, pp. 215–216.

[21] Dale C. Dahl and Winston W. Grant, eds., "Antitrust and Agriculture," Conference Proceedings, Agricultural Experiment Station Misc. Report 137, St. Paul, Minn.: University of Minnesota, 1975, pp. 2, 23, 35, 45–49.

their interests ranging from stockyard ownership to railroad terminals, refrigeration, and market news services. That ownership structure had provided them with the ability to discriminate successfully against the smaller packers.

The Cooperative Marketing Act permits farmers, agricultural associations, or federations of such associations to acquire, exchange, and disseminate a variety of price and market information.

The Robinson–Patman Act governs such activities as price discrimination between dealers of commodities and promotional allowances or services in kind for a firm's agents.

The Agricultural Marketing Agreement Act deals with marketing orders, especially important in such commodity areas as fruits, vegetables, and milk. Grower-producer agreements are permitted for fresh and processed classes of these commodities, resulting in a better control of product quality and a more orderly flow through the marketing channels.

These acts (including Capper–Volstead) are of special significance to agriculture in that they assign to the Secretary of Agriculture authority to issue cease and desist orders in cases of violation. The secretary's orders may be backed by resort to district courts for enforcement of penalties, should that become necessary.

Two economic objectives are evident in our antitrust laws as they apply to agriculture. These laws are protective devices for agriculture, because of the relative market weakness of individual producers as they face large firms handling their products. And food itself is viewed as being so essential to a healthy people and economic system that it deserves special measures to ensure an abundant supply at reasonable cost to consumers.

An agricultural bargaining association, if it is to be effective in increasing its farmer-members' incomes, must be able to control or influence those variables that determine whether a firm is a price taker or price searcher. In other words, it must either be accepted by processors as the bargaining agent for all growers of that commodity, or gain control over production and supply to the extent that processors will treat them so.

But control over total supply is very difficult to achieve and almost impossible to maintain. Seldom can all growers of a commodity be convinced that they should join the bargaining association. Unless some method is used to restrict production through grower contracts, or other legal means, it is difficult to control supply through membership maintenance. And in the absence of special power over its members, it is difficult to keep them in the group. A further difficulty is caused by their inability to control the production and supply of substitutes and imported quantities of that good.

Success in bargaining for the members of the association carries its own weakening forces. Given a low price for a product, growers may willingly become a part of an association. As product price is increased, through the actions of the group, some of its members may be tempted to "go it alone" because the original cause of group action has passed. Furthermore, the now higher price is, in itself, a divisive factor. It is a strong incentive for each producer to expand production and thus the total market supply of that product. If the higher price came about through agreements to reduce output, this would cause a loss of members and weaken the association in the very area where it had derived its market strength.

As a consequence, bargaining associations have not been very successful in the longer term. Despite this problem, they have been a balancing power and a beneficial force in improving the degree of competition in many of the markets for agricultural commodities.

# SUMMARY

A monopolist maximizes profit by producing that rate of output where $MR = MC$. For this rate of output, the monopolist will charge a price as determined by the demand curve. A monopolist misallocates resources because the firm produces less output and charges a higher price than an efficient allocation obtained where $P = MC$. As a consequence, consumers are forced to sacrifice some of the monopolist's product for other less desirable products.

Government regulation has been necessary to prevent firms from monopolizing markets. The Sherman Antitrust Act, the Clayton Act, and the Federal Trade Commission Act are examples.

Agricultural producers have formed agricultural bargaining associations or encouraged national legislation to give farmers market power. These attempts have been marginally successful because it has been difficult to maintain control of the total supply of a product, the supply of substitute products, and imports from other countries.

# CHAPTER HIGHLIGHTS

1. A pure monopolist is the only seller in a market; barriers to entry prevent competition from others.
2. The monopolist's demand curve and the market demand curve are one and the same.
3. The monopolist's marginal revenue curve lies below the demand curve because market demand curves always slope downward to the right.

4. The monopolist maximizes profit in the same manner as a purely competitive firm by producing the output at which $MR = MC$, but the monopolist's $MR$ does not equal price as it does for the firm in pure competition.

5. A monopolist's selling price always exceeds $MC$ at equilibrium output. (Prove this with a diagram.)

6. A monopoly is subject to the law of diminishing returns just as is any other firm.

7. A monopoly is inefficient because the value to society of the last unit produced by the monopolist is always greater than its opportunity cost to society.

8. The term "imperfect competition" covers all those market possibilities that do not meet the conditions of pure competition.

9. In imperfect competition one firm's optimizing depends on its competitor's price policies because the demand curves are interdependent.

10. Imperfectly competitive firms avoid price competition, whenever possible, because of demand interdependencies.

11. Public regulation of markets is intended to protect the public by preserving competition in markets.

12. Firms grow by either internal or external expansion, or both.

13. Antitrust laws are designed to prevent monopolistic firms from driving out competitors and to protect consumers from unfair pricing.

14. Agricultural producers form bargaining groups to counterbalance buyers' market power.

15. Special acts by Congress provide bargaining groups with limited immunity from the antitrust laws.

16. Agricultural bargaining groups face difficulties in maintaining continued member cooperation.

## KEY TERMS AND CONCEPTS TO REMEMBER

Agricultural bargaining
Conglomerate merger
External growth
Horizontal merger
Imperfect competition
Interdependent demand curves

Internal growth
Nonprice competition
Oligopolistic market
Oligopoly
Vertical merger

# REVIEW QUESTIONS

1. Explain why a monopolist's marginal revenue and demand curves are two different curves.
2. Does a monopolist always make excess profits? Draw a graph of this situation.
3. What is imperfect competition? How does imperfect competition differ from pure competition?
4. Why do some firms compete on "nonprice factors" rather than on price? Discuss.
5. Explain the generalized business merger periods in the United States. What happened to the structure of American industry during the first merger period?
6. What are horizontal, vertical, and conglomerate mergers?
7. Why have antitrust laws been passed? Are these laws performing their function?
8. What factors must farmers analyze if they want to form farm marketing organizations to improve their prices? Can farmers control these factors? Can the government control these factors? Explain.

# SUGGESTED READINGS

Alchian, Armen A., and William R. Allen. *Exchange and Production: Competition, Coordination, and Control,* 3rd ed., Belmont, Calif.: Wadsworth Publishing Company, 1983, Chapters 7 and 17.

Bain, Joe S. *Industrial Organization,* New York: John Wiley & Sons, 1959, Chapter 1.

Caves, Richard E. *American Industry: Structure, Conduct, Performance,* 7th ed. Englewood Cliffs, N.J.: Prentice Hall, 1992.

Dahl, Dale C., and Jerome W. Hammond. *Market and Price Analysis: The Agricultural Industries,* New York: McGraw-Hill, 1977, Chapters 13 and 14.

Moore, John R., and Richard G. Walsh, eds. *Market Structure of Agricultural Industries,* Ames, Iowa: The Iowa State University Press, 1966, Chapters 1–14.

Samuelson, Paul A., William D. Nordhaus and Michael J. Mandel. *Economics,* 15th ed., New York: McGraw-Hill, 1995, Chapters 25 and 26.

Vernon, John M. *Market Structure and Industrial Performance: A Review of Statistical Findings,* Boston: Allyn & Bacon, 1972, Chapter 6.

## AN OUTSTANDING CONTRIBUTOR

**William G. Tomek**   William Tomek is professor of agricultural economics at Cornell University where he has taught since 1961. He has served twice as the administrative officer for the graduate program in agricultural economics at Cornell. A grandson of Czechoslovakian immigrants, Tomek was a U.S. Army bandsman in 1953–55. Tomek received his B.S. degree in 1956 and an M.A. degree in agricultural economics in 1957 from the University of Nebraska. He earned his Ph.D. in agricultural economics from the University of Minnesota in 1961.

Tomek teaches graduate and undergraduate courses in agricultural marketing, econometrics, price analysis, and statistics. His research has emphasized marketing, futures contracts, and price analysis in several agricultural commodities. He has done a considerable amount of study on the effectiveness of futures markets. The results of his research have been published in numerous journal articles and chapters in books and bulletins. Tomek co-authored a popular textbook, *Agricultural Product Prices*. One of his marketing articles was selected in 1980 by the American Agricultural Association for its Outstanding Journal Article Award.

Tomek is a member of five professional associations. He served as editor of the *American Journal of Agricultural Economics* from 1975 to 1977 and as president of the American Agricultural Economics Association in 1985–86. He was a visiting scholar at the Food Research Institute at Sanford University in 1968–69, a visiting professor at the University of Connecticut in the summer of 1970, and a visiting economist with the Economic Research Service in the U.S. Department of Agriculture in 1978–79. He was the recipient of a National Science Foundation fellowship in 1965, is a member of the agricultural honorary Gamma Sigma Delta, and was selected a Master Alumnus of the University of Nebraska in 1976.

He is single, and music continues to be an avocation.

CHAPTER **9**

# Macroeconomics

$\mathbf{M}$acroeconomic conditions affect our daily lives more than most of us might realize. We cannot read a newspaper or turn on the television without reading or hearing references to the rate of inflation or the number of people that are unemployed, to changes in consumer incomes, or to the rate of increase or decrease in the levels of business inventories or investments. With family and friends, we might often discuss government tax changes, planned federal budget outlays, or mounting government deficits. What are the impacts of these changing economic conditions on our daily lives? What incentives or disincentives do they provide for workers, managers, consumers, savers, and investors? These are but a few of the many facets of aggregate economics, or macroeconomics.

When a single economic entity, such as a business firm or a household, tries to decide which of several alternative possible actions it prefers, it is using a method of analysis called *microeconomics*—our main focus in Chapters 3 through 8. *Macroeconomics*, on the other hand, deals with totals or aggregates, studying characteristics of the entire economy in such dimensions as total employment, total consumer income, and the price level of all goods and services. These totals are derived by adding the value of output for all households and business firms in the economy.

Before we can effectively analyze macroeconomic issues, we need to have some measure or indicator of the performance of an economy. The basic data that allow such performance measures to be developed are called *national income accounts*. These accounts "describe" the economy in more easily understood dollar terms. Looking at these accounts over a series of years can be useful in assessing the overall performance of the economy and in recommending policy changes.

The purpose of macroeconomic analysis is no different from that of microeconomics, or any other science, in that it attempts to explain and predict economic behavior. This is a difficult task because the world keeps changing, yet economists, like physicists, must continue to improve their theories if economics is to be of much use to society.

## THE NATIONAL INCOME ACCOUNTS _____

A simplified model of the circular flow of economic activities between households and business firms in an economy is shown in Figure 9-1. The opposite halves (or "loops") of the diagram demonstrate the flows of real goods and services, using their dollar values as the common denominator.

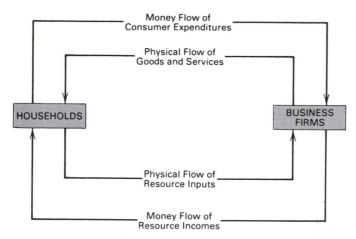

**FIGURE 9-1**  Circular flows of money and goods in a simple economy.

On the left side of the diagram are the households who own the inputs (labor and other resources) that firms use to produce all the finished products that people buy. Households are the consumers of those final[1] products (cars, food, clothing, haircuts, computer games, housing, TV sets, etc.). Households sell their labor and other resource services to business firms in return for income (wages and salaries, interest, rent, and distributed profits), which, in turn, is used to pay for the goods and services that have been produced by the firms within the system.

On the right side of the diagram are the business firms that purchase labor and other resource services from households in order to produce the final products for sale to households. The flow of income to households in the lower loop is, by definition, equal to consumer expenditures in the upper loop of the diagram. Therefore, the total value of output in an economy can be measured by the flow of dollars in either the upper or lower loop of the diagram. The total value of production is thus the total value of consumer spending for final goods and services (upper loop), or total spending by business firms for productive inputs (lower loop).

_____

[1] Rather than totaling the sales at all producing stages from the raw material supplier to the finished product, we must use *final product values*, or alternatively, tally only the *value added* at each step in the production process, so as to avoid double counting the nation's value output.

Suppose that a lawn mower fabricating plant buys all of the component parts for its mowers from other suppliers, then sells the assembled mower for $250. If we include the sales of just this firm and its immediate suppliers—mower housing and wheels ($50), motor ($150), and cutting blade ($15), as well as the finished product ($250)—our GDP would incorrectly total $465, yet the economy has produced a lawn mower worth only $250.

In this simplified model of an economy, we have assumed that households spend all their money income on goods and services and, for ease of understanding, have ignored such other economic activities as government spending and taxation, personal saving and investment, and international trade. These dimensions of an economy will be added as we expand our definition of the national income accounts.

## Gross Domestic Product

The total market value of all the finished goods and services produced within the domestic economy whether by foreign or American resources in a given period of time (a month, quarter, or year), is called the *gross domestic product* (GDP). GDP can be measured, as we have shown in Figure 9-1, by using either the flow of consumer expenditures (upper loop) or the flow of income (lower loop), since the two loops are equal in their dollar amounts.

The *expenditure method* of measuring GDP sums all the final sales in the economy over the accounting period. This total includes sales to consumers (*C*); investment (final producer goods sold to business firms, such as buildings and other facilities, machines and equipment, trucks, typewriters, and a great many other capital items), and increases or decreases in business inventories (*I*); sales to government (*G*); and sales to foreigners minus our purchases from foreigners, or net exports (*NE*).[2] Thus, on the expenditure side, $GDP = C + I + G + NE$.

GDP for 1995 (using the expenditure method)

| | | |
|---|---|---|
| (C) | Final sales to consumers | $4,924.3 billion |
| (I) | Final sales to businesses plus inventory changes | 1,065.3 |
| (G) | Final sales to government | 1,358.5 |
| (NE) | Final sales to foreigners (exports minus imports) | −102.3 |
| | Total GDP (Expenditure Method) | $7,245.8 billion |

GDP in nominal (i.e., actual or current year) dollars was $7.2 trillion in 1995. As these figures show, most of our economic efforts are devoted to producing goods and services for domestic consumption. Our international accounts amounted to $804.5 billion exported (almost 11.1 percent of GDP) and $906.2 billion imported, accounting for the negative trade balance of $101.7 billion in 1995.

Another approach to measuring GDP is the *income method*. Whereas the upper loop measures final-goods spending by consumers, the lower

[2] In order to measure the total output produced by the U.S. economy, we must add the value of all the goods and services exported to other countries and subtract our imports of goods and services that were produced overseas and sold in the United States.

loop measures spending by business firms for all the resources used in production and thus is income to households in the economy. The dollar value of GDP by this method is equal to the cost of production plus any profit earned by businesses. Costs of production are the wages and salaries paid to employees for their labor and managerial services plus proprietors' income; interest paid for the use of borrowed funds; depreciation (the dollar amount of plant and equipment worn out during the year in producing our goods and services); business taxes (sales, excise, and property taxes paid to federal, state, and local governments); rental income (payments made for the use of real property); and corporate profits (the surplus of revenue over all costs of doing business).

GDP for 1995 (using the income method)

| | |
|---|---|
| Payments to employees and managers | $4,209.1 billion |
| Net interest (interest paid minus interest earned) | 401.0 |
| Depreciation | 825.9 |
| Business taxes | 612.4 |
| Net rental income | 122.2 |
| Profits and proprietor's income | 1,066.9 |
| GNP | $7,237.5 billion |
| Plus: net foreign factor income earned in the United States | 8.3 |
| Total GDP (Income Method) | $7,245.8 billion |

Whether we add all final sales (C + I + G + NE) or incomes earned (wages and salaries + net interest + depreciation + business taxes + net rental income + profits), the totals are the same.

As we have defined it, GDP is the sum of the values of all the goods and services produced in an economy that move through the market system. But many goods and services that greatly affect our lives are excluded from the GDP computation because they do not entail market transactions. For instance, the value of work done in the home by homemakers—services worth uncounted billions of dollars—is not counted in the GDP tally. Such services are included in GDP, however, if another person is hired to perform those duties. If you spend $100 on a dinner for the family at a restaurant, that $100 is counted in GDP; if the homemaker prepares the identical meal in the home, its value added is not counted in GDP.

Undesirable side effects of production activities are not accounted for in GDP either. Factory wastes that pollute our rivers and streams and the air we breathe are ignored costs, yet expenditures that we make to clean up those problems are counted. Similarly, GDP includes the value of guns, tanks, and explosives for making war, but we know that we are worse off in war than in peace. Nor do we "net out" the effects of natural or other disasters. The millions of dollars paid to firefighters to combat the rash of forest fires in 1985 were counted in our GDP; the values of burned timber stands, homes, and other destroyed property were ignored, yet, just as for assets worn out in producing goods and services (and depreciated in GDP accounting), these assets are no longer available for future use.

We know that GDP is a measure of market-oriented activity values, but it tells us little about the economic welfare or well-being of the population. It tells us only the value of all marketed goods and services and, as we compare different time periods, whether that value is increasing or decreasing.

## Other National Income Concepts

From GDP, other useful income concepts can be derived. *Gross national product* is equal to gross domestic product adjusted for the net income which Americans earn overseas. *Net national product* is found by subtracting depreciation (the dollar value of plant and equipment consumed during production) from GNP. *National income* is net national product minus indirect taxes that were paid during the year. National income is thus the total of all wages, salaries, proprietors' incomes, net interest, net rents, and corporate profits. This measure is a general term used to denote the income of the total economy, or nation. *Personal income*, the income received by individuals, is determined by subtracting from national income all payroll taxes, corporate profit taxes, and retained earnings (earnings kept by business firms); and adding transfer payments (gifts and other expenditures for which no goods or services are received in exchange). Finally, in order to calculate how much money consumers are able to spend on goods and services, personal taxes must be subtracted from personal income. The result is termed *disposable personal income*—an important income concept, because it is a measure of our spendable, take-home pay.

Following is a flow diagram of these income accounts for 1995:

| Gross Domestic Product |
| :---: |
| $7,245.8 billion |

– Foreign Factor Income
  Earned in the United States
+ American Income Earned Overseas

```
┌─────────────────────────────────────┐
│         Gross National Product       │
│            $7,237.5 billion          │
└─────────────────────────────────────┘
```

– Depreciation

```
┌─────────────────────────────────────┐
│         Net National Product         │
│            $6,411.6 billion          │
└─────────────────────────────────────┘
```

– Indirect Business Taxes

```
┌─────────────────────────────────────┐
│           National Income            │
│            $5,799.2 billion          │
└─────────────────────────────────────┘
```

– Payroll Taxes,
  Corporate Profit Taxes, and
  Retained Earnings
+ Transfer Payments

```
┌─────────────────────────────────────┐
│           Personal Income            │
│            $6,101.7 billion          │
└─────────────────────────────────────┘
```

– Personal Taxes

```
┌─────────────────────────────────────┐
│       Disposable Personal Income     │
│            $5,307.4 billion          │
└─────────────────────────────────────┘
```

## Nominal Values and Real Values

*Nominal values* are the dollar values of goods and services at their current prices. The nominal value of a candy bar, for instance, is its current market price—the amount you will pay for it at the check-out counter.

In periods of inflation or deflation, the purchasing power of the dollar changes, causing nominal values to be of little use in making comparisons over time. Suppose the current price of the candy bar just mentioned is $0.50, and that its price last year was $0.40. Assume, further, that you bought 10 candy bars in each of those years. Our current year GDP would show an increase of 25 percent in your candy bar purchases, yet there has been no change in the quantity of candy bars. We must, therefore, distinguish between an increase in GDP caused by higher prices and an increase in GDP that results from producing a greater quantity of goods and services. Such a distinction is necessary, because, with income unchanged, a price rise worsens the economic well being of people whereas the latter improves their material well being.

To compare values over time, the nominal values of goods must be "deflated" by an appropriate index in order to determine their real values. Any time you see the word *real* used in an economic context, it means that the values are being expressed in "constant" dollars—dollars that have been adjusted to remove the effect of inflation or deflation.

In 1988, the U.S. GDP was $4,900.4 billion in current dollars; in 1989, our GDP was $5,250.8 billion. In that one-year period, GDP increased 7.2 percent in *nominal* terms. But the value of the dollars used to measure GDP also changed from 1988 to 1989. To adjust GDP for the effect of a change in the value of the dollar, we use indexes such as the Consumer Price Index (CPI) and the GDP Price Deflator Index. These indexes measure the change in prices of specific quantities of goods and services over time. From 1988 to 1989, the prices of all goods and services, as reflected in the GDP Price Deflator Index, increased from 103.9 to 108.5—a decline of 4.4 percent in the purchasing power (or value) of money. Real 1989 GDP increased (in 1987 dollars, as the reference base year for the GDP Price Deflator Index is 1987) from [$4,900.4/1.039 =] $4716.5 billion in 1988 to [$5,250.8/1.085 =] $4839.4 billion in 1989. Real GDP, the actual output of the economy, increased from 1988 to 1989 by $122.9 billion, or 2.6 percent. The difference between the 7.2 percent increase in nominal GDP and the 2.6 percent real GDP increase was the result of increased prices.

# TOOLS OF MACROECONOMICS— AGGREGATE DEMAND AND AGGREGATE SUPPLY _____

Demand and supply curves are defined in macroeconomics in a much different manner than in the microeconomic approach to decision making. In microeconomics, the relative price or cost in dollars per unit of a par-

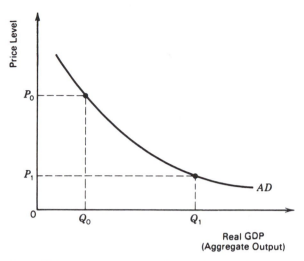

**FIGURE 9-2**   An aggregate demand curve showing the inverse relationship between the price level and aggregate real domestic output.

ticular commodity is measured along the vertical axis, and the quantity of the commodity is measured along the horizontal axis.

In macroeconomics, the **aggregate demand curve** (*AD*) shows the amounts of real goods and services that households, businesses, and government are willing to purchase at various price levels[3] (Figure 9-2). To demonstrate that relationship graphically, we plot the price level on the vertical axis and real GDP along the horizontal axis. Because we are interested in the specific relationship between the price level and aggregate demand, all other factors that may influence the purchases of goods and services must be held constant. We are thus concentrating on movements *along* the *AD* curve, postponing until later in this chapter any consideration of those factors that will cause the *AD* curve to shift its position in the diagram.

As shown in the figure, at any high price level such as $P_0$, the aggregate real domestic output purchased is $Q_0$; with a lower price level $P_1$, a greater aggregate real domestic output, $Q_1$, is purchased. Notice the inverse relationship between the price level and real output. The *AD* curve slopes downward to the right for at least the following reasons.[4] First, as the price level is increased, more money will be demanded to purchase a given amount of goods and services at higher prices. If the supply of

---

[3] In our use of the term *price level*, we are referring to an average of the prices for all goods and services, rather than the price of any single good or service.
[4] William J. Baumol and Alan S. Blinder, *Economics: Principles and Policy*, 4th ed., San Francisco: Harcourt, Brace, Jovanovich, 1988.

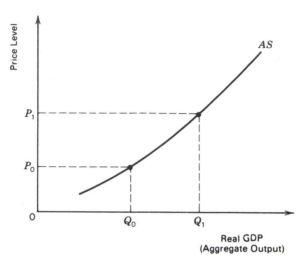

FIGURE 9-3   An aggregate supply curve showing the direct relationship between the price level and aggregate real domestic output.

money remains constant while the quantity of money demanded rises, the interest rate must increase. A higher interest rate will reduce investment and, therefore, will reduce aggregate demand. A second and less important reason that the *AD* curve slopes downward to the right is that higher price levels, *ceteris paribus*, reduce the purchasing power of consumer assets (money, stocks, and bonds). A lower real value of assets reduces consumption spending.[5]

The *aggregate supply curve* (*AS*) in macroeconomics has the same appearance as its supply counterpart in microeconomics, with the same important differences in axis labels as for the micro- and macroeconomic demand curves. Using the same labels on the axes as for the *AD* curve, the *AS* curve shows the aggregate output supplied by producers at various price levels during a given period of time (Figure 9-3). At a price level of $P_0$, the aggregate quantity supplied is only $Q_0$; at a higher price level, such as at $P_1$, the quantity supplied will be $Q_1$.

Note the positive slope of the *AS* curve—a reflection of the direct relationship between the price level and quantity supplied.[6] As the price level rises, selling prices increase relatively more than do input costs, so producers are encouraged by their short-run profitability to increase their

---

[5] Real income is held constant when comparing consumer expenditures at different price levels. *Ibid.*, p. 147.
[6] We will postpone until later in this chapter (under the heading "The Shape of an Aggregate Supply Curve") consideration of some important deviations from this generalized shape of the *AS* curve.

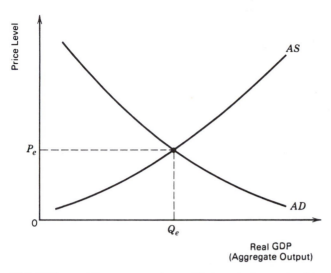

**FIGURE 9-4**   The economy's equilibrium is determined by the intersection of the aggregate demand and supply curves.

output of goods and services. Conversely, a reduced aggregate output accompanies a decline in the price level.

## Equilibrium Output

An equilibrium in any physical or other system occurs when a balance between opposing forces within the system is established. That system will remain in such a "state of rest" until one or more of the forces in the system is changed. *Macroeconomic equilibrium* occurs when there are no forces in the economy acting to either increase or decrease real GDP. As shown in Figure 9-4, the equilibrium real GDP and price level for a particular period of time, holding everything else constant, is determined at the point where the aggregate demand curve intersects the aggregate supply curve. When $AD = AS$, the aggregate quantity of goods and services demanded is equal to the aggregate quantity supplied. This equilibrium condition is shown in Figure 9-4 at the price level $P_e$ and real output level $Q_e$.

## Shifts in Aggregate Demand

When the $AD$ shifts to the right (an increase in aggregate demand), the entire curve moves to the right, from $AD_0$ to $AD_1$, as shown in Figure 9-5. With $AS$ given, the impact of the increase in aggregate demand is to increase the price level from $P_0$ to $P_1$ and to increase aggregate real output

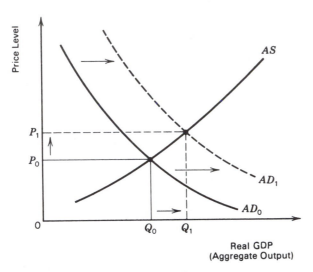

**FIGURE 9-5**   An increase in aggregate demand shifts the demand curve to the right.

from $Q_0$ to $Q_1$. A decrease in aggregate demand would shift the $AD$ curve to the left from $AD_1$ to $AD_0$ in the diagram. As a result, the general price level would fall from $P_1$ to $P_0$, and aggregate real output would fall from $Q_1$ to $Q_0$.

These shifts in aggregate demand, with their corresponding influence on the price level, have significant effects on people in the economy. We know the severe effect of price level increases (inflation) on people with fixed incomes. Implicit within the concept of aggregate output is the employment of people and other resources. A short-run increase in output is brought about by increased employment of labor and other resources. And when output falls, employment also declines. When an economy is not fully utilizing all its resources, society loses the output that could have been produced by those idled resources.

Changes in aggregate demand (shifts of the $AD$ curve) occur because of spending changes in the economy. An increase in spending by government, consumers, businesses, or foreign buyers will increase aggregate demand. Some examples of increased aggregate demand are an increase in government defense spending, reduced personal taxes that result in an increase in disposable income, increased business investment spending, or an increase in the nominal money supply. All these factors tend to raise output, employment, and the price level.

Increased business investment spending, additional government spending, or tax cuts increase personal incomes in the economy and encourage more short-run spending by consumers, with a resulting shift of the *AD* curve to the right.

Aggregate demand will decrease (shift the *AD* curve to the left) and reduce the price level, output, and employment when there is a drop in government and investment spending, an increase in taxes, and a decrease in the money supply. Other examples of factors that can cause a decrease in aggregate demand are changes in expectations that cause consumers to increase their rate of saving, reduced exports or increased imports, or reduced Social Security payments.

Total spending changes resulting from any or all of the factors mentioned here will cause the entire *AD* curve to *shift* its position in the quadrant space. Any movement *along* a given *AD* curve, however, will be in response to a change in the general price level.

## Shifts in Aggregate Supply

Shifts or changes in the position of the *AS* curve result from influences that impact on production costs, thus changing the total output of goods and services at any given price level. Aggregate supply shifts are caused by changes in technology, labor productivity, input prices, and other external factors affecting production, such as climatic changes. New and improved technology that increases output per unit of input will increase aggregate supply (move the entire *AS* curve to the right, as in Figure 9-6, from $AS_0$ to $AS_1$). Lower prices for inputs or more favorable growing weather will also shift the *AS* curve to the right. Given *AD*, an increase

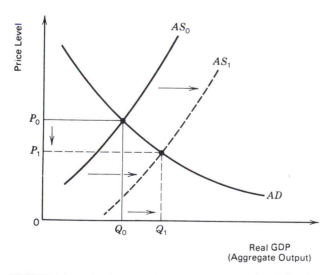

**FIGURE 9-6** An increase in aggregate supply shifts the *AS* curve to the right.

in aggregate supply will reduce the price level from $P_0$ to $P_1$ and increase aggregate real output from $Q_0$ to $Q_1$.

A decrease in aggregate supply can result from a decline in labor productivity, or a shift in worker preferences for more leisure time. Such changes will shift the *AS* curve to the left (as from $AS_1$ to $AS_0$ in the diagram), thus increasing the general price level and reducing aggregate real output and employment in the economy.

Any movement *along* a given *AS* curve is a response to a change in the general price level, while an increase or decrease in aggregate supply, due to changes in any or all of the factors just discussed, will cause a *shift* of the *AS* curve to the right or left in the diagram.

## The Shape of an Aggregate Supply Curve

There is little, if any, disagreement among economists about the upward slope of the short-run aggregate supply curve (as we have drawn it in this chapter). Under "normal" conditions, the *AS* curve reflects a positive relationship between the price level and aggregate real output. If this relationship were true without exception, however, we should never expect an increased price level without a corresponding increased real output in the economy, or an increase in real output without an accompanying increase in the general price level, both of which have been experienced in the past by the U.S. economy.

Such occurrences suggest three separately identifiable segments of the *AS* curve, as illustrated in Figure 9-7. Beginning from a situation of high

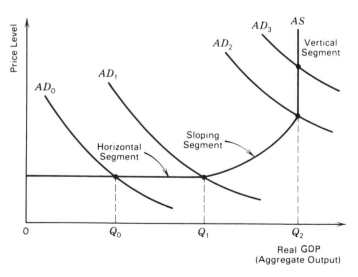

**FIGURE 9-7**   The three segments of an aggregate supply curve.

unemployment in the United States in the 1930s (and other similar periods, especially in earlier years), substantial growth of real GDP was produced without seeing the general price level also increase. Such a possibility is shown by the horizontal segment of the *AS* curve. Real output can be increased to output $Q_1$ without a necessary increase in the price level because unemployed labor and underutilized plant and equipment can be put to productive use. Prices do not rise in such a situation because these newly employed labor and capital resources are not being bid away from alternative goods production. Without an upward pressure on input prices (thus no rise in production costs), the greater spending of $AD_1$ (as compared with $AD_0$) can be accomplished without causing prices to rise.

An opposite extreme is one in which the price level rises, without an accompanying increase in real output in the economy. Such a condition is represented by the vertical segment of the *AS* curve. Were the labor and capital resources of the economy to be fully employed, an increase in aggregate demand (as from $AD_2$ to $AD_3$) would simply inflate prices with no increase in real output because the economy is already producing at capacity output.

Between output levels $Q_1$ and $Q_2$ (the segment of the *AS* curve that slopes upward to the right), prices of some resources are bid up in order to obtain the increased output that a movement from $Q_1$ to $Q_2$ represents. Higher input prices and the more limited short-run physical availability of inputs causes higher production costs, thus higher prices for all goods and services. These costs will normally increase with increased output until full employment is reached (the vertical segment of the *AS* curve).

## Full Employment

The U.S. government formally committed itself to a policy of full employment in the Employment Act of 1946.[7] Although we may speak of "full employment," we do not mean *zero* unemployment but rather a rate of unemployment that is due to the normal functioning of the labor market. Full employment would currently be defined as an unemployment rate of about 6 percent, whereas about 4 percent was regarded as full employment as recently as the 1960s.

Assume that full-employment GDP is at level $Q_f$ in Figure 9-8. Will the aggregate real domestic output always be at full employment GDP? The answer to this important question depends on whether wages are sufficiently flexible to move the economy to full employment, or whether

---

[7] In 1933, the Roosevelt Administration adopted full employment as a policy objective with the implementation of its New Deal, but it was not until the Act of 1946 that the detailed procedures to achieve full employment were specified.

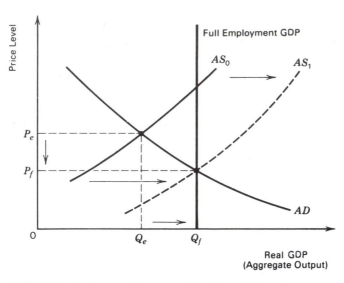

**FIGURE 9-8**   An equilibrium at less than full employment.

there are rigidities in the economy that will cause the system to be in equilibrium at output levels other than full employment GDP. In Figure 9-8, equilibrium GDP ($Q_e$) is below the nation's full employment GDP ($Q_f$).

Most economists would argue that full employment GDP can be reached automatically if wages and prices are free to adjust. With unemployed resources, there will be downward pressure on resource prices that will reduce the cost of production and shift the aggregate supply curve to the right (from $AS_0$ to $AS_1$). Because of the increased aggregate real output, the price level falls and, as a consequence, increased consumer spending can be relied on to move the economy to full employment. Although this automatic correction may occur—eventually— many policy analysts and politicians are frequently unwilling to wait for the length of time it might take while the economy moves toward full employment.

Some economists maintain that wages and prices are not sufficiently flexible because of a number of factors that restrict short-term adjustments within the economy, making it possible to have an equilibrium for a period of time at a level below full employment. Their solution to the dilemma is to rely on fiscal policies to stimulate aggregate demand, emphasizing increased government spending, reduced taxes, and increased money supply. Such policies result in greater real output and higher employment levels through a shift of the aggregate demand curve. According to the Keynsians, demand-stimulating policies can achieve the de-

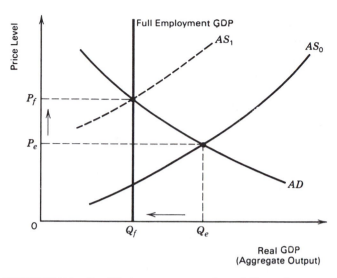

**FIGURE 9-9**   Equilibrium at greater than full employment GDP.

sired results much more rapidly than can reliance on the slower adjustments of the price system.[8]

If, as shown in Figure 9-9, equilibrium real GDP ($Q_e$) is above full employment GDP ($Q_f$), the shortage of resources and products causes inflationary pressures in the economy. The demand for labor is high, causing wage rates to increase. Rising labor and other costs will shift the aggregate supply curve to the left (from $AS_0$ to $AS_1$). As a result, the price level will rise and cause consumer spending to decline until full employment equilibrium is reached, at $Q_f$.

## Simultaneous Inflation and Unemployment

During the late 1970s and the early 1980s, the economy experienced a situation of both high rates of inflation (12 to 13 percent in 1979 and 1980) and high levels of unemployment (rising to near 10 percent by 1981 and 1982). The word *stagflation* was coined to describe that situation. In general, it is the view that the aggregate supply curve had shifted to the left, in response to higher production costs, and had caused higher prices for goods and services with a lower real output and employment level. This

---

[8] A school of thought (following John M. Keynes) that relies heavily on active government fiscal policies to stimulate demand and reach full employment GDP. Monetarists, another school of thought, emphasize the money supply and its changes as a crucial variable in reaching full employment GDP.

leftward shift of the aggregate supply curve was, in part, a result of higher petroleum and other input prices.

In an effort to relieve the problem, policies established by the Reagan administration and continued by the Bush administration attempted to influence both the demand and supply sides of the equation. The Federal Reserve was asked to maintain a tight rein on the money supply and Congress was asked to cut the rate of increase in federal spending. These steps were intended to reduce the rate of growth in aggregate demand. On the supply side, the administration attempted to provide incentives to workers and businesses by cutting both personal and business tax rates, expecting to shift the aggregate supply curve to the right and thereby hold down prices while increasing employment in the economy. As a result, interest rates declined steadily to a prime rate[9] low of 6.25 percent. Unemployment fell in each year from its peak to a rate of 5.2 percent in 1989 and had risen to 7.3 percent in early 1992. The number of unemployed people declined from 10.7 million in 1982 and 1983 to 6.6 million by the end of 1989 and back to 9.4 million by 1992.

Although per capita real income grew by less than $75 per year in the 10 years ending in 1982, the period during which stagflation also became most serious, real income has grown an average of $240 per year up to the 1990 recession. But these favorable results have not been achieved without other undesirable changes. Federal budget deficits and the public debt both grew rapidly, resulting in the Gramm–Rudman–Hollings Act, passed by the Congress to force tighter control over federal spending. Similarly, the nation's balance of trade turned very negative in 1983. That imbalance worsened until in 1987 when the net export of goods and services reached –$143.1 billion, improving since then to –$42 billion in 1992, but expanding to –102.3 billion in 1995.

The American economy fell into a recession in the second half of 1990 and has been in a slow recovery since March 1991. Factors contributing to this most recent recession were the oil price shock near the end of 1990, and various structural relationships and imbalances including defense downsizing, a boom-bust cycle in commercial real estate aggravated by wide swings in the tax laws and bank regulations, credit market constraints due to the problems within financial institutions, a buildup of corporate debt relative to profits, and excessive household debt relative to income. Economic performance has been very sluggish, and there has been a general slowdown in the international economy that is also restricting the U.S. recovery. The economy has continued to expand, however, since March 1991. Inflation and unemployment have remained low and stable. Recent monetary policy has involved open market operations

---

[9] The prime rate is the interest rate that large New York banks charge their better borrowers for short-term, low-risk loans; it is used as a benchmark for setting local interest rates.

to adjust interest rates. Fiscal policy has been used indirectly by reducing the budget deficit and the federal workforce. This contractionary fiscal policy has reduced interest rates and stimulated private investment. In the next few years, the economy is expected to remain sound with a growth rate slightly over 2 percent per year; inflation is expected to remain about 3 percent; unemployment about 5.7 percent; and interest rates to fall about 1 to 1½ percentage points.

# MACROECONOMIC LINKAGES TO AGRICULTURE _____

Because agriculture is only one of many sectors of the general economy, and as interdependent as all others, changes in macroeconomic policies will influence incomes in agriculture through changes in taxes, government spending, and the money supply.

As mentioned in Chapter 2, *monetary policy* deals with the money supply and credit conditions: *fiscal policy*, on the other hand, deals with federal government spending and taxation. Changes in monetary and fiscal policies affect farmers as both producers and consumers. *Expansionary monetary policy*, such as the Federal Reserve purchasing bonds, reducing the reserve requirement, or lowering the discount rate, will increase the money supply, thus making more money available for credit and loans. An increase in the supply of loanable funds, *ceteris paribus*, causes the interest rate to decline, which should increase investment spending and aggregate demand. As purchasers, agricultural firms are then able to buy more goods and services, which will cause their prices to rise. Expansionary policy, because of its impact on the real interest rate (the nominal rate minus the rate of inflation), will reduce production costs for agriculture, as elsewhere in the economy, but the higher prices that farmers pay for their inputs could offset this advantage to some degree.

Expansionary monetary policies also influence the amount of international trade in agricultural commodities. Under a floating exchange rate, when the U.S. interest rate falls relative to interest rates in foreign countries, a flow of funds out of this economy results as financiers increase the foreign holdings in their portfolios. This outflow of funds is an increase in the demand for foreign currencies, which depreciates the value of the dollar and, in turn, increases the exports of U.S. agricultural commodities. U.S. exports increase because the depreciated value of the dollar makes American goods cheaper for foreign buyers.

Farmers benefited from the depreciated value of the dollar during the 1970s, especially in the volume of agricultural exports to nations that allowed the exchange rate to float. However, not all countries allow their currencies to float on international money markets. Although the major trading nations allow the value of their currencies to fluctuate with changing demand and supply of currencies, it is often a "managed" float. As such, they intervene in their money markets to offset the effects of a depreciating dollar by having their central banks purchase the increased supply of dollars. Consequently, the dollar may actually appreciate against those nations' currencies and cause a decline in U.S. exports to those nations. Many developing nations, on the other hand, often overvalue their currencies, causing their food imports to be greater than would have occurred with a market-determined exchange rate for their currencies.

When the government increases its spending or decreases taxes, it is engaging in *expansionary fiscal policy*. Higher government expenditures are made possible by borrowing from the public. In this case, the rise in government spending increases aggregate demand, and prices and incomes may also increase temporarily. This increase in prices and incomes (or output) increases the U.S. demand for loanable funds, thus increasing the interest rate. With a higher interest rate, domestic investment expenditures decline and domestic aggregate demand falls. Internationally, the rise in U.S. interest rates causes loanable funds to flow into the United States, as foreigners add to their U.S. investments. This increase in the demand for the dollar causes the value of the dollar to rise against foreign currencies. With the increased value of the dollar, U.S. goods become more expensive for foreigners, and our exports decline. The net effect of such domestic and foreign adjustments is for a decline of aggregate demand in the United States.

Because the United States does not often intervene directly in the foreign exchange markets, most foreign central exchange banks will most likely let the dollar appreciate because of the expanded U.S. imports of those nations' products.

The general impacts of short-run changes in monetary and fiscal policies, under flexible exchange rates, can be summarized as follows:

| | Monetary Policy (Changes in Money Supply) | | Fiscal Policy (Changes in Government Spending) | |
|---|---|---|---|---|
| | *Expansionary* | *Contractionary* | *Expansionary* | *Contractionary* |
| Domestic farm prices | increase | decrease | increase | decrease |
| Export farm prices | increase | decrease | decrease | increase |
| Farm input prices | increase | decrease | increase | decrease |
| Real interest rate | decrease | increase | increase | decrease |

Expansionary monetary policies during the 1970s led to increasingly high inflation rates, lower real interest rates, and a depreciated dollar in the international money markets, which helped spur foreign demand for U.S. agricultural commodities. High inflation rates and increased prices for farm inputs increased farm production costs, but export demand expansion kept farm prices and net income rising. Farm debt also rose over this period as many farmers borrowed against their rising land values to add to their land holdings, and borrowed other funds to increase their production as well.

Beginning in late 1979, monetary policy was abruptly switched to a contractionary policy designed to lower the rate of inflation. As a result, interest rates reached record levels in the 1980–82 period. At the same time, foreigners began to view the United States as a more favorable place to invest, causing the value of the dollar to increase by 50 percent between 1980 and early 1985. Foreign demand for U.S. agricultural products fell both because of the dollar's appreciation (which made U.S. goods more expensive for foreigners to buy) and because many food-importing countries suffered economic reverses that limited their purchasing power. United States agriculture was severely hit by the combination of lower product demand and higher production costs—especially the much larger interest payments required to service a sharply increased farm debt.

The contractionary monetary policy was successful in reducing the rate of inflation from more than 10 percent in the early 1980s to about 3 percent in 1984–85. Nominal interest rates began to decline in 1984, in part because of lowered expectations of inflation. Current interest rates have declined to the lowest level in 25 years. The exchange value of the dollar peaked in early 1985, and by 1995 had declined to its 1980 level. These changes, along with more emphasis on trade liberalization, were more favorable to U.S. agriculture. Foreign demand for U.S. agricultural products has improved with the economic recovery in many other countries. While macroeconomic changes can (and do) have important impacts on agriculture, microeconomic factors also play a crucial role.

# SUMMARY

Macroeconomics is the study of the entire economy in the dimensions of the aggregate price level and aggregate income, employment, and real output of goods and services. The total market value of all finished goods and services produced within the economy in a given period of time is called the gross domestic product. Gross domestic product can be measured either from the expenditure side as $(C + I + G + NE)$ or by the income side as the sum of payments made to employees and managers plus proprietors' income, net interest payments, depreciation, business taxes paid, net rental income, and profits. Other income accounts can be derived from GDP. Real gross domestic product is nominal GDP adjusted for the change in the value of the dollar. To calculate real GDP, nominal GDP is divided by either the Consumer Price Index or the GDP Price Deflator Index.

Aggregate demand and aggregate supply determine the equilibrium price level and real domestic output rate for the economy. Aggregate demand will shift if spending increases or decreases. The aggregate demand curve will shift to the right with increased spending by consumers, the government, businesses, or foreign buyers of U.S. goods; it will shift to the left with decreased spending by any of those entities. Also, aggregate demand will increase with a cut in personal taxes or an increase in the money supply; it will decrease with an opposite change in those policy variables.

Aggregate supply will increase if a change in technology increases output per unit of input, and with lowered input prices and other changes such as more favorable weather conditions during the growing season.

Equilibrium real GDP is possible either above or below full employment GDP. Government policies are initiated as efforts to move equilibrium real GDP to full employment GDP, but some economists argue that there is no need to do this because it aggravates the movement to equilibrium. They argue that the automatic mechanisms involved in any economic situation will move the system to full employment if given enough time.

Monetary and fiscal policies influence every sector of the economy, including agriculture. Expansionary monetary policy can increase domestic and export farm prices and decrease the interest rate. These factors will affect every aspect of farm life, including farm production costs, farm prices, purchases of consumption goods, investments, and net income from farming. Expansionary fiscal policy can increase domestic farm prices and reduce export prices, increase farm input prices, and increase real interest rates. Changes in these factors will affect consumption and investment decisions, as well as farm prices, production costs, and net farm income.

# CHAPTER HIGHLIGHTS

1. Macroeconomics deals with totals or aggregates, studying characteristics of the entire economy in such dimensions as total employment, consumer incomes, and the general price level.
2. Gross domestic product (GDP) is the total market value of all finished goods and services produced within the domestic economy in a given period of time.

3. GDP can be measured by either the expenditure method or the income method.

4. Other income concepts, beside GDP, include gross national product, net national product, national income, personal income, and disposable personal income.

5. Nominal values are the dollar values of goods and services at their current market prices.

6. Real values are the dollar values of goods and services expressed in constant dollars. Real values are adjusted values that have removed the effects of inflation or deflation.

7. An aggregate demand (*AD*) curve shows the amounts of goods and services that households, businesses, and government are willing to purchase at various price levels during a given period of time.

8. An aggregate supply (*AS*) curve shows the amounts of goods and services that will be supplied by producers at various price levels during a given period of time.

9. Macroeconomic equilibrium occurs when there are no forces in the economy acting to either increase or decrease real GDP. This equilibrium occurs where *AD* = *AS* and determines the general price level and real GDP.

10. Shifts in aggregate demand are caused by changes in spending by consumers, businesses, the government, or foreign buyers.

11. Shifts in aggregate supply are caused by changes in technology, labor productivity, input prices, and other external factors (such as climatic changes) that affect production.

12. The economy can be in equilibrium at less than or more than full employment GDP.

13. Monetary policy deals with the money supply and credit conditions in the economy.

14. Fiscal policy deals with federal government spending and taxing.

15. Changes in monetary and fiscal policies shift the aggregate demand curve and the aggregate supply curve.

16. Changes in macroeconomic policies affect all sectors of the economy, including agriculture.

17. Expansionary monetary policy should increase domestic farm prices, export prices, and input prices, but decrease interest rates. Contractionary monetary policies should have the opposite effects.

18. Expansionary fiscal policies should increase domestic farm prices, input prices, and the interest rate and reduce farm export prices. Contractionary fiscal policies should result in opposite effects.

19. Macroeconomic policies of other nations can offset the intended effects of U.S. macroeconomic policies.

## KEY TERMS AND CONCEPTS TO REMEMBER

Aggregate demand

Aggregate supply

Contractionary fiscal policy

Contractionary monetary policy

Disposable personal income

Expansionary fiscal policy

Expansionary monetary policy

Expenditure method

Fiscal policy

Gross domestic product

Gross national product

Income method

Macroeconomic equilibrium

Macroeconomics

Microeconomics

Monetary policy

National income

National income accounts

Net national product

Nominal values

Personal income

Stagflation

Value added

## REVIEW QUESTIONS

1. Microeconomics and macroeconomics differ in many ways. Explain the major differences.
2. Define gross domestic product. What does GDP measure? Is GDP an indicator of the well-being of society? Explain.
3. Explain the meaning of nominal values and real values. In comparing GDP over time, why must nominal values be converted to real values?
4. What do we mean by aggregate demand? What factors can cause the aggregate demand curve to shift? Do monetary or fiscal policy changes cause the aggregate demand curve to shift? Why, or why not?
5. Review the shape of an aggregate supply curve. How, if at all, does the shape of the aggregate supply curve differ in the short run from its long-run shape?
6. Why should any person be concerned with monetary and fiscal policies? Can these macroeconomic policies be used to move the economy to "full employment"? Why not just let the economy find its own level of equilibrium without federal intervention?
7. Construct a diagram showing the impact of contractionary monetary policy (given flexible exchange rates) on the money supply, supply of credit, interest rates, investment spending, aggregate demand, dollar

flows into or out of the United States, demand for foreign currencies, the exchange value of the dollar, and U.S. net exports.

8. Macroeconomic policy changes influence agriculture. Explain the general effects on agriculture of a contractionary fiscal policy.

9. Discuss the relative impacts of macroeconomic policies on U.S. grain prices in recent years, versus the price effects of increased U.S. production and U.S. and foreign agricultural programs.

## SUGGESTED READINGS

Baumol, William J., and Alan S. Blinder. *Economics: Principles and Policy*, 5th ed., New York: Harcourt, Brace, Jovanovich, 1991, Chapters 5, 7-10.

Gwartney James D., and Richard Stroup. *Economics, Private and Public Choices*, 6th ed., Fort Worth: Dryden Press, 1992, Chapter 14.

Mansfield, Edwin. *Principles of Macroeconomics*, 7th ed., New York: WW Norton, 1992, Chapters 9 and 20.

Samuelson, Paul A., William D. Nordhaus and Michael J. Mande. *Economics*, 15th ed., New York: McGraw-Hill, 1995, Chapters 23 and 24.

Schiller, Bradley R. *Essentials of Economics*, 2nd ed., New York: Random House, 1996, Chapters 11, 12, and 14.

## AN OUTSTANDING CONTRIBUTOR

**Gordon C. Rausser**   Gordon Rausser is Robert Gordon Sproul Distinguished Professor at the University of California at Berkeley. He has been president of the Institute for Policy Reform since 1990. He received the B.S. degree from Fresno State University in 1965, his M.S. in 1968 from the University of California at Davis, and a Ph.D. from the University of California at Davis in 1971.

Rausser was assistant, and associate professor, at the University of California at Davis, 1971–74; professor of economics and statistics at Iowa State University, 1974–75; professor of business administration and managerial economics at Harvard University, 1975–78; and has been professor of agricultural and resource economics at the University of California at Berkeley since 1979. He was chairman of the Giannini Foundation 1982–84. He served as chairman of his department 1979–85 and was renamed chairman in 1993. In addition to his distinguished professorship, he served as Chief Economist with the Agency for International Development in Washington, D.C.

He has been a visiting professor at Hebrew University and at Ben-Gurion University, Israel; a Fulbright Scholar in Australia, and a Ford Foundation Scholar in Argentina. Rausser was presented Harvard University's Outstanding Teaching Award in 1978. At Berkeley, Professor Rausser has been dissertation advisor to more than 61 Ph.D. candidates.

Professor Rausser has developed excellent research and instruction programs in environmental and natural resource analysis, applied econometrics, public policy for food and agricultural sectors, futures markets, antitrust, industrial organization, asset evaluations and portfolio analyses, political economics of policy reform, statistical decision analysis, and trade and development economics. In each of these fields, Rausser and his associates have provided professional leadership and made major advancements; as a result, in 1971, 1976, 1978, 1980, 1982, 1986, 1987, 1992, and 1993, he and his collaborators were selected to receive outstanding published research awards from national and international associations. He has been selected to present the keynote or major plenary session at over 40 major professional society meetings or conferences.

Dr. Rausser has published extensively in a wide variety of academic journals in agricultural economics, economics, and statistics. His major articles have been in econometrics, natural resources, and applied economics. He received the American Agricultural Economics Association award for his published research in 1976, 1980, and again in 1986, and the association's Outstanding Journal Article award in 1982, and his articles have been a finalist for this award on numerous occasions. In 1993 the association awarded him their Publication of Enduring Quality Award for contributions to environmental economics, statistical decision theory, and natural resource analysis; he also received the AAEA Distinguished Policy Contribution Award for econometric analysis of public policies.

Rausser was on the Editorial Board of the *American Journal of Agricultural Economics* from 1977–80 and served as Editor of that journal from 1983–86. He also served as associate editor for both the *Journal of the American Statistical Association* (1973–77) and the *Journal of Economic Dynamics and Control* (1978–82).

Dr. Rausser has worked with several organizations in consulting and non-academic assignments. These include the U.S. Office of Saline Water, U.S. Bureau of Mines, U.S. Federal Trade Commission, U.S. Department of Agriculture, U.S. Office of Management and Budget, U.S. Agency for International Development, the U.S. Department of State, Chicago Mercantile Exchange, Chicago Board of Trade, and the Farm Credit Administration. He was Senior Staff Economist and Special Consultant to the Council of Economic Advisors, 1986–87, and is cofounder of the Institute for Policy Reform and the Research Fellow Program (Washington, D.C.) and cofounder and principal of the Law & Economics Consulting Group (Berkeley, CA; Washington, DC; and Chicago, IL).

In 1991 Dr. Rausser was named a Fellow of the American Statistical Association; in 1990 he was named a Fellow of the American Agricultural Economics Association; and in 1987 he was named a Fulbright Scholar.

# CHAPTER 10

# Financial Picture of Agriculture

$\mathbf{T}$his chapter discusses the financial position of agriculture, sources of farm credit, the banking system, and how to compute simple interest rates. The vitality of agriculture depends on managers who understand finance and can apply it to the farm and ranch business.

A balance sheet gives you some idea of the present financial position of an individual or business. It is the result of all past transactions. A balance sheet is divided into *assets, liabilities,* and *net worth.* Assets are items of money value *owned* by a business. Liabilities are items of money value *owed* by a business and represent creditors' claims against the assets. Net worth is the excess of assets over liabilities and represents the owner's residual claim to assets.

The value of all farm assets increased from about $839 billion in 1990 to $957 billion in 1995. 1n 1995, real estate (land and buildings) accounted for $726 billion, nonreal estate assets (equipment, livestock, and crops) totaled $183 billion, and financial assets amounted to $48 billion. Debt in agriculture was about $151 billion, or 16 percent of assets. Most of this debt is split between real estate and nonreal estate loans (Table 10-1). Farmers' equity or ownership is $806 billion.

The average farm in the United States has real estate worth $350,217, nonreal estate assets worth $88,085, and financial assets amounting to $23,155 (Table 10-2). Total assets increased in nominal terms from $21,530 in 1950 to $461,457 in 1995. Over this same time period, claims on those assets increased from $1,930 to $72,697 per farm. Thus, proprietor's equity increased from $19,600 to $388,760 per farm. As can be seen, farmers are in relatively good financial condition *on the average.* However, these data do not give direct information on the profitability of farming, nor do they reveal the financial difficulties a great many farmers have experienced especially in the mid- to late 1980s. The debt crisis experienced by farmers in mid-1980s caused financial problems for the Farm Credit System (FCS), a federally chartered network of cooperatively owned lending associations and banks. To strengthen the FCS, Congress established the Farm Credit System Insurance Corporation (FCSIC), required increased control over access to the federally sponsored agency funds market, and reorganized the Farm Credit Administration (FCA) as an independent agency that regulates the Farm Credit System.

## SOURCES OF FARM CREDIT _____

Most farm credit has been used to finance farm expansion and higher production cost items such as farm machinery and motor vehicles. 1n 1995,

248

TABLE 10-1 Balance Sheet of Farming Sector, December 31, Selected Years[a]

| Item | 1950 | 1960 | 1970 | 1980 | 1985 | 1990 | 1995 |
|---|---|---|---|---|---|---|---|
| Assets | | | (Billion Dollars) | | | | |
| Physical Assets[b] | | | | | | | |
| Real estate | 75.4 | 123.3 | 202.4 | 782.8 | 586.2 | 618.4 | 726.0 |
| Nonreal estate | | | | | | | |
| Livestock and poultry | 17.1 | 15.6 | 23.7 | 60.6 | 46.3 | 70.9 | 68.3 |
| Machinery and motor vehicles | 12.3 | 19.1 | 30.4 | 80.3 | 82.9 | 85.4 | 87.2 |
| Crops stored on and off farms[c] | 7.1 | 6.2 | 8.5 | 32.7 | 22.9 | 23.0 | 23.1 |
| Purchased inputs[d] | — | — | — | — | 1.2 | 2.8 | 4.0 |
| Financial Assets | | | | | | | |
| Investments in cooperatives | 2.7 | 4.2 | 7.2 | 19.3 | 24.3 | 27.5 | 33.3 |
| Other[e] | 7.0 | 5.8 | 6.5 | 7.4 | 9.0 | 10.9 | 14.7 |
| Total | 121.6 | 174.2 | 278.7 | 983.2 | 772.7 | 838.8 | 956.6 |
| Claims | | | | | | | |
| Liabilities | | | | | | | |
| Real estate debt | 5.2 | 11.3 | 27.5 | 89.7 | 100.1 | 74.7 | 78.7 |
| Nonreal estate debt | 5.7 | 11.7 | 21.2 | 77.1 | 77.5 | 63.2 | 72.0 |
| Total Liabilities | 10.9 | 23.0 | 48.7 | 166.8 | 177.6 | 137.9 | 150.7 |
| Proprietor's Equity | 110.7 | 151.7 | 229.9 | 816.4 | 595.1 | 700.9 | 805.9 |
| Total | 121.6 | 174.2 | 278.7 | 983.2 | 772.7 | 838.8 | 956.6 |
| Debt to Asset Ratio | 9.8% | 15.2% | 21.2% | 20.4% | 29.8% | 16.4% | 15.8% |

SOURCE: U.S. Government Printing Office, Economic Report of the President, 1996 and USDA/ERS Agricultural Outlook, June 1996.
[a] Data for 50 states beginning with 1960.
[b] Does not include household equipment and furnishings.
[c] All crops held on farms including crops under loan to Commodity Credit Corporation (CCC) and crops held off farms as security for CCC loans.
[d] Includes fertilizer, chemicals, fuels, and supplies.
[e] Includes currency and demand deposits.

**TABLE 10-2**  Balance Sheet of the Farming Sector: Average per Farm, December 31, Selected Years[a]

| Item | 1950 | 1960 | 1970 | 1980 | 1985 | 1990 | 1995 |
|---|---|---|---|---|---|---|---|
| | | | | (Dollars) | | | |
| **Assets** | | | | | | | |
| Physical Assets[b] | | | | | | | |
| Real estate | 13,350 | 31,121 | 51,580 | 320,884 | 255,700 | 288,915 | 350,217 |
| Nonreal estate | | | | | | | |
| Livestock and poultry | 3,028 | 3,937 | 8,105 | 24,841 | 20,196 | 33,124 | 32,947 |
| Machinery and motor vehicles | 2,178 | 4,821 | 10,397 | 32,916 | 36,162 | 39,899 | 42,065 |
| Crops stored on and off farms[c] | 1,257 | 1,565 | 2,907 | 13,404 | 9,989 | 10,746 | 11,143 |
| Purchased inputs[d] | — | — | — | — | 523 | 1,308 | 1,930 |
| Financial Assets | | | | | | | |
| Investments in cooperatives | 478 | 1,060 | 2,462 | 7,911 | 10,600 | 12,848 | 16,064 |
| Other[e] | 1,239 | 1,464 | 2,223 | 3,033 | 3,926 | 5,092 | 7,091 |
| Total | 21,530 | 43,968 | 95,315 | 403,032 | 337,051 | 391,886 | 461,457 |
| **Claims** | | | | | | | |
| Liabilities | | | | | | | |
| Real estate debt | 921 | 2,852 | 9,405 | 36,770 | 43,664 | 34,900 | 37,964 |
| Nonreal estate debt | 1,009 | 2,953 | 7,250 | 31,605 | 33,805 | 29,527 | 34,732 |
| Total Liabilities | 1,930 | 5,805 | 16,655 | 68,374 | 77,469 | 64,427 | 72,697 |
| Proprietor's Equity | 19,600 | 38,289 | 78,625 | 334,657 | 259,582 | 327,459 | 388,760 |
| Total | 21,530 | 43,968 | 95,315 | 403,032 | 337,051 | 391,886 | 461,457 |
| Debt to Asset Ratio | 9.8% | 15.2% | 21.2% | 20.4% | 29.8% | 16.4% | 15.8% |

SOURCE: U.S. Government Printing Office, Economic Report of the President, 1996 and USDA/ERS Agricultural Outlook, June 1996.

[a] Data for 50 states beginning with 1960.

[b] Does not include household equipment and furnishings.

[c] All crops held on farms including crops under loan to Commodity Credit Corporation (CCC) and crops held off farms as security for CCC loans.

[d] Includes fertilizer, chemicals, fuels, and supplies.

[e] Includes currency and demand deposits.

outstanding nonreal estate loans secured by farm assets totaled $72 billion. Commercial banks supplied 50 percent, the Farm Credit System 17 percent, the Farm Service Agency 7 percent, the Commodity Credit Corporation supplied 5 percent, and individuals and others 21 percent.

The Farm Credit System supplied 31 percent of all outstanding real estate debt, life insurance companies 12 percent, commercial banks 28 percent, the Farm Service Agency 6 percent, and individuals 23 percent. Historically, individuals have been the major source of funds for land transfers. The Farm Credit System is the largest lender involved in the land mortgage field. The total real estate debt outstanding as of December 31, 1995, was $78.7 billion.

# THE FARM CREDIT SYSTEM OVER THE YEARS _____

Following the experience gained from earlier Federal Land Bank loans to agriculture, the Farm Credit System was established by Congress to make more adequate credit available to the nation's farmers and their related business sector. From a modest beginning with $9 million "seed money" from the U.S. government to help the Federal Land Bank get started, it has evolved into a complex system that by 1984 had $65 billion in outstanding farm loans.

Until its recent restructuring (some institutional elements of which still remain in place), the Farm Credit System included federal land banks, federal intermediate credit banks, production credit associations, and banks for cooperatives supervised by the Farm Credit Administration in Washington, D.C. The Farm Credit Administration continues as an independent regulatory agency of the federal government. The Federal Farm Credit Board was composed of 13 members, 12 of whom were appointed by the president of the United States (based on recommendations from the farm credit system) and one member who represented the secretary of Agriculture. This Board set policy for the Farm Credit Administration and appointed the governor or chief executive of the Farm Credit Administration. The underlying principle of operation was one of providing short-, intermediate-, and long-term credit to producers on a cooperative basis.

# FARM CREDIT DISTRICTS _____

Historically, there have been 12 Farm Credit Districts in the United States, including Alaska, Hawaii, and Puerto Rico. Within each district, there was a district farm credit board and three banks. These banks were the Fed-

eral Land Bank and its Federal Land Bank associations, the Federal Intermediate Credit Bank and its related Production Credit Associations,
and a Bank for Cooperatives. The District Farm Credit Board served as
the board of directors for all three banks in the district. This board was
composed of seven members, two borrowers from each of the three banks
and one individual who was appointed by the governor of the Farm
Credit Administration.

The farm credit districts have been changed recently as a result of the
1987 Agriculture Credit Act.

## Federal Land Banks

The Federal Land Banks were organized under federal charter established by the Federal Farm Loan Act of 1916, with 12 Federal Land Banks
in the United States, one in each Farm Credit District. These banks made
long-term loans secured by first mortgages on real estate through more
than 490 local Federal Land Bank Associations. The federal government
provided initial financial support to these banks ($9 million in 1916 that
was repaid in 1932, plus $189 million obtained in 1933–37). By 1947, the
federal support was repaid and the banks became entirely owned by their
borrowers.

The Federal Land Bank in each district supervised the Federal Land
Bank Associations in their district. The producer actually borrowed funds
from the Federal Land Bank Association, but the loan was transferred directly from the Federal Land Bank to the borrower. The Federal Land Bank
delegated the authority to the Association to make and service all loans,
but supervised the activities of the associations. Voting members of these
associations selected a board of directors and employed a manager.

When farmers borrowed from the local association, they were required
to purchase capital stock or participation certificates in the local association equal to 5 percent of the loan. The local association then bought the
same amount of capital stock or certificates in the district Federal Land
Bank. District bank and local association stock and certificates were re-
funded to the borrower when the loan was repaid.

The Federal Land Bank obtained funds through the sale of farm credit
system bonds in the national money market, with the size of any bond
issue being dependent on the estimated money needs of the farm credit
system. In addition to bond issues, the banks borrowed from other financial institutions with the Farm Credit Administration's approval.

The Federal Land Banks made loans with a range of 5 to 40 years for
maturity. Originally, they could provide loans limited to 50 percent of the
value of the land used as collateral. That was raised to 75 percent in 1933,
lowered to 65 percent in 1947, then raised higher yet to 85 percent in 1967,
where it remained until the financial crisis of the 1980s. Loans could not

exceed 85 percent of the appraised value of borrowers' collateral, except when loans were guaranteed by a governmental agency.[1]

Interest rates were determined by the bank boards subject to the Federal Credit Administration's approval. All banks used a variable interest rate in which interest rates could rise or fall over the length of the loan depending on the cost of money to the farm credit system.

The federal land banks, after providing for reserves and operating expenses, could distribute earnings to the local associations in the form of dividends. These local associations, in turn, could decide to pass their dividends on to their member patrons.

## Production Credit

The production credit system was established under federal laws in 1923 and 1933 to provide short-term and intermediate-term loans to farmers, ranchers, harvesters of aquatic products, and rural residents. The federal intermediate credit banks were first authorized by the Farm Credit Act of 1923. These 12 banks were intended only to discount short-term notes that farmers had given to various financial institutions. Farmers and ranchers did not make use of the credit banks, so short and intermediate credit remained a problem. Therefore, in 1933, Congress authorized local production credit associations.

The relationship between the Federal Land Bank and the federal land bank associations was similar to the relationship between the federal intermediate credit banks and the production credit associations. The 12 Federal Intermediate Credit Banks discounted loans for, and made loans to, the production credit associations. They also supervised some of the production credit associations' operations, which usually were educational in nature, including such functions as helping on problem loans and developing credit standards.

The federal intermediate credit banks obtained funds in the same way as the Federal Land Bank, by selling systemwide bonds in the national money market. These bonds provided the production credit associations with a dependable source of credit at current interest rates.

By 1986, there were more than 420 production credit associations throughout the United States with 1500 full-time offices. Members with voting stock in the production credit associations elected a board of directors from its members and employed a staff to conduct their business affairs.

Ownership of the federal intermediate credit banks was by the local production credit associations through purchases of capital stock and cer-

---

[1] The Federal Land Bank was able to make loans up to 97 percent of the borrower's collateral when loans were guaranteed by a federal agency or a state government.

tificates issued by the bank. Every production credit association borrower purchased stock in the association equal to 5 percent of the loan and could be required to own as much as 10 percent. When these loans were made, the amount to be repaid included the cost of the stock. As the loan was repaid, the amount of stock owned by the borrower was reduced proportionately, so that when the loan was repaid the borrower did not own stock in the production credit association.

Loans to borrowers were made by the individual production credit association, with producers' notes and mortgages then used as collateral to borrow from the district Federal Intermediate Credit Bank. In 1971, the Farm Credit Act was passed that provided a participation agreement between federal intermediate credit banks and production credit associations in making and servicing loans. This provision facilitated the making of large loans, as producers become fewer and larger, by spreading risks.

The interest rate policy was much the same as for the federal land banks. Many production credit associations used the "fixed-interest spread," adding their costs of making and servicing loans and requirements for reserves to the amount they must pay for money from the Federal Intermediate Credit Bank.

## Banks for Cooperatives

The 12 district banks for cooperatives serviced the credit needs of agricultural cooperatives. Also there was a Central Bank for Cooperatives in Denver, Colorado, that participated with district banks on large loans. The Farm Credit Act of 1933 established the organization and initial capitalization for these 13 banks.

The board of directors of the Central Bank for Cooperatives was composed of one director elected from each district farm credit board and one additional member appointed by the governor. The banks for cooperatives were owned by current and former borrowing cooperatives. A cooperative's equity in its district bank was acquired by: (1) purchasing shares of stock when a loan was made; (2) purchasing additional shares of stock in proportion to interest paid on borrowings; and (3) net savings of the banks, which could be distributed to the cooperative in stock as patronage refunds or allocated surplus.

The banks for cooperatives obtained most of their loan funds through the sale of systemwide debenture bonds backed by borrower collateral. These bonds were sold by a Fiscal Agency in New York City. Other funds were obtained by borrowing from systemwide notes, commercial banks, and other financial institutions.

Any association of farmers, ranchers, or producers or harvesters of aquatic products, or any federation of such associations, was eligible to borrow from a bank for cooperatives. To be eligible to borrow, a cooperative had to have at least 80 percent of its voting control with agricultural producers, and the cooperative had to do at least 50 percent of its business with its members.[2] No member could have more than one vote, and dividends on stock or membership capital were restricted to a fixed percentage rate per year.

Loans to eligible borrowers were made to finance long-term assets or working capital. Most loans were made for constructing, remodeling, or expanding facilities, or purchasing land, buildings, or equipment.

Interest rates varied from district to district, depending on the type and length of the loan. Short-term loans usually carried a lower rate of interest than longer term loans. As the cost of money in the money market rose or fell, the rate to cooperatives also rose and fell.

Until recently, the farm credit system in the United States was very successful. All original equity capital supplied by the government to establish the farm credit system was repaid by 1968. This producer owned and controlled organization provided access to the major money market for all producers. Therefore, producers had the advantage of maintaining competition with local commercial banks in obtaining loanable funds at the lowest possible rates. By 1985, the system had $65 billion loaned to more than one million producers. Loans outstanding had increased sixfold since 1969, and the farm credit system had become the major source of agricultural credit for farmers.

# FARMERS HOME ADMINISTRATION _____

The Farmers Home Administration (FmHA) was begun in 1935 as an independent government agency called the Resettlement Administration, established under President Franklin D. Roosevelt's Executive Order 7027. This agency provided supervised loans to help needy farmers become self-supporting. In the first 2 years of its operations, the conviction grew that supervised agricultural credit held the solution to the worsening economic conditions on U.S. farms, and especially for tenant farmers. By 1937, the rate of tenancy had increased to 40 percent of all farms in the country.

As an expansion of federal aid to agriculture, the Farm Security Administration was authorized to succeed the Resettlement Administration in 1937, taking over all assets and functions of its predecessor. Farm rehabilitation loans and farm ownership loans were made to farmers who

---

[2] In 1980, the farmer membership requirement to be eligible to borrow from a bank for cooperatives was lowered to 60 percent for rural electric, telephone, and some local supply cooperatives.

were unable to obtain credit from existing sources. Over its 10-year life, this agency developed numerous rural-area projects having both social and economic objectives. As the critical financial problems in agriculture eased through the mid-1940s, some older assistance programs were no longer needed, others could be improved, and yet other new program needs were anticipated. Thus, the Farmers Home Administration Act was passed in 1946. As the official successor to the Resettlement Administration, the Farmers Home Administration was established within the U.S. Department of Agriculture.[3]

This agency had 46 state offices covering all 50 states plus Puerto Rico and the Virgin Islands. All rural counties were served by more than 1700 field offices normally located in county seat towns. This organization was established to assist beginning producers and other farmers and ranchers with limited resources who are unable to obtain credit from the farm credit system or commercial lenders. These credit programs were expanded to include help in providing new employment and business opportunities, improve the environment, assist in acquiring homes, and improve the quality of life in rural America.

The Farmers Home Administration primarily provided two types of loans. One was a guaranteed loan handled by a private lender. Farmers Home Administration guaranteed to limit the loss on the loan to a specified percentage. The second type was a direct loan by the Farmers Home Administration. Loan funds were obtained from insured notes backed by the government. Also, Congress authorized the Farmers Home Administration to provide grants for rural development planning, pollution abatement, and control, and to facilitate business and industry.

Consolidations of the functions of other rural-area agencies and a series of Congressional acts broadened the scope of the Farmers Home Administration's activities. By the mid-1980s, loans made by this agency were of the following general types:

1. Farm ownership
2. Farm operating
3. Farm emergency
4. Irrigation and drainage
5. Grazing association
6. Resource conservation and development
7. Watershed
8. Indian land acquisition
9. Soil and water conservation

---

[3] Farmers Home Administration, *A Brief History of Farmers Home Administration*, U.S. Department of Agriculture, Washington, D.C., February 1990.

10. Recreation enterprise
11. Youth
12. Community facility
13. Business and industrial
14. Individual homeownership
15. Repair and rehabilitation housing
16. Rental and cooperative housing
17. Farm labor housing
18. Homesite development
19. Self-help technical assistance grants
20. Aquaculture

The maximum amount that could be borrowed, the repayment schedules, and the terms of the loan differed according to the type of loan.

As a separate, federally funded credit agency, and independent of the Farm Credit System, the Farmers Home Administration was not included in the structural changes required under the Agricultural Credit Act of 1987. Changes in their lending regulations, however, have required some new operational procedures.

Being a lender of last resort, thus at higher risk in both short- and long-term farm loans, this agency experienced a greater proportion of loan delinquencies and defaults in its loan portfolio. Between 1988 and 1995, the number of delinquent loans decreased from 50 percent to 39 percent, and delinquent direct farmer loans decreased from $12.5 billion to $4.5 billion over the same period. Loan losses, however, averaged about $1.7 billion per year in the last 5 years.

The FmHA ceased to operate with the signing of the Federal Crop Insurance Reform and Department of Agriculture Reorganization Act of 1994. Farm loans are currently handled by the Consolidated Farm Service Agency (FSA), including both direct and guaranteed loans.

The nonfarm portion of FmHA, Rural Development Administration, Rural Electrification Administration, and Agricultural Cooperative Service were merged into a new USDA organization in 1994 called the Rural Economic and Community Development (RECD) Mission Area. RECD programs are administered through three rural development services: the Rural Utilities Service (RUS), the Rural Business and Cooperative Development Service (RBCDS), and the Rural Housing and Community Development Service (RHCDS). RECD programs and services are provided through State, District, and County offices. Former FmHA services continue to be provided by the RHCDS, including the financing of new and improved housing for over 65,000 moderate to low-income families each year. RHCDS financed rental housing requires that low-income tenants should pay no more than 30 percent of their income for rent. Loans are also provided to improve community facilities. RECD has been recently renamed Rural Development.

# RESTRUCTURING THE FARM CREDIT SYSTEM _____

The nation's agricultural industry experienced serious financial difficulties through much of the 1980s. After many years of generally rising farmland values, with farmers expanding farm size by taking on more real estate debt at rising interest rates, the market for agricultural products dropped significantly. U.S. farm products lost some of their international competitiveness, and increased domestic production caused farm-product prices to decline. As a result, farmers were less able to repay their debts.

The value of farmland is closely tied to the value of its products, so as farm product prices fell land prices also fell. As collateral for real estate mortgages, land was also of declining value to creditors. Consequently, the nation's credit delivery system began having its own financial problems.

As its only charge is to provide credit to farm and farm-related businesses, the financial condition of the Farm Credit System is closely linked to that of agriculture. The boom and bust of the 1970s and 1980s in agriculture was reflected by its credit institutions. Loan losses began mounting in 1983. By 1985 the Farm Credit System reported that it had $13 billion in loans that might be in default over the next few years, so the agency set up loan loss allowances totaling $3.9 billion in 1985 and 1986. For those two years, the agencies operating within the System reported combined net income losses that totaled $4.6 billion. The Federal Land Banks lost almost two-thirds of this total, the Federal Intermediate Credit Banks and Production Credit Associations one–third, and the Bank For Cooperatives about 1 percent.

The severe financial problems of the Farm Credit System prompted Congress to pass the Agricultural Credit Act of 1987. In addition to reorganizing the System, the Act also provided up to $4 billion of assistance to the Farm Credit System through the sale of 15-year bonds guaranteed by the U.S. Treasury. The interest payments on those bonds are to be paid by the Treasury for the first 5 years, one-half each by the Treasury and the Farm Credit System in the second 5 years, and by the Farm Credit System alone during the remaining 5 years. At the end of 15 years, however, the Farm Credit System must repay the principal on the bonds and the interest paid by the U.S. Treasury.

The Act restructured the System by authorizing the establishment of (1) the Farm Credit Assistance Corporation, (2) the Farm Credit Assistance

Board, (3) the Farm Credit Insurance Corporation and its Insurance Fund, and (4) the Federal Agricultural Mortgage Corporation.

The Farm Credit Assistance Corporation is responsible for approving and providing financial help to institutions within the Farm Credit System that are experiencing difficulties. This agency can issue bonds to purchase preferred stock of financially troubled institutions.

In its supervisory capacity, the Assistance Board may request the Farm Credit Administration to approve or require a merger or consolidation, appoint a receiver, or exercise enforcement powers authorized under the Act. The Insurance Corporation is required to ensure timely payment of principal and interest on notes, bonds, debentures, and other financial obligations of the institutions within the Farm Credit System.

The 1987 Act also required the Federal Land Banks and Intermediate Credit Banks within the 12 Credit Districts to merge. There now is a single institution in each District, called the Farm Credit Bank (Figure 10-1), which handles both long- and short-term credit needs of farmers. These mergers resulted in the establishment of seven Federal Credit System districts, including an Agricultural Credit Bank (ACB), five district Farm Credit Banks (FCBs), and their related local lending associations (Figure 10-2). The 1987 Act also provided for Federal Land Bank Associations (FLBAs) to form Federal Land Credit Associations (FLCAs) to make long term loans. The PCAs and FLBAs were allowed to merge into Consolidated Agricultural Credit Associations (ACAs) with direct short, intermediate, and long-term lending authority. The local Federal Land Bank Associations and Production Credit Associations were required to consolidate their operations wherever their offices served overlapping territories. Some Federal Land Banks and Production Credit Associations have combined to form Agricultural Credit Associations. Also, the Agricultural Credit Act authorized the formation of Federal Land Credit Associations. These associations derive their lending authority from the Farm Credit Banks. It has been estimated that such mergers should eliminate about half of the local offices of the Farm Credit System. In 1980, there were 915 local associations, and by 1996 mergers and consolidations reduced the number of associations to 228.

Ten district Banks for Cooperatives have merged with the Central Bank for Cooperatives to form a single institution named CoBank, headquartered in Denver, Colorado. The former separate banks are now operated as branches of the consolidated institution. The Bank for Cooperatives in two districts (St. Paul, Minnesota and Springfield, Massachusetts) voted to remain independent and are authorized to make loans anywhere in the nation.

The Farm Credit Administration is no longer a part of the Farm Credit System. Its new role is to regulate the Farm Credit System and enforce the specific regulations of the 1987 Act.

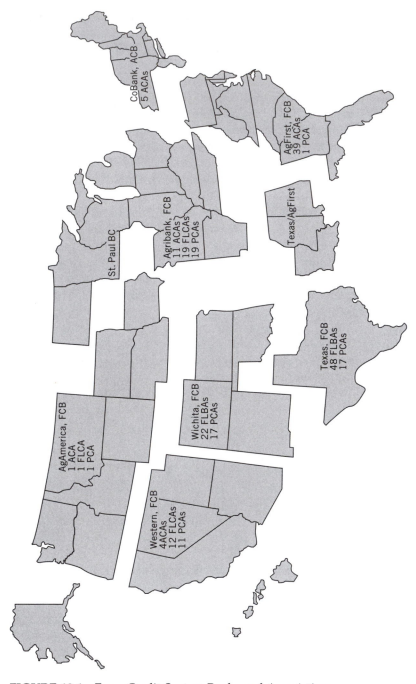

**FIGURE 10-1**   Farm Credit System Banks and Associations

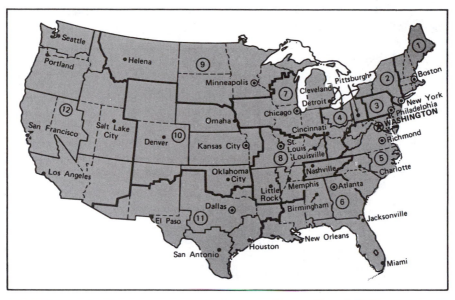

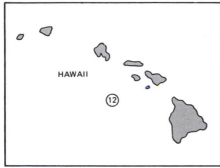

(*Source:* Board of Governors, Federal Reserve System.)

LEGEND

| | |
|---|---|
| ——— Boundaries of Federal Reserve Districts | ⊙ Federal Reserve Bank Cities |
| ——— Boundaries of Federal Reserve Branch Territories | ● Federal Reserve Branch Cities |
| ✪ Board of Governors of the Federal Reserve System | ● Federal Reserve Bank Facility |

**FIGURE 10-2** Districts in the Federal Reserve System.

The Farm Credit Administration's new authority includes the power to issue cease-and-desist orders to prevent or stop unsafe or unsound practices or violations of federal laws or regulations; the power to penalize violators; and the power to remove or suspend officers and directors of FCS institutions for cause as in the case of the Federal Deposit Insurance Corporation, the Federal Reserve System, the Comptroller of the Treasury, and the state banking authorities that regulate commercial banks.

The lending authority of the Farm Credit Banks and affiliated associations is limited to providing credit to farmers, ranchers, and some businesses that supply on-farm services to agricultural producers. Only cooperatives dealing in farm or aquatic products can borrow from the Banks for Cooperatives. The Federal Farm Credit Banks Funding Corporation is responsible for raising funds to support the Farm Credit Banks by selling securities in national financial markets. Three types of securities include: Federal Farm Credit Banks Consolidated System-wide Bonds, medium-term notes, and discount notes that may be purchased by both private and public investors.

## A New Secondary Market

The Federal Agricultural Mortgage Corporation (given the acronym "Farmer Mac") is a relatively new agency under the supervision and regulation of the Farm Credit Administration. Farmer Mac offers buyers of farms and rural homes a broader access to financial markets. It works by guaranteeing securities issued against agricultural and real estate loans that participating lenders originate and pool. Private lenders can sell 90 percent of the value of their loans to loan "poolers" who, in turn, sell securities backed by the loans to investors. Thus, Farmer Mac guarantees the securities, but the loan poolers hold the mortgages.

Farmer Mac has been funded by the sale of stock to the Farm Credit System institutions, banks, life insurance companies, and any other lender wanting to use the secondary market. Its purpose is to free up funds of these loaning institutions and reduce their risks.

Should a loan default occur, investors were covered by a 10 percent reserve established by the lenders in cooperation with the pools. If these reserves are inadequate, Farmer Mac holds additional reserves built up by charging participants a guarantee fee. Farmer Mac can also use its own stock if the need should arise and, if that is insufficient, may borrow from the federal government by issuing securities to the U.S. Treasury.

A number of benefits to rural areas were expected from the development of a secondary market for loans. An active Farmer Mac was expected to help stabilize primary loan markets by enabling lenders to sell their agricultural loans and gain additional liquidity when that is needed. A properly functioning secondary market also should help to better allocate the flow of loanable funds between regions. That is expected to make more funds available in localities where deposits are not sufficient to meet lending needs. Small-town rural areas would benefit from such improved credit supplies. And the increased loanable funds should increase competition among lenders, reducing interest rates for borrowers in those areas.[4]

The volume of mortgages sold through Farmer Mac for high quality agricultural real estate and rural homes has been very limited to date, making Farmer Mac unprofitable. New legislation was introduced for Farmer Mac under the Farm Credit System Reform Act of 1996 in an attempt to lower costs, to cope with higher pending capital standards, and to provide guidelines for recapitalization and for an orderly liquidation of the corporation if capital becomes inadequate. The new legislation permits Farmer Mac to become a more competitive portfolio lender and to purchase loans directly from lenders which may be held in portfolio or resold as mortgage-backed securities. Farmer Mac is required to increase it's core capital up to $25 million over the next two years or risk being terminated by the FCA.

# COMMODITY CREDIT CORPORATION ———

The Commodity Credit Corporation (CCC) was formed in 1933 to support and stabilize the prices of a number of agricultural commodities. Placed under the supervision of the Agricultural Stabilization and Conservation Service of the U.S. Department of Agriculture, the primary function of the CCC involves commodity loan programs established by federal farm policy. Over the years, most of these farm programs have included nonrecourse loans to agricultural producers.

CCC nonrecourse loans to farmers use eligible agricultural commodities as collateral for their loans. When a loan expires, or at the option of the borrower, the loan terms may be fulfilled either by repaying the loan or by transferring title in the commodity to the CCC. If the market price is above the loan rate, the farmer can sell the commodity and repay the loan. The word *nonrecourse* means that if the market price is below the loan rate, the borrower may transfer title to the CCC in full payment of the loan, and the CCC absorbs the loss.

In recent years, the CCC has become a major source of short-term funds for agricultural producers. Direct loans to farmers have amounted to about $7.8 billion per year.

---

[4] Sources for recent changes in the agricultural credit system. Farmers Home Administration, "A Brief History of Farmers Home Administration," U.S. Department of Agriculture, February 1990; Merritt R. Hughes, "Recent Developments at the Farm Credit System," Agricultural Information Bulletin No. 572, U.S. Department of Agriculture, September 1989; and Gene D. Sullivan, "Changes in the Agricultural Credit Delivery System," *Economic Review*, Federal Reserve Bank of Atlanta, January/February 1990.

# THE BANKING SYSTEM _____

Agriculture and agricultural financial institutions do not operate in isolation from conditions in other sectors of the economy. The agricultural sector must compete for available funds with public and private borrowers from all segments of the economy. While both monetary and fiscal policies influence the availability of loanable funds to agriculture, monetary policy is of greater concern here.

Fiscal policy is the government policy regarding expenditures and taxation. Thus, the government can increase expenditures of such agencies as the Farm Service Agency and increase loanable funds, or restrict credit by cutting budgets. Also, of course, tax levels influence the amount of money available to producers to invest in their operations. Fiscal policy changes affect the flow of funds to financial institutions and therefore affect loanable funds. These changes also influence business activity and savings. A tax decrease tends to stimulate incomes, employment, and savings; a tax increase will have an opposite impact on the economic system.

The ability of banks to create or destroy money as they perform their usual business has a great deal to do with the performance of the economy. Given their potential impact throughout the system, the activities of the banking industry must be coordinated so as to inhibit wide swings in prices and employment levels.

Monetary policy is an effective tool for attempting to meet the national goals established by Congress and the president of maintaining price stability and full employment. To do so requires that the flow of bank credit and the supply of money maintain a certain stability because of their effects on incomes, employment, and savings. By regulating the supply of money and therefore the terms on which people borrow money, monetary authorities can stimulate or retard business activity. The federal reserve system, which was established by the Federal Reserve Act of 1913, is the principal organization that regulates monetary policy in the United States.

There are 12 Federal Reserve Bank districts serving all 50 states, with one bank in each district plus 25 branch banks (Figure 10-2). All these banks are part of the Federal Reserve System. Each bank has nine directors, six elected from member banks and three appointed by the Board of Governors of the bank system.

The Federal Reserve System has a membership of less than 5000 commercial banks, which control about 80 percent of the commercial bank assets, but only comprise about 36 percent of the approximately 13,000 banks in the United States. Although these member banks operate for profit, the system does not. In its operation the reserve system does earn money, but all income over expenses and reserves is transferred to the U.S. Treasury.

The supervision of the Federal Reserve System is handled by a board of governors. This board consists of seven members appointed by the president of the United States and confirmed by the Senate for a 14-year term. These members are full-time employees whose primary function is to formulate national monetary policy and supervise its execution.

The Federal Reserve system regulates the supply of credit through controlling the supply of money. There are four basic ways this control occurs. These are regulating bank reserve requirements, conducting open market operations, adjusting the discount rate, and using various selective controls.

1. *Bank Reserves.* The Federal Reserve has control over the volume of money and the interest rate by regulating the amount and use of bank reserves. On March 31, 1980, the Depository Institutions Deregulation and Monetary Control Act of 1980 was signed into law. This law granted additional powers to the Federal Reserve, including the power to regulate the reserves of all depository institutions including non-member banks, mutual savings banks, saving and loan associations and credit unions.[5]

Reserve requirements on transactions accounts (demand deposits, NOW accounts, etc.) are established at 3 percent for amounts under $36.9 million and 12 percent for amounts over $36.9 million. For non-personal time deposits (time deposits of businesses and corporations), the reserve requirements can be varied from 0 to 9 percent, and in addition, the rates can vary depending on the length of maturity of that deposit.

As long as banks have excess reserves, they can continue to loan or invest funds. But if the Federal Reserve increases the reserve requirement, it means that banks have less excess reserves, which restricts the amount of credit the bank can extend. For example, assume a local bank in Burns, Oregon, has deposits of $50 million. If the reserve requirement is 10 percent, that bank can make loans or investments up to a total of $45 million.[6] If the reserve requirement is increased to 14

---

[5] Jeffrey Rogers Hummel, "The Deregulation and Monetary Control Act of 1980," *Policy Report*, A Publication of the Cato Institute, Vol. II, No. 12, December 1980.

[6] An individual bank can lend or invest its excess of reserves. The entire commercial banking system, however, can create money by a multiple of its excess reserves. If the reserve requirement is 20 percent, then 80 percent of any new deposit can be loaned. So if a commercial bank receives a new deposit of $10, the banking system can create $40 in additional deposits (for a total of $50). Money creation is accomplished with a reserve requirement of 20 percent as follows: The first bank receives $10 and loans $8, which is deposited in the second bank; the second bank can loan $6.40, which is deposited in the third bank; the third bank can loan $5.12, which is deposited in the fourth, etc. This process can be repeated until the banking system has recorded new deposits of $50.00. A formula is used to calculate the total amount of new deposits in the banking system. The formula is:

percent, the maximum amount it can lend or invest is reduced to $43 million. Also, if the reserve requirement is lowered, the amount of credit available to borrowers can be increased. Changes in the reserve requirement are an infrequently used tool of the Federal Reserve System. Day to day operations of the money supply are handled through open market operations.

2. ***Open Market Operations.*** Open market transactions are controlled by the Federal Open Market Committee, which buys and sells government securities, that is, bonds, in the open market. This committee has 12 members, seven from the board of governors and five of the presidents of the Federal Reserve banks. By purchasing or selling securities, the Federal Reserve system can influence the volume of bank reserves. For example, assume the Federal Reserve system buys bonds worth $100,000 from commercial banks. The system receives the bonds, and the banks get dollars in the form of deposits. The deposits increase reserves, so more loans can be made. Remember, bankers are in business to make money, and they don't want to lose interest on money by leaving it idle. Thus, by increasing the supply of loanable funds, there is a tendency for the interest rate to fall, and this lower rate may lead to additional borrowing.

   The reverse occurs if the open market committee decides to decrease reserves. In this case, the government sells bonds in the open market. People or businesses buying these bonds reduce their bank demand deposits by paying for the bonds, and therefore bank reserves are reduced. This policy tightens credit, tends to increase interest rates, and causes reduced borrowing.

3. ***The Discount Rate.*** When a bank runs short of money, it can borrow from its district Federal Reserve bank just as we borrow for an automobile when we lack cash to pay for it. The rate of interest charged by the Federal Reserve bank is called the discount rate. It is termed the discount rate because when a bank borrows money, it must have collateral that is usually loan notes from customers. These notes then are discounted at the "Fed window" at a set discount rate.

   By raising the discount rate, the Fed discourages borrowing by banks and their customers because the cost of borrowing money is increased. On the other hand, by lowering the discount rate, the Fed encourages bank borrowings. However, the discount rate is relatively unimportant in controlling the volume of money, because the Fed does not loan funds except for short periods of time.

---

$$\text{Total new deposit} = \frac{\text{New deposit}}{\text{Reserve requirements}}$$

In this example, the total new deposits = $10/0.20 = $50.

4. **Selective Controls.** The Fed also uses many selective controls and moral suasion to induce banks to follow bank policy. For example, it regulates the amount of interest banks can pay on various types of savings programs. It controls margin requirements on stocks and bonds. Also, from time to time, the Fed has used its authority to regulate consumer installment credit. Such controls may deal with down payment requirements, the maximum length of loan, and the rate of interest to be paid.

Federal Reserve Bank activities affect the federal farm credit system because bonds sold by the system must be purchased by commercial banks and other investors. If the Fed wants to reduce the volume of funds available, it means that commercial banks have less reserves to purchase farm bonds, and producers must pay a higher rate of interest on borrowings.

## Money Market

There are demand and supply schedules for all commodities, and money is no exception. As has been explained earlier, the supply of money is largely determined by the Federal Reserve System, while the demand for money is determined by individuals, business firms, and governments. These groups demand money for three basic reasons. First, they desire money for *transactions purposes*; second, for *unexpected happenings*; and third, to *guard against economic losses*.

Individuals need currency or checking accounts to pay their monthly bills. We carry currency around daily to pay for lunch, newspapers, and small-item spending. Our checking accounts are normally used to pay for larger or regularly occurring expenses such as the monthly house payment, weekly groceries, or clothing. The transactions demand for money depends on our spending habits, which are related to our income.

The precautionary demand for money is motivated by our desire to hold money in case of an unforeseen event. Unexpected events may or may not occur, but people will hold some amount of cash to cover such an eventuality.

The third demand for money is the speculative demand. It relates the demand for money to the interest rate. When the interest rate is high, the quantity of money demanded is low because the opportunity cost of money (what that money could earn in alternative investments) is relatively high. When the interest rate is low, the quantity of money demanded is relatively high, because the opportunity cost of money is low (the loss of interest income is minimal) and, in addition, the interest rate may increase in the future. The demand for money is shown as the relationship between the quantity of money demanded and the interest rate; likewise, the supply of money is the relationship between the quantity of

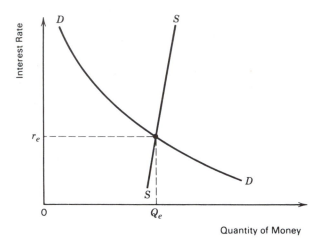

**FIGURE 10-3**   Equilibrium in the money market.

money supplied and the interest rate. The money supply and demand curves are diagrammed in Figure 10-3.

The demand for money (the *DD* curve) and the supply of money (the *SS* curve) determine the price of money—the equilibrium interest rate $r_e$ and the equilibrium quantity of money demanded $Q_e$. An increase in the money supply diagrams as a rightward shift of the money supply curve reduces the interest rate; a decrease in the money supply will increase the interest rate, determined by the point of intersection of the money supply and demand curves. An interest rate below the equilibrium rate results in an excess demand for money and a "bidding up" of the interest rate to $r_e$. An interest rate above the equilibrium rate results in an excess supply of money and a fall in the interest rate to its equilibrium, at $r_e$.

With this brief background description of the economics of demand and supply in the market for money, we can follow through with the logic behind decisions regarding the nation's monetary policy. If the Federal Reserve acts to increase the money supply, the interest rate will fall, which will cause business investment to increase because the cost of investment borrowing has been reduced, thus increasing aggregate demand ($C + I + G + NE$) and the nation's GDP. A decrease in the money supply will cause the interest rate to increase, resulting in decreased investment spending and reduced aggregate demand and a lower GDP.

The Federal Reserve can achieve its objectives by either controlling the money supply or changing the interest rate. From time to time, the Fed has emphasized controlling the interest rate; at other times, it has chosen to change the money supply. The particular policy choice is dictated by the Fed's perception of general economic conditions and the appropriate policy to achieve its specific macroeconomic goals; this has led to conflicts with the U.S. Congress.

Congress is occasionally at odds with the Fed because the Fed often makes "trade-offs" between the money supply and the interest rate. Congress has been concerned in recent years that the rate of growth in the money supply has been too severely restricted, so that the interest rate has been too high, with consequent reductions in investment, aggregate demand, output, and employment. At other times, Congress gets somewhat upset with the Fed because the Fed has held fast on a particular interest rate and has varied the money supply, which has caused the price level to fluctuate. Congress, therefore, requires the Fed to report its money supply targets to Congress, and in subsequent reports, to tell Congress how well the Federal Reserve has met its goals. These reports give Congress a better understanding of the Fed's national monetary policy so that Congress may decide whether the Fed's monetary policy is consistent with Congress' own fiscal policy objectives.

# COMMERCIAL BANKS _____

There are about 13,000 commercial banks in the United States providing the major credit source for farmers. These banks are corporations chartered under federal or state law. They are stock organizations that are owned by the people who have invested in them. The stockholders elect a board of directors that determine bank policy and employ a staff to operate the bank. Thus, a commercial bank is a private business, but it is supervised by many federal and state (public) agencies. This regulation is required to ensure the safety of depositors' funds and that banks follow the national regulation of the Federal Reserve System.

Most commercial bank loans to agricultural producers are of a short-term nature for production expenses. These expenses are for fertilizer, cattle, feed, seed, fuel, labor, and so forth. Loans of longer duration are made for machinery and equipment, but little mortgage money of a long-term nature is available from commercial banks. As of 1995, commercial banks provided about 52 percent ($38 billion) in nonreal estate loans to agriculture and about 28 percent ($22 billion) in real estate loans. In 1995 they increased their market share to 40 percent of total farm debt.

Interest rates commercial banks charge depend on the demand for loans, policy of the Fed, rates charged by competitors, and their excess reserves. Also important is the risk involved and the management capacity of the borrower. Usually banks are competitive with production credit associations and other lending institutions. But it is necessary for producers to shop for credit like shopping for any other commodity. The interest rate is the price of money, and large savings are often possible if one can borrow at lower rates.

Many banks have agribusiness or agricultural departments that specialize in agricultural credit needs. They employ skilled agricultural representatives who understand agriculture. Most of these specialists can analyze the farm business and provide information and guidance in assisting producers so they use capital wisely and are able to meet their repayment schedules.

# LIFE INSURANCE COMPANIES _____

Life insurance companies finance about 7 percent of the total farm debt and about 13 percent of the real estate loans, or about $10 billion. In 1950, life insurance companies provided more than 25 percent of the mortgage money to agriculture. In recent years, however, commercial banks have become more competitive.

Life insurance companies are in business to sell life insurance. As people pay on their policies, these companies accumulate large sums of money. Such funds are used in all types of investments to produce income for their policyholders. Most of their investments are made in the nonfarm economy, but insurance companies do diversify their portfolios and invest in long-term credit to agriculture.

There are two types of insurance companies: stock companies and mutual companies. Mutual life insurance companies are cooperatives. Those who purchase insurance from these companies own the assets of the company. The company provides insurance at cost. If the premiums paid by policyholders cover all expenses with reserves left at the end of the year, these reserves are credited to policyholder accounts as dividends. Mutual companies issue more than one-half of all life insurance sold. Stock life insurance companies are organized as private stock ventures. They are controlled by stockholders who operate the business for the profit of investors. Therefore, the stockholders share in the gains or losses from the enterprise.

Most life insurance companies that are fairly large have field representatives located in branch offices around the country. They hire trained agriculturalists to procure and service loans. Farm mortgage loans fit the nature of the insurance business. Most policyholders have held their insurance policies for more than 40 years. Also, most farm mortgages are repaid over 20–25 years, although some of their loans may be for shorter time periods. In terms of dollar amounts, insurance companies have been making much larger loans than commercial banks or the Farm Credit system.

The shares of different creditors in the farm loan market included 40 percent for commercial banks, 25 percent for the Farm Credit System, 22 percent for individuals, 7 percent for the Farm Service Agency and 6 percent for life insurance companies.

## Interest Rates

An *interest rate* is the price of borrowed money. A producer who wants to borrow money must be able to compare interest rates and charges made by lenders, because the method used to compute interest charges affects the final cost of a loan. The 1969 Truth in Lending Act is a recognition of the difficulty borrowers long have had in understanding loan charges. The act requires that lenders inform their borrowers of the total finance charges and the annual percentage rate of interest (APR).[7]

Three different methods are used by lenders to determine the interest charges on loans. These methods are (1) the *remaining balance method*, (2) the *add-on method*, and (3) the *discount method*. The choice of which method is used will affect the annual percentage rate in most cases.

When the remaining balance method is used, the interest payment is calculated by multiplying the contractual interest rate by the remaining loan balance outstanding at the beginning of the current period.

Equal periodic interest payments result when the add-on method of computing interest charges is used. The periodic interest payment is determined by multiplying the contractual interest rate by the original amount of the loan, then dividing this product by the number of loan payments to be made. Thus, although the loan balance outstanding declines over time, the interest charges remain constant.

The third method discounts the amount of the loan before the borrower receives the loan. Under the discount method, the interest charge equals the contractual rate of interest times the loan amount. The result of that computation is then subtracted from the loan amount. Thus, the loan is *discounted* by the amount of interest.

The example in Table 10-3 shows total interest payments and APR for the three different methods, given the same loan and contractual rate of interest.

The remaining balance method of determining interest payments has become a common practice among agricultural lenders. The add-on method, with its constant interest payments, makes it easier for the bor-

---

[7] The following simplified formula will yield a reasonably close estimate of APR: $APR = 2I/t(B + a)$, where $I$ = interest amount; $t$ = loan term in years; $B$ = beginning principal loan amount; and $a$ = the amount of each periodic payment. *Source:* Warren F. Lee, Michael D. Boehlje, Aaron G. Nelson, and William G. Murray, *Agricultural Finance,* 7th ed. Ames, Iowa: The Iowa State University Press, 1980, p. 128.

**TABLE 10-3**  Comparison of Principal and Interest Payments, and Annual Percentage Rate of Interest for a $100 loan with a Contractual Interest Rate of 10 Percent

| Year | Remaining Balance Method | | Add-on Method | | Discount Method | |
|------|-----------|----------|-----------|----------|-----------|----------|
| | Principal | Interest | Principal | Interest | Principal | Interest |
| 0 | — | — | — | — | — | $50.00[a] |
| 1 | $ 16.38 | $10.00 | $ 20.00 | $10.00 | $ 20.00 | |
| 2 | 18.02 | 8.36 | 20.00 | 10.00 | 20.00 | |
| 3 | 19.82 | 6.56 | 20.00 | 10.00 | 20.00 | — |
| 4 | 21.80 | 4.58 | 20.00 | 10.00 | 20.00 | — |
| 5 | 23.98 | 2.40 | 20.00 | 10.00 | 20.00 | — |
| Total | $100.00 | $31.90 | $100.00 | $50.00 | $100.00 | $50.00 |
| APR | 10.0% | | 15.239% | | 28.65% | |

[a] Deducted from loan proceeds at the time the loan is made.

rower to keep track of interest payments as expenses for income tax purposes. Personal loans are most commonly made using the discount method of calculating interest payments.

One needs to shop for money to borrow just as you would for a tractor, because there often is a large difference in finance charges among lending institutions. The interest paid on a loan will be influenced by the cost of funds to the lending institution, length of loan, risk of default on the loan, cost of servicing the loan, and applicable state and federal regulations.

# SUMMARY

The general financial picture of agriculture is shown by the balance sheet. This financial statement shows all assets, liabilities, and net worth. Agriculture's assets totaled $957 billion by 1995. Claims against these assets amounted to only $151 billion. Thus, farmers have substantial amounts of assets that they own. The debt-to-asset ratio per farm was 15.8 percent by 1995.

Agricultural producers borrow production funds from commercial banks, life insurance companies, individuals, the Farm Service Agency, and the farm credit system.

The farm credit system is composed of six Farm Credit Banks, one District Federal Intermediate Credit Bank, and two Banks for Cooperatives. Also included in this system are 228 local associations that handle farm loans. This farm credit system is farmer owned and controlled. They make long-term, intermediate, and short-term loans.

The Farmers Home Administration of the U.S. Department of Agriculture was established as a lender of last resort. It was an agency to help finance beginning producers with limited resources. Subsequently this agency was involved in financing rural development projects and rural housing. Its functions were taken over by the Farm Service Agency and three rural development services.

The Federal Reserve System was established by the Federal Reserve Act of 1913 and is the principle organization that regulates monetary policy in the United States. This system has less than 5000 member banks that control about 80 percent of the commercial bank assets. It regulates the supply of money available by controlling (1) bank reserves, (2) open market operations, (3) discount rates, and (4) selective controls.

A market exists for money, with its demand and supply curves, just as for any other good or service. Changes in money supply and the interest rate affect the amounts of investment and other borrowing, causing changes in aggregate demand and the nation's GDP.

The interest rate is the price of borrowing money. The 1969 Truth in Lending Act requires that lenders inform borrowers of all finance charges and the annual percentage interest rate they would pay. Interest rates do vary, thus, it is important to be able to compute simple interest rates so they can be compared among alternative lenders. Quoted interest rates may differ because lending institutions obtain loanable funds at different costs, the risks of loss vary among borrowers, and the costs of serving loans differ.

## CHAPTER HIGHLIGHTS

1. Assets, liabilities, and net worth are shown on a balance sheet.
2. Farm assets amounted to $957 billion by 1995. Land is the major asset owned by the farm business.
3. Agricultural debt is $151 billion, about 16 percent of assets.
4. Farmers and ranchers own most of their farms. Net worth was $806 billion in 1995.
5. Farm credit is available from the farm credit system, life insurance companies, commercial banks, Farm Service Agency, and individuals.
6. The federal farm credit system includes six Farm Credit Banks, and two Banks for Cooperatives and 228 local associations.
7. The farm credit system is farmer owned, although it started with government assistance.

8. The Farm Service Agency assists farmers with credit if they meet specific regulations or are unable to attain funds from the farm credit system or other commercial lenders.
9. National monetary policy is established by the Federal Reserve System.
10. The Federal Reserve System influences the national money supply by (1) regulating bank reserve requirements, (2) conducting open market operations, (3) adjusting the discount rate, and (4) using selective credit controls.
11. The Fed can cause changes in aggregate demand and GDP by its ability to regulate the money supply and the interest rate.
12. Commercial banks provide most of the nonreal estate loans to agriculture.
13. Commercial banks are the major source of funds for purchasing real estate.
14. An interest rate is the price of borrowing money.
15. The simple or actuarial interest rate is based on the average loan balance over the period of the loan.

## KEY TERMS AND CONCEPTS TO REMEMBER

| | |
|---|---|
| Assets | Liabilities |
| Bank reserves | Net worth |
| Discount | Open market operations |
| Discount rate | Selective controls |
| Interest rate | |

## REVIEW QUESTIONS

1. What is the general financial condition of agriculture? Does the balance sheet give a complete financial picture of agriculture?
2. What are the sources of farm credit? Which of these sources provide short-term credit?
3. The farm credit system was established as a cooperative credit system initially funded by the U.S. government. At present, who owns the farm credit system?
4. Where does the farm credit system borrow the money it loans to farmers and ranchers?

5. Are farm loans granted by the Farm Service Agency? What type of loans does it handle?
6. Explain the difference between government monetary and fiscal policy.
7. How does the Federal Reserve System regulate the supply of credit available by controlling the money supply?
8. Explain why a change in the interest rate should have any effect on the nation's GDP.
9. Will a change in the interest rate have a different effect on your demand for a new house as compared to your demand for a new bowling ball? Explain your answer.
10. Why should you "shop" for farm credit? How can you compare the costs of borrowing money from different financial institutions?

## SUGGESTED READINGS

Barry, Peter J., Paul N. Ellinger, John A. Hopkin, and C. B. Baker. *Financial Management in Agriculture*, 5th ed., Danville, Ill.: The Interstate Printers and Publishers, 1995, Chapters 4 and 5.

Hoag, Gifford W. *The Farm Credit System,* Danville, Ill.: The Interstate Printers and Publishers, 1976.

Lee, Warren F, Michael D. Boehlje, Aaron G. Nelson, and William G. Murray. *Agricultural Finance*, 8th ed., Ames, Iowa: The Iowa State University Press, 1988, Chapters 19–24.

Penson, John B. Jr., and David Lins. *Agricultural Finance*, Englewood Cliffs, N.J.: Prentice Hall, 1980, Chapters 8, 19–22.

Stokes Jr., W. N. *Credit to Farmers,* Washington, D.C.: The Federal Intermediate Credit Banks, Farm Credit Administration, 1973.

Willis, James F., and Martin L. Primack. *Explorations in Economics*, 3rd ed., Redding, CA: CAT Publishing Co., 1990, Chapter 12.

## AN OUTSTANDING CONTRIBUTOR

**Peter J. Barry**   Peter Barry was born in Toronto, Ontario, Canada. He grew up in Illinois and New Jersey and spent part of his life on a grain-hog farm in central Illinois.

Dr. Barry received his B.S., M.S., and Ph.D. degrees from the University of Illinois, although he attended Purdue University for 2 years. While he was a student and for a time after earning his B.S. degree, he worked as an assistant county agent primarily in 4-H programs.

His first position after completing graduate school was as an assistant professor at the University of Guelph from 1963 to 1971. He served at Texas A&M University as assistant professor from 1971 to 1973, associate professor from 1973 to 1979, and was promoted to full professor in 1979. In that year he moved to the University of Illinois as a professor of agricultural finance.

Barry has taught a variety of agricultural economics classes including an introduction to agricultural economics, farm management, production economics, agricultural finance, and risk theory and capital markets.

Barry's research program has involved analyses of the financial conditions of farming. He has been concerned with financing the farm business, farm growth, risk management, estate planning, tax management, farm credit, the farm credit system, and financial stress in agriculture.

He has received publication awards from the Western Agricultural Economics Association in 1978, the International Confederation of Agricultural Credit in 1987, and Texas A&M University in 1975. Five of his students have received outstanding Master's thesis and Ph.D. dissertation awards. These awards were from the WAEA in 1976, the AAEA in 1976 and 1983, and two awards from the University of Illinois in 1983.

He has edited and co-authored four textbooks. His co-authored book, *Financial Management in Agriculture,* is widely used and is now in its fourth edition.

Dr. Barry is well known for his excellence in journal editing. He was editor of the *Western Journal of Agricultural Economics* from 1976 to 1979, associate editor of the *American Journal of Agricultural Economics* from 1984 to 1986, and editor of the *AJAE* from 1987 to 1991.

Barry has been a consultant to the American Bankers Association; state banking associations of Illinois, Texas, and Iowa; the Farm Credit Bank of St. Louis and the Farm Credit Bank of Texas; the U.S. Department of Agriculture; the National Planning Association, Office of Technology Assessment; and the CANFARM Data System.

He is married to the former Jane Dague, and they have two children.

CORN

| MAR | MAY | JUL | DEC | NOV | PW JAN | MAR | DEC | WHEAT MAR | MAY | JUL | SEP | NOV |
|---|---|---|---|---|---|---|---|---|---|---|---|---|
| | | | | | | | | | | | | 3850 |
| | | | | | 875 | 890 | 4920 | 4830 | 4620 | 4090 | | 3865 |
| 2580 | 2604 | 2614 | 2204 | 885 | | | 4890 | 4810 | | 4084 | | |
| | 2614 | | | 888 | | | | | | | | 3935 |
| | | | | | | | | | | | | 3750 |
| 2584 | 2614 | 2614 | 2204 | 888 | 875 | 890 | 4920 | 4830 | 4620 | 4090 | | |
| 2482 | 2520 | 2540 | 2180 | 866 | 855 | 873 | 4714 | 4654 | 4430 | 3890 | | 835 |
| | | | | | | | | | | | | 30 |
| 514 | 544 | 560 | 194 | 874 | 866 | 880 | 714' | 654' | 440 | 900 | | 25 |
| | 40 | 54 | 90 | 80 | 65 | 75 | 14 | 54 | 30 | 90 | | |
| 12 | 44 | 50 | 84 | 74 | 61 | 80 | 14' | 54' | 40 | 10 | | |

| MAR | MAY | JUL | DEC | NOV | JAN | MAR | DEC | MAR | MAY | JUL | SEP | NOV |
|---|---|---|---|---|---|---|---|---|---|---|---|---|
| 56 | 259 | 260 | 221 | 880 | 870 | 876 | 4910 | 4850 | 4620 | 4080 | | 3810 |
| 6 | 9 | 0 | | 84 | 74 | | 2 | 60 | 40 | 10 | | 20 |
| 66 | 69 | 70 | 31 | 952 | 942 | 946 | 11 | 05 | 830 | 290 | | 015 |
| 46 | 49 | 50 | 11 | 812 | 802 | 806 | 71 | 65 | 43 | 89 | | 615 |
| 42 | 145 | 146 | 138 | 1863 | 1612 | 1418 | 2107 | 2127 | 2101 | 1833 | | 2557 |

# CHAPTER 11

# Agricultural Price and Income Policies

$\mathbf{F}$ederal government price and income programs in agriculture have been in existence in one form or another since 1929. The programs have been changed, revised, and extended from time to time and continue to be important to most agricultural producers of grains, cotton, rice, soybeans, tobacco, peanuts, sheep and lambs, milk, and sugar. Before we get into a discussion of agricultural policies, let us first look at the basis for the formulation of policy—the goals and values of people.

Hathaway has defined *public policy* as a specific type of group action designed to achieve certain aspirations held by members of society.[1] For any public policy to come into existence, national policymakers must have knowledge of the objectives of the group being represented. The policy must be consistent with many of the goals of society, the most basic of which is the *quality of life*.

Quality of life includes such general goals as peace, security, freedom, and justice. We all know that peace is a worldwide aspiration, but we also know that there have been very few periods in which confrontations have not been taking place. Peace has a high priority in this country; so much so that vast sums (18 percent of total federal spending) are spent on national defense in an attempt to deter foreign aggression.

When we speak of security, we include economic, political, and social stability. With the price level increasing at a 3 to 4 percent annual rate during the 1990's, there still is concern over the loss of economic security due to the decline in the purchasing power of the dollar. As far as political stability in the United States is concerned, we are fortunate that our governing system has stability built into it with the sharing of power between the executive, legislative, and judicial branches of government. Our great desire for personal security is evidenced in the fact that we have employed such policies as retirement programs, health insurance, unemployment insurance, social security, and medicare.

Freedom is basic to the United States' origin, but we must enforce some restrictions in order that one person's freedom does not deny another person's rights. When we speak of justice, we must include not only the legal protection of life and property, but economic and social justice as well. Women's rights and civil rights movements have been based to a large extent on securing economic (e.g., the attaining of equal pay for equal work) and social justice (for example, the removal of discrimination because of race or sex).

Each of these goals of peace, security, freedom, and justice is based on society's values. *Values* are the principles that guide human action and

---

[1] Dale E. Hathaway, *Government and Agriculture*, New York: Macmillan, 1963, p. 3.

are basic to our concepts of what is good or bad. Values are therefore a reflection to our cultural heritage. There can, however, be a conflict of values within an individual and more often within a group. Policymakers must consider trade-offs between the attainment of one goal or another. As Heady has said, economists can suggest several different ways to solve U.S. farm problems, but any solution is dependent on the resolution of conflicts in goals and values.[2]

Because of the many basic aspirations involved, no one policy will assure that all goals will be realized. One of the major tasks of policy formation is recognizing and weighing competing goals and values of individuals and groups to arrive at a policy that will provide the maximum quality of life to those people involved or affected.

## VALUES OF FARM PEOPLE _____

Until recently, agricultural fundamentalism had an important impact on agricultural policy. These agrarian values are based on "laissez-faire" economics, Jeffersonian Democracy, and the French physiocratic philosophy. These basic ideas were summarized by Paarlberg in what he called the Agricultural Creed. The creed's articles are:[3]

1. Farmers are good citizens, and a high percentage of our population should be on farms.
2. Farming is not only a business, but a way of life.
3. Farming should be a family enterprise.
4. The land should be owned by the person who tills it.
5. It is good to make two blades of grass grow where one grew before.
6. Anyone who wants to farm should be free to do so.
7. A farmer should be his/her own boss.

It is indeed true that a high proportion of farmers are good citizens, but it is also true that a high proportion of urban people are good citizens. One would be hard pressed to say that one group was better than the other and thus support the statement that a high percentage of our population should be on farms. In 1890, 65 percent of the population lived in rural areas as compared to 20 percent now. In 1920, about 30 percent of the population lived on farms as compared to the present 2 percent.

---

[2] Earl O. Heady, *Goals and Values in Agricultural Policy*, Center for Agricultural and Economic Development. Ames: Iowa State University Press, 1961, pp. v-vi.

[3] Don Paarlberg, *American Farm Policy*, New York: John Wiley & Sons, 1964, p. 3.

Farming has been a way of life, but it is being transformed into a business. Computerization of farm records, growth of the farm firm, and modern scientific advances have forced producers to take a more business-like attitude toward farming. Farming in the United States is primarily a family enterprise and will continue to be in the foreseeable future. Farms have grown in size and complexity, but technology has made it feasible for a producer and family to remain a viable economic unit. The investment that producers have in the farm firm is much larger than for most businesses found in rural America. Modern agribusiness methods are a must for today's producers.

Most agricultural producers own their own farms, but it has been necessary for farm size to increase in order to maintain a competitive position. This expansion in farm size has been done to a large extent through the purchasing of land on credit. Agriculture, like any other business, has found the wise use of credit to be a necessity. From 1900 to 1992 those producers who owned their farms free and clear of any mortgages increased from 56 percent to 58 percent. Over the same period, part-owners increased from 8 percent to 31 percent of all agricultural producers.

The implementation of article 5 of the agricultural creed has virtually eliminated the impact of much of this historic creed as a value basis in determining agricultural policy. Policymakers have been forced to determine agricultural policy on values outside this agricultural creed. The scientific revolution in agriculture, coupled with the competitive structure and the physical nature of agriculture, has increased agricultural production much faster than the demand for agricultural products has increased. As a consequence, farm product prices have declined relative to other prices in the economy, making agricultural policy decisions more important than ever to some agricultural producers.

Over the years, government policies have provided opportunities for people to farm. Since colonial times, the government has provided incentives to encourage land settlement throughout the United States. The major public land policy was the Homestead Act of 1862. Under this act, a settler could pay a small registration fee and then if the settler resided on and worked 160 acres of land for 5 years, the settler would gain title to the land; or the settler could reside on the land 6 months and pay $1.25 per acre and thus gain title. Between 1868 and 1879, 70 million acres of land were made available to farmers under this act.

At present, however, anyone desiring to farm has a limited opportunity to do so because of the capital requirements that are necessary to obtain an economic-sized farm. It is almost essential to either inherit a family farm or have one's relations give the land or provide some substantial financial assistance. Stam has shown that the probability of a farm youth taking over an economic-sized unit is only about 1 in 12.[4]

Most producers who are financially sound are free to make production and marketing decisions, and thus are their own bosses. Most farmers who own their farms have maintained their managerial control even though they may have had to give up some managerial freedom because of credit requirements, vertical coordination, and governmental policies.

Property rights give producers the benefit of using their resources as they see fit. With population increasing, some discussion has been taking place as to the advisability of exclusive control over a resource. People are indirectly affected by property rights, and some believe that these rights should be amended to restrict the owner's control over the use of a resource so as to maintain its use in the public interest.

Even though some of these fundamental attitudes may have changed, they still influence the feelings with regard to agriculture. With this in mind, an update to the creed might include the following:

1. Farming should be a family business.
2. Country living has many virtues and it should be made available to urban residents (parks, greenbelts, etc.).
3. Agriculture should provide adequate food and fiber at reasonable prices to producers and consumers.
4. Society should assist farm youth in providing farming and other agribusiness opportunities.
5. A farmer should be his/her own boss.
6. The land should be controlled by the person who owns it.
7. Agriculture is vital to mankind and therefore should receive priority in national goals.

These articles will continue to provide the basis of our future agricultural legislation. Out of these articles one can derive policy goals that will form the basis of future policies.

A *policy goal* is defined as a desirable end being sought that is consistent with the values of the group proposing it.[5] To date, most farm policy has revolved around the concepts of the "family farm," "parity," and "equality of bargaining power."

---

[4] Jerome M. Stam, "Farming Opportunities for Rural Farm Youth in the North Central Region," Economic Study Report No. 569–3. Department of Agricultural Economics. University of Minnesota, St. Paul, July 1969, p. 23.

[5] Hathaway, op. cit., p. 61.

## Family Farms

Much discussion has taken place regarding the family farm. In most areas, the family farm is considered to be consistent with the values of freedom, political and social stability, and economic justice and will therefore influence agricultural policy for some time. In fact, many states have introduced legislation to restrict corporations from operating agricultural firms in order that the family farm might remain as the basic economic unit in agriculture. However, it is difficult to define a family farm. By a *family farm* some people mean a farm that is operated by one family and possibly some small amount of hired help. Others try to define a family farm in terms of size, such as acreage. Such size measurements are not very precise because one family may be able to operate only a few acres in a fruit, nut, nursery or vegetable farm, but thousands of acres in a wheat or cattle ranch.

## Parity

Because of past problems of low incomes in agriculture, all major agricultural legislation has attempted to promote parity in some way. Some of the ways have been in terms of parity prices, parity incomes, or a fair return on factors of production used in agriculture. Parity price is the concept most discussed and presented in newspapers. *Parity* means equality of value. It means prices that will give a unit of an agricultural commodity the same purchasing power as that unit had in some previous period. So if a bushel of corn purchased a shirt during the base period, then parity means that one bushel of corn today should buy one shirt at current market prices. The historic base period for calculating parity for many commodities was August 1909 to July 1914. Full parity, therefore, would give agricultural commodities the same purchasing power as they had in the 1909–14 base period.

The definition of the parity concept, however, causes problems because it does not take into account changes in demand, supply, and resource productivity over time. To allow for these changes and for product substitution, the calculation of parity price was modified in the Agricultural Act of 1948. This act changed the base period for calculating parity prices for individual farm products to the most recent 120-month moving average. An example of the modification of parity price for wheat is as follows:

$A$ = average price of wheat for the last 120 months = \$3.93

$B$ = average index of farm prices received for the last 120 months = 633

$C$ = current farm prices paid index = 1241

Computation of parity price of wheat:

$$\frac{A}{B} \times 100 = \text{Adjusted price}$$

$$\frac{\text{Adjusted price} \times C}{100} = \$7.71$$

Although this formula may appear to be complicated, it is not. All this formula does is take the current price of wheat and divide it by the index (weighted average) of prices received by farmers in order to adjust for increases in inflation. This gives a "real" or adjusted price that takes out the changes in the value of the dollar and also takes out the change in demand and supply for wheat relative to other farm commodities. This real price is then multiplied by the index of prices paid to inflate the price of wheat to reflect the increase in the costs of producing agricultural products.

Producers are now interested in other measures of parity because, at certain time periods, the formula just shown would not give them parity incomes. Presently, these parity prices are far above market prices and would be very expensive for society to maintain. The new concept is one that is based on cost of production or a fair return on investment in land, labor, management, and capital. The measure then would be what these factors could earn in the nonfarm economy. This means that cost of production studies need to be conducted. One would have to impute the cost of labor, land, management, and capital and divide the sum of these costs by output to arrive at a cost per unit or parity price per unit. Because costs of production differ widely between farms and production regions for a particular crop, a sizable representative sample of firms on which to run cost studies would have to be selected. Even then the problems associated with imputing a value for land are severe.

Most of the problems associated with parity price are contained in the fair return concept. However, parity is consistent with the values of economic justice and social and economic stability, security, and freedom, and will continue to be a goal for a long time.

## Equality of Bargaining Power

The central theme of the National Farmers Organization (NFO) formed in 1955 in Iowa is bargaining power for agriculture. The NFO believes that agricultural producers are being taken advantage of by the large processors, wholesalers, and retailers in the food and fiber system. The NFO states that "if producers want to price their products, they must go to the marketplace with equal or greater strength than those who buy their products." Thus, they are attempting to increase their muscle in the marketplace by forming a quasimonopoly organization that could hold products off the market for higher prices and use contracts to market agricultural products in an orderly manner (see Chapter 12).

Although this method of increasing agriculture's bargaining power is used by the NFO, other farm organizations are using farmer-owned and controlled cooperatives such as the National Farmers Union, National Grange, and American Farm Bureau Federation.

Bargaining power for agriculture has been with us since the 1920s and is consistent with farmer values of economic justice, economic stability, and security. This goal will also be around for some time to come.

# FARM PROBLEMS _____

In general, farmers have not prospered as the rest of the economy has, primarily because the supply of agricultural products has been increasing faster than the demand for them. Until recently, agricultural production has been increasing at about 2 percent per year, while demand has been increasing at a slightly less rapid rate. Hence, agricultural prices and incomes have been relatively low. The major historical problem has been one of chronic overproduction, but other problems also exist. They include the variability of agricultural prices with changes in demand and supply and very low incomes for the producers who are marginal farmers.

## Overcapacity

Many countries around the world have difficulties because they cannot adequately feed their people—this is not the problem in the United States. Rather, it is one of plenty, not scarcity. Excess production has not been especially large in American agriculture in absolute terms. It has been estimated at from 6 to 12 percent of total production; however, its economic impact is large. This overproduction has caused price and income troubles for many agricultural products. We refer here to excess production in the aggregate, yet there have been lean and abundant years for individual products.

Tweeten estimates the aggregate price elasticity of demand for total farm output for domestic use to be − 0.20 in the short run and − 0.40 in the long run.[6] This elasticity means that a 1 percent increase in prices will decrease quantity demanded by 0.20 percent in the short run and 0.40 per-

_____

[6] Luther Tweeten, "Economic Instability in Agriculture: The Contributions of Prices, Government Programs and Exports," *American Journal of Agricultural Economics*, Vol. 65, No. 5, December 1983, p. 923.

cent in the long run. If we include foreign demand for U.S. products, this increases the price elasticities of demand to $-0.25$ in the short run to $-0.63$ in the long run.[7] Aggregate price elasticities are increasing over time, because greater amounts of products have been exported in recent years. Thus, agricultural prices should fluctuate greatly with small changes in agricultural output,with price fluctuations being lessened because the foreign market increases the price elasticity of demand.

Income elasticity of demand shows the relationship between food consumption and consumers' incomes. Tweeten estimates this to be 0.15 for all farm products; if consumer disposable income increases 1 percent, the demand for farm products will increase by only 0.15 percent.

Two major factors that affect food consumption are population and incomes. With the American population growing at 1.05 percent a year, and per capita real income at only 1.01 percent, domestic demand for farm commodities should grow by a total of 1.20 percent per year [1.05 percent for population, plus $(0.15 \times 1.01 =)$ 0.15 percent for incomes]. Thus, if farm output is capable of increasing more than 1.20 percent per year, excess capacity will exist unless land continues to be retired and export demand stays at its current high levels.

U.S. agricultural capacity to produce is immense. American farmers have the potential to vastly increase their output of agricultural products if prices are favorable, if there are no restrictions on land use, if inputs are adequate, and if growing conditions are normal (Table 11-1). Under these conditions, compared to 1980, the United States could increase its 2000 feed grain production by 56 percent, soybean production by 50 percent, cotton by 22 percent, and wheat by 33 percent.

## Instability of Farm Prices

One of the basic aims of government farm programs has been to reduce the year-to-year variation in farm prices and incomes. These programs included the granting of low interest loans to producers to build on-farm storage and the establishment of the Commodity Credit Corporation in 1933 to take delivery and sell agricultural products.

Farm prices are influenced by farm program changes as well as by weather on the supply side, and by foreign demand on the demand side. The inelasticity of demand along with these other factors sharply reduces farm prices and incomes as production increases. As shown in Figure 11-1, a slight increase in production from $Q_0$ to $Q_1$ forces the price of wheat from $P_0$ to $P_1$. Also, total revenue drops from $O$-$P_0$-$A$-$Q_0$ to $O$-$P_1$-$B$-$Q_1$ be-

---

[7] Ibid:, p. 924.

**TABLE 11-1**   Actual and Potential Production of Selected Crops, United States

| Production | Average | | | Potential[b] | Actual |
|---|---|---|---|---|---|
| | 1969-71 | 1980[a] | 1985 | 2000 | 1995 |
| **Corn** | | | | | |
| Harvested acres (mil) | 58.7 | 73.1 | 75.2 | 75.0 | 65.0 |
| Yield (bu/ac) | 82.2 | 91.0 | 118.0 | 131.4 | 113.5 |
| Production (bil bu) | 4.8 | 6.6 | 8.9 | 9.9 | 7.4 |
| **Soybeans** | | | | | |
| Harvested acres (mil) | 42.1 | 67.9 | 61.6 | 71.0 | 61.6 |
| Yield (bu/ac) | 27.4 | 26.8 | 34.1 | 38.0 | 34.9 |
| Production (bil bu) | 1.2 | 1.8 | 2.1 | 2.7 | 2.2 |
| **Feed grains** | | | | | |
| Harvested acres (rnil) | 100.4 | 101.6 | 111.8 | 107 | 82.6 |
| Yield (tons/ac) | 1.81 | 1.95 | 2.45 | 2.88 | 2.82 |
| Production (mil tons) | 182 | 198 | 274.4 | 308 | 233.2 |
| **Wheat** | | | | | |
| Harvested acres (mil) | 46.1 | 70.9 | 64.7 | 81 | 61.0 |
| Yield (bu/ac) | 31.9 | 33.4 | 37.5 | 39.4 | 35.8 |
| Production (bil bu) | 1.5 | 2.4 | 2.4 | 3.2 | 2.2 |
| **Cotton** | | | | | |
| Harvested acres (mil) | 11.2 | 13.0 | 10.2 | 13.8 | 16.0 |
| Yield (lbs/ac) | 437 | 411 | 630 | 700 | 540 |
| Production (mil bales) | 10.2 | 11.1 | 13.4 | 20.1 | 18.0 |
| **Total acreage of crops harvested (mil)** | 292 | 341 | 334 | 356 | 286 |

SOURCE: David W. Culver and Milton H. Ericksen, "American Agriculture, Its Capacity to Produce," *The Farm Index*, Commodity Economics Division, USDA, Special Report, December 1973.

[a] 1980 was a dry year with reduced yields and production.
[b] Authors' estimate based on 1982 acreages and historical yields.

cause the percentage decline in price is greater than the percentage increase in production.

Changes (shifts) in demand as shown in Figure 11-2 can also cause large changes in farm prices. In this case, the demand curve shifts from $DD$ to $D'D'$, so prices fall from $P_0$ to $P_1$ and total revenue falls from $O$-$P_0$-$A$-$S$ to $O$-$P_1$-$B$-$S$. These price changes result in income variations causing many family hardships, especially for the marginal producers.

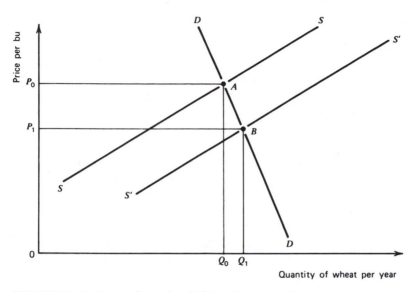

**FIGURE 11-1**   Price effect of a shift in wheat supply.

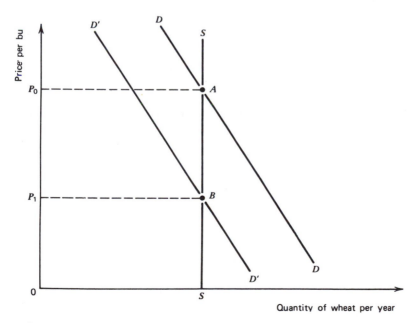

**FIGURE 11-2**   Price effect of a shift in wheat demand.

## Poverty

In a study published in 1967, there were approximately 33.7 million people living in poverty in the United States.[8] In 1994, there were 38.1 million poor. Of these poor, 9.2 million lived in rural America. Twenty percent of the total population lived in rural areas, yet about 24 percent of the nation's poor lived there. In that year, it is estimated that there were about 0.6 million poor people who lived on farms comprising 12.5 percent of the total farm population.

Poverty is a relative concept and difficult to define. *Poverty* for our purpose is defined as "a lack of access to respected positions and the lack of power to do anything about it; insecurity and unstable homes; and a wretched existence that tends to perpetuate itself from one generation to the next." The symptoms of poverty include low levels of formal education, high unemployment rates, high dependency rate on those of working age, and dilapidated housing. The official *poverty level* is established as "an income level below that needed to provide the kind of living that our society considers a basic right."[9] The poverty income level changes with the above conditions and is updated each year to adjust for inflation. By 1995, the poverty level for an urban family of four was increased to $15,570 per year.

Past U.S. agricultural programs have done little to help farm poverty. The reason for this is that government farm payments are based on production, and poor farms have little to sell. Sixty percent or 1,246,000 farms have cash receipts of less than $20,000 annually. Subtracting production costs from such amounts received leaves little if any net agricultural income. The question then arises as to whether or not these farms have been able to supplement farm income with more adequate nonfarm income. In general, there has been substantial substitution of nonfarm income for farm income, but there still remain some low-income problems.[10]

In 1994, disposable income per capita for the farm population was only 53 percent of nonfarm income, $10,141 compared with nonfarm of $19,253. However, when off-farm income is included, per capita income for the farm population is about $27,231. The income from off-farm sources by those farms that had farm sales of less than $20,000 was $38,842, and for those with farm sales between $20,000 and $99,999 it was $34,606. On the average, these farm families had farm and nonfarm incomes exceeding

---

[8] *The People Left Behind*, A Report by the President's National Advisory Commission on Rural Poverty, Washington, D.C., September 1976, p. 3.

[9] W. W. McPherson, "An Economic Critique of the National Advisory Commission Report on Rural Poverty," *American Journal of Agricultural Economics*, Vol. 50, No. 5, December 1968, p. 1363.

[10] E. J. R. Booth, "The Economic Dimensions of Rural Poverty," *American Journal of Agricultural Economics*, Vol. 51, No. 2, May 1969.

the poverty level and enjoyed many benefits of life desired by most Americans, but these data tend to hide poverty problems that remain in agriculture.

# FARM PROGRAMS THROUGH THE YEARS _____

Agricultural policy in the United States can be broken down into three basic types of programs. These are (1) two-price plans, (2) land retirement programs, and (3) direct payment programs.

## Two-Price Plans

Two-price programs are designed to take advantage of the different elasticities of demand in the domestic and foreign markets so as to increase total revenue to agricultural producers. The relatively inelastic domestic demand and the relatively more elastic demand in the foreign market can be used to the producers' advantage. The price elasticity of demand in the foreign market is more elastic for most agricultural products because other countries' products are substitutes for many U.S. agricultural exports. The more substitutes there are for a product, the more elastic the demand curve becomes. As an illustration, consider the domestic and foreign demand for U.S. wheat. U.S. producers are usually the sole supplier to the domestic market partly because of import quotas restricting the amount of imports. The domestic demand for wheat is very inelastic (about − .20), which means that consumption does not vary much with price changes. In the foreign (or world) market, the United States competes for customers primarily with the European Community, Canada, Australia, and Argentina.

As shown in Figure 11-3, assume the United States has 1.5 billion bushels of wheat to sell. Now, by restricting domestic marketings to 500 million bushels, a $4.00 per bushel price is attained for a total domestic revenue of $2.0 billion. If the other 1 billion bushels is sold at the foreign market price of $3.00 per bushel, total foreign revenue of $3 billion is attainable. Total revenue from both markets is $5 billion. If we were to market more wheat in the domestic market (e.g., 800 million bushels at a price of $2.00 per bushel), domestic revenue would be $1.6 billion. Now if the remaining 700 million bushels is sold overseas, then foreign revenue is $2.45 billion, resulting in a total revenue of only $4.05 billion. This market allocation reduces total revenue by $950,000. Therefore, the two-price plans were enacted in an attempt to increase revenue to agricultural producers by restricting output sold in the domestic market, which has an inelastic demand, and selling the balance in the foreign market, which has

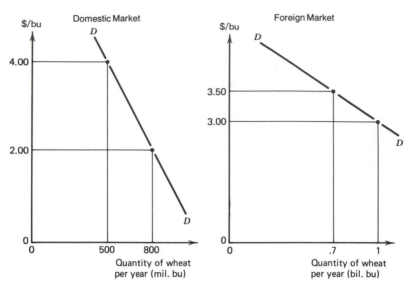

**FIGURE 11-3**   Economic effect of two-price plans.

a more elastic demand. In this illustration (disregarding costs of production and allocation costs between markets), total revenue is maximized when marginal revenue in the two markets is equalized.

These two-price plans were used to support the domestic prices of agricultural commodities above equilibrium prices and then merchandise the surplus output in foreign markets. This market segmentation is shown in Figure 11-4. In a free market, given demand curve $DD$ and supply curve $SS$, the equilibrium price and quantity produced and sold in the market would be $P_0$ and $Q_0$, respectively. If the support price is above equilibrium, say at $P_1$, then consumers will take quantity $Q_1$ but production will be at $Q_2$. Therefore a surplus ($Q_1$ to $Q_2$) remains that must be either stored or exported.

In the past, the Commodity Credit Corporation (CCC) has accumulated these surpluses through the use of nonrecourse loans. A *nonrecourse loan* is one where a loan is made to a producer participating in the program, with the borrower having the option of repaying the loan or turning over ownership of the commodity without further financial responsibility. If the market price rises above the loan rate, the producer can sell the product on the open market and repay the loan plus interest. If the loan rate is above the open market price, the producer can relinquish the product to CCC in full payment of the loan. Therefore, during periods of excess production, CCC took delivery of large amounts of some agricultural commodities. CCC then had the responsibility of disposing

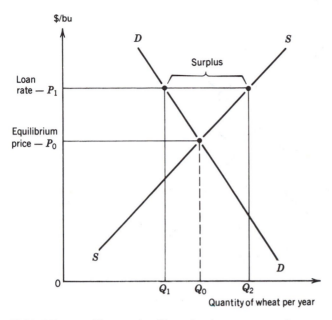

**FIGURE 11-4**  Economic effect of price supports above equilibrium.

of these surpluses in domestic aid programs and in foreign markets at prices in relation to the domestic loan rate as specified by Congress.

Two-price plans have been used extensively for many agricultural commodities. The government has used two-price plans since 1929, with the primary purpose of these programs being to increase the incomes of the agricultural producers.

## Land Retirement

Land retirement has been used in conjunction with two-price programs to restrict production and increase agricultural income. One of the most massive land retirement programs was the soil bank program established under the Agricultural Act of 1956. The act established an acreage reserve and a conservation reserve. The *acreage reserve* was a short-term land retirement program where producers were paid to divert part of their allotted acreage from current use. The *conservation reserve* was a long-term program to divert all or part of a producer's land from crop production to soil-conserving uses.

The purpose of the land retirement was to reduce supply by limiting the land input. Remember, a part of the basic problem was that the supply curve was continuing to shift to the right and a land retirement pro-

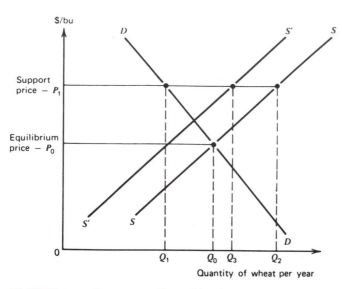

**FIGURE 11-5** Economic effect of land retirement programs.

gram attempted to reduce the amount by which the supply curve could shift or even reverse it in the shorter run.

Figure 11-5 shows the open market equilibrium price and quantity as $P_0$ and $Q_0$ given demand curve $DD$ and supply curve $SS$. At a support price of $P_1$, consumers want only quantity $Q_1$ and producers supply quantity $Q_2$ so a surplus ($Q_1$ to $Q_2$) develops. In order to reduce this surplus, land retirement was used to take land out of production, shifting the supply curve from $SS$ to $S'S'$. This movement of the supply curve reduces the surplus from ($Q_1$ to $Q_2$) to ($Q_1$ to $Q_3$). Payments to producers to restrict cropland use should reduce both surpluses and government storage costs.

In 1957, under the acreage reserve, 21.4 million acres were taken out of production. The conservation reserve program increased to a maximum of 28.7 million acres in 1960. The total acreage taken out of agricultural production under the conservation reserve and other acreage diversion programs reached a maximum of 65 million acres in 1962. These were the largest land retirement programs until the 1983 and 1988 federal farm programs reduced cropland harvested by 78 million acres. Current land retirement is about 55 million acres.

## Direct Payment Programs

Direct payment programs were first proposed for agriculture in 1949, but not used until 1973. They were included in the Agriculture and Con-

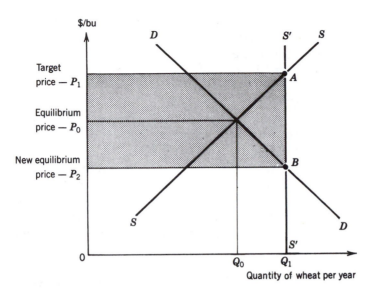

**FIGURE 11-6**   Economic effects of direct payment programs.

sumer Protection Act of 1973 for wheat, corn, rice, cotton, and barley and in the Food and Agriculture Act of 1977, the Agriculture and Food Act of 1981, the Food Security Act of 1985, and the Food, Agriculture, Conservation, and Trade Act of 1990.

Figure 11-6 illustrates the use of a direct payment program. It shows the equilibrium price and quantity as $P_0$ and $Q_0$ given $DD$ and $SS$. If the target price is $P_1$, producers will supply quantity $Q_1$. Quantity $Q_1$ now represents the amount produced shown by $S'S'$. To get consumers to take quantity $Q_1$ the price in the market must drop to $P_2$ (where $S'S'$, and $DD$ intersect). Thus under a direct payment program, the price would fall to the market clearing level of $P_2$. The U.S. Treasury would make up the difference between the target price $P_1$ and the free market price $P_2$. The supported income to producers would be the area $O\text{-}P_1\text{-}A\text{-}Q_1$; consumers would pay an amount $O\text{-}P_2\text{-}B\text{-}Q_1$, with the shaded area being the amount contributed by the U.S. Treasury. Thus, consumers have a large amount of the product at a low price, but tax revenues must be used to pay the amount represented by the shaded area.

## Farmer-Owned Reserves

In April 1977, a farmer-owned grain reserve (FOR) was established to provide a means for farmers to keep grain supplies off the market until grain prices reach specified levels. The purpose of the reserve program is to stabilize prices and provide assurance of dependable grain supplies.

To be eligible for the reserve, commodities must be held for 9 months under price support loan by the CCC. Agreements with producers are for 27 months of storage from the date the original loan expires; however, the Secretary of Agriculture could extend the loan for an additional six months. Reserve grain can be released from the reserve at any time at the discretion of the producer by repaying the loan.

Annual storage payments are made quarterly to producers to offset storage costs. Producers do not earn storage payments if the national average market price of the commodity equals or exceeds 95 percent of the commodity's current target price and for any 90-day period following the last day the price was at such a level. Interest on reserve grain loans may be charged whenever prices equal or exceed 105 percent of the target price for that year. Once interest is levied, it will continue to accrue for 90 days from the last day the market price equaled or exceeded 105 percent of the target price.

The Secretary of Agriculture must specify the maximum quantity in the FOR to be between 300 and 450 million bushels of wheat and between 600 and 900 million bushels of feed grains. There is no minimum quantity, which must be maintained in the FOR. The Secretary may permit crops to enter storage in FOR under two conditions: I) the average market price is 80 percent of loan rate (for 90 days before December 15 for wheat and for 90 days before March 15 for corn), or 2) the projected stocks-to-use ratio is greater than 37.5 percent for wheat or 22.5 percent for corn. If both of these conditions are met, the Secretary must permit grain to enter the FOR. However, no direct entry in the FOR is permitted. Producers can only enter grain after an original 9-month loan. As mentioned, a producer may exit the FOR at any time by repaying the loan.

The reserve acts to maintain minimum grain prices. On the downside, producers can take out loans at the prevailing loan rate.

Under the 1996 farm bill authority for the FOR was suspended through the 2002 crop year. This change in legislation means that most grain stocks will be privately held rather than publicly held.

## Conservation Reserve

Large carry-over stocks of grain, record crop production, and reduced exports forced the Reagan administration to establish a conservation reserve, payment-in-kind (PIK) program in 1983. The program reduced acreages of wheat, cotton, rice, corn, and grain sorghum by about 78 million acres. Wheat was reduced by 30 million acres, feed grains 39 million, upland cotton 7 million, and rice 2 million acres.

The Food, Agriculture, Conservation, and Trade (FACT) Act of 1990 was signed into law on November 28, 1990. This act covered 5 crop years from 1990 through 1995 and continued the past method of supporting agriculture through target prices, loan rates, and deficiency payments. Target prices were frozen under the 1990 program. Minimum target prices for wheat were set at $4.00 per bushel. Target prices for corn were left at $2.75 per bushel.

The minimum loan rate was set at $2.44 per bushel for wheat and at $1.76 per bushel for corn. The basic loan rate is calculated as 85 percent of the simple average of prices received by producers for the preceding 5 marketing years, dropping the high and low prices from the calculation. The Secretary of Agriculture can reduce the price support rate by up to 10 percent based on projected ending stocks-to-use ratios for the current marketing year.

Loan levels can be reduced further by the Secretary of Agriculture up to an additional 10 percent to ensure that U.S. commodities are competitive in world markets. However, if the Secretary of Agriculture adjusts the basic price support rate to maintain a competitive position internationally, producers are paid emergency compensation to provide the same total return. This compensation is known as Findley payments.

Deficiency payments were made if the market price is lower than the target price. The deficiency payment rate is the difference between the target price and either the national weighted average market price (for the first 5 months of a marketing year) or the basic price support level prior to any adjustments, whichever is higher. For 1994-95, deficiency payment calculations were changed to the difference between the target price and the lower of either a 12-month national weighted average marketing year price or the 5-month marketing year price plus some small amount per bushel (wheat, 10¢ per bushel; corn, 7¢ per bushel).

The Agricultural Reconciliation Act of 1990 continued to provide the Secretary of Agriculture with authority to require reductions in acreages planted to participate in the FACT program. The maximum required acreage reduction for wheat is 15 percent and the minimum acreage reduction for corn is 7.5 percent. The Secretary of Agriculture has the authority to expand the marketing loan concept (now covering rice and cotton) to wheat and feed grain producers. Under a marketing loan, when the market price is below the price support rate, producers can repay their loan at the higher of the Secretary's determined world market price or 10 percent of the basic price support rate. The marketing loan tends to reduce loan forfeitures, reduces accumulation of stocks, and makes commodity prices competitive internationally.

Cropland diverted from production under the federal support program was 78 million acres in 1983/84. In addition, a conservation reserve program (CRP) was established to reduce planting on highly erodible cropland. Producers could implement a plan approved by the local conservation district to place highly erodible cropland into grasses, trees, and other acceptable vegetative covers for 10 to 15 years. Annual rental payments were on a per-acre bid basis for the 10-year contracts. About 36 million acres have been idled under current Conservation Reserve Program contracts.

A new farm bill, the Federal Agricultural Improvement and Reform (FAIR) Act, became law in 1996. The new act removes the link between income support payments and farm prices by providing for seven annual fixed but declining "production flexibility contract payments." This new bill marks a major change from past farm programs where deficiency payments were dependent on farm prices and were designed to help stabilize farm income. Another important change is that farmers now have much greater flexibility in planting, including the elimination of annual acreage idling programs, and will rely more heavily on the market rather than government programs in making production decisions. The "production flexibility contracts" that will expire with the 2002 crop will require producers to comply only with existing conservation plans and wetland provisions, with no limitations on planting except for fruits and vegetables. Land eligible for contract acreage for the new FAIR Act is equal to the farm's base acreage for 1996 calculated under the previous farm program plus any returning CRP base and less any continuing CRP enrollment. Payments on production flexibility contracts are limited to $40,000 per person.

Features of former programs continued in the new 1996 legislation include the floor price protection of nonrecourse loans with marketing loan provisions where loan rates are generally set at 85 percent of prior 5-year average farm prices (excluding high and low years) for most commodities. The Farmer-Owned Reserve concept has also been retained but authority is suspended through the 2002 crop year. Authority to enter into new P.L. 480 Program agreements, Food for Progress agreements, and Export Credit Guarantee Programs is extended through 2002. Other former conservation programs are extended, e.g. the national CRP enrollment can be maintained at a maximum of 36.4 million acres. New enrollment of environmentally sensitive land to CRP is permitted to replace land taken out of the program. The Food Stamp Program is reauthorized for two years while congress considers comprehensive welfare reform. Permanent provisions of the 1938 Agricultural Adjustment Act and the 1949 Agricultural Act are continued after 2002. These permanent provisions authorize marketing quotas, marketing certificates, acreage allotments, and parity-based price support for wheat, feed grains, cotton, and sugar.

In comparison with recent annual total deficiency payments for all crops, ranging from a high of $8.6 billion in 1993 to $3.9 billion in 1995, the total outlay for production flexibility contract payments will range from $5.4 to $5.8 billion to 1999 and then drop off to $4.0 billion in 2002 when they are terminated. The general formula for allocating the annual contract payments to different crops is 46.2 percent to corn, 7.4 percent to other feed grains, 26.3 percent to wheat, 11.6 percent to upland cotton, and 8.5 percent to rice. The program payment rate for each commodity would be calculated from the annual total allocation for each commodity divided by the sum of all individual commodity contract quantities for the year. An individual farm's payment quantity equals the farm's program payment yield multiplied by 85 percent of the farm's commodity contract acreage.

# HAVE AGRICULTURAL PROGRAMS INCREASED AGRICULTURE INCOME? _____

Although the three types of programs discussed above were intended to increase agricultural incomes, they have been unable to do so in the long run. Income benefits are capitalized into land values and therefore become production costs. Higher prices for agricultural commodities mean that agricultural land produces greater income, so people bid up its price because of its increased profitability. In the long run, benefits accrue to present landowners rather than to later generations of producers.

With greater reliance on the export market for wheat, feed grains, and soybeans, farm incomes are subjected to wider swings because of fluctuating demand in the foreign market.

Because of the increasing share of exports to total farm product sales, the elasticities of demand for farm products has been increasing over the past 25 years. The individual short-run price elasticities of wheat, feed grains, and soybeans are less than – 0.5, whereas their long-term elasticities are approaching or slightly exceed – 1.0. The implication of this situation is "that permanent supply controls will not raise real farm receipts markedly in the long run."[11]

Government programs have provided other benefits to farmers, however. Past federal farm programs and off-farm income have stabilized farm incomes and have reduced some of the risk and uncertainty in farming. Farm programs have also made it possible for some of the excess labor in agriculture to adjust to nonfarm employment. In addition, these programs have provided food stocks for national and international emergencies. More importantly, however, these programs have undoubtedly provided for a greater degree of political and social stability in the United States than would have existed without them; and agricultural producers have been more able to lead the life they desire, while providing more stable and higher quality food supplies to urban dwellers.

---

[11] Luther Tweeten, "Economic Instability in Agriculture: The Contributions of Prices, Government Programs and Exports," *American Journal of Agricultural Economics*, Vol. 65, No. 5, December 1983, pp. 923-924.

# ALTERNATIVE PROGRAMS _____

Many alternative programs could be employed to solve some of agriculture's problems. The alternative selected would depend on the criteria used to judge a program. For example, if the only criterion is to minimize treasury cost, a move to a free market would accomplish this objective, whereas a direct payment program would not do so. If economic efficiency is the criterion, a free market would be best, and a support program would be less desirable. Some of the most popular alternative programs are (1) demand expansion, (2) free market, (3) land retirement, and (4) market quotas. These can be short-run programs. Long-run programs to move the labor resource from agriculture would include a national employment service, education, alternative skill training programs, to make surplus agricultural labor more competitive in the urban labor market, and the relocation of industry to rural areas.[12]

## Demand Expansion

Increasing the domestic and foreign demand for products is often suggested as a method to increase farm prices and incomes. The United States carried on many demand-increasing programs. These are the food stamp program; the special supplemental food program for women, infants, and children; the school breakfast and lunch programs; and direct food distribution programs. In 1995, these programs cost some $38.8 billion. The food stamp program is of major assistance to low-income people, as well as to agriculture. It increases the income of low-income people and thus increases their consumption, expanding the market for agricultural commodities. In 1994, about 27 million people participated in this program. Those who participated received food stamps that had a value of $24.5 billion in the grocery store. The government thereby increased the participants' purchasing power by $24.5 billion or about $891 per person.

The food stamp program can be analyzed using indifference curve analysis, as in Figure 11-7. Budget constraint I shows what a consumer could purchase without food stamps. The budget constraint is tangent to indifference curve $I_0$. In equilibrium (at point $A$) this consumer would buy $C_0$ of other commodities and $F_0$ of food.

Budget constraint II shows the effect of an increase in income from the voucher food plan. The voucher is a "gift" of income spendable only on the specific commodity."[13] The relevant part of the budget line is the segment $E$ to $H$, as $FE$ represents reduced food consumption, which the

---

[12] Tweeten, op. cit., Chapter 11.
[13] Jack Hirschleifer, *Price Theory and Applications*, 4th ed. Englewood Cliffs, N.J.: Prentice Hall, 1988, pp. 108–110.

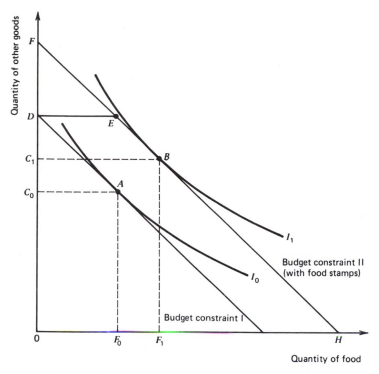

**FIGURE 11-7**   Effect of a food stamp plan on consumption.

voucher system is attempting to prevent. With the voucher increasing spendable income, the consumer is able to reach a new equilibrium at point $B$ and a higher level of satisfaction. Although food purchases increase to $F_1$, the consumer is able to increase the consumption of other commodities to $C_1$. Whether there actually is an increase in the consumption of other goods depends on the consumer's tastes and preferences for food versus other commodities, which would be revealed in the slope of the indifference curves.

In foreign demand expansion, the United States has relied primarily on Public Law 480 (Food for Peace Program) to increase the demand for agricultural products. Under this program America sells commodities for foreign currency or long-term credit, or sends products abroad for famine relief or barter. In 1995, the United States sent $400 million worth of commodities overseas under this and similar programs.

The U.S. government spent $409 million in export subsidies in 1973 to move agricultural products abroad. Most of that total was used for wheat subsidies. These export subsidies were eliminated in 1975, but credit guarantees were instituted in 1979 to increase the export of agricultural commodities. General Sales Manager Program 102 (GSM-102), and 103

(GSM-103) the two credit guarantee programs are funded at about $5.5 billion per year for 1996 through 2002. The GSM-102 program provides assistance to exporters by guaranteeing payment to commercial banks up to 98 percent of the principal and up to 4.5 percent of the interest on export credit loans made through foreign banks at commercial rates. GSM-103, provides intermediate credit guarantees from more than 3 years to 10 years duration.

The 1990 Act reauthorized the Direct Credit Sales Program and the Export Credit Guarantee Program. The Short-Term Direct Credit Sales Program allows CCC to offer direct financing up to 3 years for export sales while the Intermediate-Term Direct Credit Sales Program authorizes credit terms of 3–10 years. The CCC determines the amount of funding for these programs. The 1990 Act also provided $1 billion for fiscal years 1991–95 for credit guarantees to promote agricultural exports to emerging democracies. A portion of these guarantees can be used to establish or improve facilities in emerging democracies to promote U.S. agricultural exports. All export promotion programs in the 1990 Act have been extended in the FAIR Act through 2002.

Economists believe that there has been some success in expanding the domestic demand for food. All income support programs help increase food demand because the income elasticity of demand for food is greater for low income groups. But this effort offers little long-run hope for solving agriculture's excess capacity problems. Food aid abroad could solve the capacity problem if the United States were willing to make the growing financial commitment that would be necessary to feed a larger portion of the world's population.

## Free Market

Agriculture could return to a free market if the government decided to eliminate production controls and price supports on agricultural commodities and other food and credit programs. In the short run, Tweeten estimates that for each 1 percent increase in farm output placed in the market, prices on the average would likely be depressed by 2 percent, gross cash receipts by 1 percent, and net farm income by almost 3 percent.[14] Therefore, if the excess capacity was from 6 to 12 percent, and we use the low estimate of a 6 percent increase in farm output, farm products prices would drop 12 percent, cash receipts would fall 6 percent, and net farm income would be reduced by 18 percent. Producers would also lose their government support payments. The total loss in the short run would amount to a decrease in net farm income of 18 percent plus the forfeited government payments. This loss would cause considerable hard-

---

[14] Tweeten, op. cit., p. 325.

ship, unless the transition were buffered with a world shortage situation, such as those that developed in 1972 and 1980. After a 4-year adjustment period, Heady and Tweeten estimated that farm prices, gross income, and net income would have recovered to 90, 93, and 88 percent, respectively, of the level that existed before the release of 6 percent more commodities on the market.[15]

The major concern over returning to a free market is whether, because of its basic structure, agriculture can remain relatively stable, or whether it always will be feast-or-famine in terms of agricultural prices and incomes. Also, producers are concerned that they may not have the political persuasion to get another farm bill passed.

## Land Retirement

Most proposals for land retirement are either "part-farm" or "whole-farm" retirement programs that divert farm land to soil-conserving uses on a bid basis. Past programs to take agricultural land out of production on a part-farm basis have not been successful. The primary reason is that land is just one input in the production process. When the land input is restricted, producers tend to substitute other inputs such as capital equipment, agricultural chemicals and fertilizers. Consequently, rather than reduce output, it may be increased. For this reason, many economists prefer long-term land retirement programs on a whole-farm basis. Although this type of program cures the problem of substituting other inputs for land because all inputs are removed (land, equipment, labor, etc.), it does have other effects. One of these is the rural community problem. This problem develops because as producers rent their land to the government there is no need to purchase input items from the local feed, equipment, and other dealers. Many of these dealers will sell less of their goods and services, causing financial difficulties that also spread to other businesses, and therefore may increase the migration of people from rural America. As expected, farm supply firms put pressure on Congress not to pass this type of legislation.

## Market Quotas

Marketing quotas limit the amount of product a producer can sell or produce without penalty. Past marketing quota programs have been successful in reducing agricultural production, but this type of program has problems similar to the land retirement program in its effect on rural communities. This similarity is especially true if marketing quotas are

---

[15] Tweeten, op. cit., p. 325.

transferable. If farmers are allowed to sell their quotas, then these quotas will move to the most efficient farms. In many cases marginal land is concentrated and would affect the rural communities in those areas. If quotas are not transferable, then food prices would be higher than necessary largely as a result of fixing past production relationships into current production patterns.

This summary of some of the alternative farm programs that could be initiated is for perspective only. It should, however, give some idea of the advantages and disadvantages of each. Again, it is important to point out that different groups of people judge farm policy from different viewpoints. Farmers want a program that gives them maximum managerial freedom and the highest net income. Consumers want a plentiful and high quality food supply at a reasonable cost. Taxpayers want low government payments and low administrative burdens. Because basic conflicts of wants and values exist among these groups, any legislation must reflect a compromise among these conflicting forces—a compromise that changes as groups change their values or as political power shifts.

# COMMON AGRICULTURAL POLICY IN THE EUROPEAN COMMUNITY _____

All developed countries attempt to protect their agricultural producers from the vagaries of the world market, and the European Community (EC) is no exception. The intended protection is both political and economic. It is political, both in that the EC is attempting to attain a high degree of self-sufficiency and that government officials act in a manner conducive to reelection. It is economic in that they wish to maintain their farming culture.

Even though the EC was established in 1958, it was not until 1962 that the first Common Agricultural Policy (CAP) regulations were introduced. The first CAP regulations covered grain, poultry, pork, eggs, fruits, and vegetables. Regulations have been developed since that time for beef, milk, rice, fats and oils, sugar, tobacco, hops, seed, flax, silk, and fish. The three fundamental components of CAP are common pricing, community preference, and common financing.[16]

Common pricing in the EC can be explained using grains as an example. To promote intracommunity sales a *target price* is fixed in the major deficit consuming areas. For grains, the location used is Duisburg, Germany. Around this target price, small variations are permitted so that the open market forces can operate. These variations around the target

---

[16] John F. Hudson, "The Common Agricultural Policy of the European Community," Foreign Agricultural Service, USDA, FAS M-255, November 1973.

prices are called *intervention prices*. The government maintains the market floor by purchasing grain offered to them at the intervention price. A market ceiling price is maintained by the government through the selling of grain stocks. For example, assume an August 1 target price of $6.00 per bushel for wheat and a minimum intervention price of $4.90. The *threshold price* would be the floor intervention price minus the cost of transportation from the port of entry to the major consuming areas. If this cost was 30 cents per bushel the threshold price would be $4.60 per bushel. This is the price of wheat at Rotterdam (the EC central grain market) plus transportation from there to Duisburg. This price is the minimum import price the EC could offer and maintain the intervention price in the major consuming areas. The difference between the import price (cost, insurance, and freight—CIF) and the threshold price is the *variable levy* that is used to protect the intervention price. This variable levy is a tax that varies with changes in the CIF import price or the target price of grains. In this way, the EC eliminates both price and quality competition from other grain exporting countries.

Community preference is accomplished through the use of variable levies and export subsidies. Because the variable levies provide a wall of protection for EC producers, imports can be restricted to any amount and quality desired. Since there are no production restrictions in the CAP, surpluses are produced and grains and dairy products are subsidized so that they move into the world market. Also, subsidies are used on rice, olive oil, sugar, beef, pork, poultry, fruits and vegetables, and processed foods.

All countries in the EC are responsible for the financing of the CAP. The cost of agricultural support payments are met through the European Agricultural Guidance and Guarantee Fund that was established in 1962. The expenditures from this fund account for most of the EC budget. At present most of the revenue for agricultural support payments is coming from direct contributions from member states, custom duties, and levies on agricultural imports.

# THE PROCESS OF POLICY DEVELOPMENT _____

Anyone attempting to solve a problem systematically will tend to use the same method, which we call the **scientific method**. This process can be broken down into at least six steps: (1) *recognizing and defining the problem,* (2) *outlining the issues,* (3) *developing alternative solutions,* (4) *choosing a policy solution,* (5) *putting the policy into effect,* and (6) *appraising its effectiveness.*

Before one tries to solve a problem it is necessary to evaluate the facts so that everyone has the same feeling for the situation. The historical perspective gives the discussant some idea of what the problem is all about. A problem exists when people are troubled by a situation or when circumstances deviate from a norm or standard. These norms or standards may be viewed in terms of one's own values or goals. For example, a problem in the U.S. dairy industry is the large amount of dairy imports from the EC. These imports increase the supply of dairy products in the United States and reduce prices and incomes to U.S. dairy producers.

Once the problem is understood it is imperative to define what is involved in the controversy. This is called outlining the issues. Some of the issues in the dairy problem cited above are (1) the EC is merchandising many of its products in the North Atlantic market, (2) they export these products by using export subsidies in order to compete, and (3) export subsidies violate the General Agreement on Tariffs and Trade.

In order for people to have a choice, more than one alternative solution must be developed and presented. Problems can be solved in many ways. The method selected will depend on the solution that fits the beliefs and values of the person or group adopting it.

Some alternative solutions to the dairy problem might be: (1) impose a tariff (tax) on EC dairy products so they will sell for the same price as U.S. dairy products, (2) put a quota limiting the quantity that can be imported so that the amount by which prices of dairy products decrease in the North Atlantic market is reduced, or (3) allow dairy imports into the United States, but subsidize domestic producers.

No solution is going to fit everyone's needs. Usually a compromise is reached after all groups involved have been consulted. The solution picked may be a single solution or a combination of several. Policy solutions are adopted in the political process. This open process places a responsibility on interested persons to express ideas to friends, special interest pressure groups, and elected officials. Pressure groups are people with common interests working together and thus presenting common aims with a stronger voice.

Once a solution to the problem is selected it must be put into effect. Normally, agricultural policy is directed to the Agricultural Stabilization and Conservation Service, USDA, for implementation. The service performs the administrative task of interpreting the legislation and informing producers of specific conditions of the legislation.

Finally, the policy must be appraised as to its effectiveness. Just as the world changes, so does the need to modify and adapt to new situations. What was needed yesterday may not be needed tomorrow. Therefore, policies must be reviewed periodically to ensure that they meet the conditions for which they were intended.

# SUMMARY

Federal government programs in U.S. agriculture have been in effect since 1929. They have been used to support prices and incomes of farmers producing grain, cotton, tobacco, peanuts, sheep and lambs, milk, and sugar.

The major policy goals of agriculture include maintaining the family farm, parity, and bargaining power for agriculture. Almost all agricultural legislation mentions these goals. These policy goals are consistent with farm values.

Overcapacity, price instability, and poverty are the primary agricultural problems. Agricultural policies have been enacted to deal with these problems. These policies have included programs that expanded demand for agricultural products, retired land from production, or reduced the amount of product that could be produced and sold under marketing quotas. The Food, Agriculture, Conservation, and Trade Act of 1990 is a policy that uses direct deficiency payments to producers with provisions for some acreage set-aside and a land retirement program. The Federal Agricultural Improvement and Reform (FAIR) Act adopted in 1996 has discontinued large price and deficiency payments, but has continued loan protection and will provide seven years of contract payments to ease the burden of adjusting to reduced price protection in the new farm program.

Agricultural programs have had only short-run impacts on increasing farm incomes. In the longer run, these income benefits are capitalized into land values. Thus, policy benefits in the long run are captured by present landowners.

The process of policy development is a decision-making process. The steps involved are (1) recognizing and defining the problem, (2) outlining the issues, (3) developing alternative solutions, (4) choosing a policy solution, (5) putting the policy into effect; and (6) appraising its effectiveness.

# CHAPTER HIGHLIGHTS

1. Government farm programs have been in existence since 1929.
2. Enactment of any public policy requires the resolution of conflicts in basic goals and values.
3. Agricultural fundamentalism has provided the impetus for much U.S. agricultural policy in the past.
4. Basic policy goals of agriculture are (1) family farms, (2) parity, and (3) equality of bargaining power.

5. Agriculture has at least three farm problems (1) overcapacity, (2) instability of farm prices, and (3) poverty.

6. The demand for many agriculture commodities is inelastic, but is becoming more elastic as the export market grows.

7. Historically, agricultural policy has revolved around three types of economic programs: (1) two-price plans, (2) land retirement programs, and (3) direct payment programs. All these programs were enacted to increase farm incomes.

8. Farm programs have not increased farm income in the long run because government support payments become capitalized into land values.

9. Benefits of past farm programs have included (1) reducing the variability of farm income, (2) easing the outmigration of excess labor, and (3) providing food stocks for emergencies.

10. The U.S. has many programs to encourage domestic food consumption. The Food Stamp Program is the major program, and 27 million Americans participate in it.

11. The primary program to increase foreign demand for U.S. agricultural commodities is Public Law 480 that was passed in 1954.

12. There is an infinite number of possible ways to solve our farm problems. Conflicts in goals and values keep us from adopting an effective long-run solution.

13. All countries of the world have some sort of farm programs to protect their farmers from very low prices and incomes.

14. The European Community's agricultural policy is a variable levy system designed to support domestic farm prices and control imports of farm products.

15. The scientific method is useful in finding a solution to any problem.

16. Agricultural policies may need to change from time to time to meet the cyclical nature of agricultural production.

# KEY TERMS AND CONCEPTS TO REMEMBER

| | |
|---|---|
| Parity | Quality of life |
| Poverty level | Scientific method |
| Public policy | Values |

# REVIEW QUESTIONS

1. What are values? Do the values of farm people influence agricultural policy?
2. Is the agricultural creed relevant today? What changes would you make in the creed in order to update it?
3. Define parity. Is parity price a useful device for measuring the "well being" of farmers and ranchers?
4. What are three of American agriculture's major problems? How have we attempted to solve these problems?
5. What is a two-price plan? Does this type of program take advantage of different elasticities of demand in the domestic and foreign market so as to increase total revenue to agricultural producers? Explain.
6. Explain a direct payments program. Does the price elasticity of demand for the commodity affect the amount paid by consumers and the government?
7. Have agricultural programs increased farm income? What has happened to land values? Discuss.
8. What is the Common Agricultural Policy? How does it differ from a direct payments type of program?
9. Explain the steps in the policy development process.

# SUGGESTED READINGS

*Agricultural Outlook,* Provisions of the 1996 Farm Bill, Washington, D.C.: Economic Research Service U.S. Department of Agriculture, April 1996.

*Food and Agricultural Policy,* Washington, D.C.: American Enterprise Institute for Public Policy Research, 1977.

Halcrow, Harold G. *Food Policy for America,* New York: McGraw-Hill, 1977, Chapters 5, 8, and 11.

Hathaway, Dale E. *Government and Agriculture: Public Policy in a Democratic Society,* New York: Macmillan, 1963, Chapters 1–3.

Office of Publishing and Visual Communication, "1990 Farm Bill—Proposal of the Administration," U.S. Department of Agriculture, Washington, D.C.

*Provisions of the Food, Agriculture, Conservation, and Trade Act of 1990,* Washington, D.C.: Economic Research Service, U.S. Department of Agriculture.

Schickele, Rainer. *Agricultural Policy: Farm Programs and National Welfare,* Lincoln, Neb.: University of Nebraska Press, 1954, Chapters 1–5.

Shepherd, Geoffrey S. *Farm Policy: New Directions,* Ames, Iowa: The Iowa State University Press, 1964, Chapters 7–13.

Spitze, R. G. F., and Marshall A. Martin (eds.). "Analysis of Food and Agricultural Policies for the Eighties," North Central Regional Research Publication No. 271, Urbana-Champaign, Ill.: University of Illinois, November 1980.

Tweeten, Luther G. *Foundations of Farm Policy*, 2nd ed. Lincoln, Neb.: University of Nebraska Press, 1979, Chapters 3, 6, 10, and 11.

# AN OUTSTANDING CONTRIBUTOR

**Luther G. Tweeten**   Luther Tweeten is currently Anderson Professor of Agricultural Economics in the Department of Agricultural Economics at Ohio State University, an appointment he has held since January 1987.

He was born in Las Vegas, New Mexico, and grew up on a farm in northern Iowa. After attending Waldorf College in Forest City, Iowa, for 2 years, he entered Iowa State University and received a B.S. degree in Agricultural Education in 1954. Following two years of service as Special Agent in the Army Counter-Intelligence Corps, he entered Oklahoma State University where he earned his M.S. degree in 1958. He returned to Iowa State University where he received a Ph.D. degree in Agricultural Economics in 1962.

Dr. Tweeten was appointed assistant professor of Agricultural Economics at Oklahoma State University in 1962. He advanced to associate professor in 1963, to professor in 1965, and regents professor of Agricultural Economics in 1972. He was visiting professor at Stanford University in the 1966–67 academic year and at the Institute for Research on Poverty at the University of Wisconsin in 1972-73.

Tweeten's research contributions include empirical estimates of the elasticity of supply of farm output based on input demand functions, the elasticity of demand for farm output including export demand, as well as estimates of inflation pass-through and the marginal utility of income. Examples of his conceptual contributions include a general theory of economic stagnation, the decreasing-cost and cash-flow theories of chronic low rates of return to farming industry resources, the role of attitudes and income distribution in economic welfare, and the implications for the agricultural industry of inflation.

Tweeten has taught courses in research methodology, mathematical economics, rural development, and farm policy. He is a member of six professional societies and three honorary fraternities.

In 1964, Tweeten was granted an award by the American Agricultural Economics Association to attend the International Agricultural Economics Association meeting at Lyon, France. He received the Distinguished Alumni Award from Waldorf College in 1970. The Western Agricultural Economics Association rewarded him with their Outstanding Published Research Award in 1972. He served as president (1981) of the American Agricultural Economics Association and was elected a fellow by that association in 1983. Tweeten has been called to serve as an expert witness before committees of the U.S. Congress numerous times.

He is married to the former Eloyce Hugelen, and they have four children.

# CHAPTER 12

# Marketing Agricultural Commodities

$\mathbf{A}$gricultural marketing is a large and important discipline in agricultural economics. In 1994, the cost of marketing U.S. domestic food amounted to $401 billion, while the cost of producing farm food products was $110 billion. Of the $511 billion spent by consumers for food products, less than one-fourth was returned to producers while over three-fourths of that amount was for marketing costs.

Marketing costs include those associated with assembly, transportation, processing, and distribution of farm food to consumers. The cost components of marketing (sometimes called the "marketing bill" or "marketing margin") are shown in Figure 12-1 as percentages of the total spent for food. The figure indicates that the major components are labor (accounting for 37 percent of the total), farm value (accounting for 21.5 percent), packaging (accounting for 8.2 percent), and transportation (amounting to 4.3 percent). Increases in the marketing bill have been due primarily to two factors: (1) greater volume of food marketed, and (2) increases in the cost of providing these services. Over the last decade the cost of marketing services has accounted for most of the rise in the marketing bill. The rise in the marketing bill is caused by increased labor costs and by higher prices for inputs purchased by the food industry from nonfarm sources.

## WHAT IS MARKETING? _____

The term *marketing* has a variety of meanings. To some shoppers it means purchasing groceries and all other household needs. From the point of view of farmers or ranchers it means selling their commodities. From the perspective of the handler of a commodity, it means storing the commodity, transforming the product into a form that consumers want, shipping it to retail outlets, and promoting its sale. All these activities are part of the marketing process.

The American Marketing Association has defined marketing as the performance of business activities that direct the flow of goods and services from producer to consumer or final user. In agricultural marketing, the point of production (the farm or ranch) is the basic source of supply. The marketing process begins at that point and continues until a consumer buys the product at the retail counter or until it is purchased as a raw material for another production phase. However, marketing also includes input supply firms that serve the farms and ranches. Thus, marketing consists of those efforts that effect *transfer of ownership* and that create *time, place,* and *form utility* to commodities. Time utility is added to

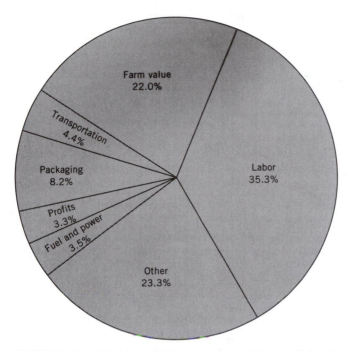

**FIGURE 12-1** Estimated components of 1994 retail food prices. (*Source*: Economic Research Service, USDA.)

commodities by storage. Place utility is added to commodities through transportation services. Finally, form utility is added to a commodity through the processing function. By the creation of these utilities, marketers are productive and add value to raw agricultural commodities that consumers want.

Because consumption is the purpose and end result of production and marketing activities, it is necessary for marketers to focus their activities toward satisfying consumer wants and needs. It is difficult to successfully market something consumers do not desire, even with massive promotional endeavors.

# HOW MARKETING DEVELOPED _____

The existence of marketing is a direct result of specialization of production in our economy. Initially, most families were self-sufficient, or nearly so. They produced most of the products they needed on their small acreages. They ground their own flour and baked their own bread, spun their fibers, butchered their meat, then cured and stored it for later use. As time passed, people discovered that their different resource endowments and

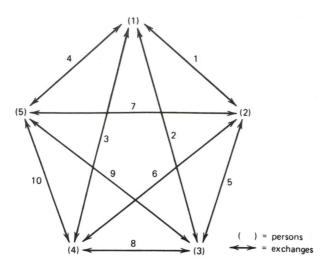

**FIGURE 12-2**   Bartering transactions among five specialized producers.

talents allowed them to produce some things better than others. Along with this realization came increasing demands for various goods and services as population grew and society became more affluent. Increased demands made specialization profitable. As the individual farmer specialized, a surplus was produced that could be exchanged for goods and services that no longer were produced or provided on the farm. As the *law of comparative advantage* states, it is beneficial to the producer to specialize in the production of the good that person can produce more cheaply and then exchange the surplus output for the surplus output of other producers. In this way, producers and society as a whole benefit because more commodities are available at lower costs.

In the early days, money was not used as a medium of exchange. Barter—exchanging goods for other goods—served society well. For example, a farmer might exchange a sack of potatoes for a wheelbarrow of coal. As people specialized and produced fewer products, it became more difficult to meet with other producers of each of the goods needed; bartering became too costly. Let's illustrate the problems of a barter system using an example of five persons, each specializing in producing different goods needed by each of the others. Ten separate exchanges must be made in order that each of the five producers obtains the five products involved (Figure 12-2). If 20 people exchange 20 different goods, then 190 separate exchanges are necessary. Thus, increased specialization brings about money as a medium of exchange in order to simplify the exchange process. Marketing is the direct result of specialization and trade.

# MARKETS AND THE MARKET ECONOMY _____

In any economic system, regardless of the type of political or social structure, there are four basic decisions that must be made. The system must somehow determine (1) what goods and services are to be produced and in what quantities, (2) how to allocate available resources (the inputs of land, labor, capital, and management) to obtain the largest output or national product, (3) what production methods should be used, and (4) how national output should be divided among the population. In most capitalist countries these decisions are made through an intricate system of market prices that are reflected through the marketing system from consumers to producers. Before going into detail on how this is accomplished, a definition of a market is essential.

A *market* consists of buyers and sellers with facilities to communicate with each other. It need not be a specific place, although some people refer to markets in this sense, such as commodity markets and auction markets. Markets may be local, regional, national, or international. The only requirement is that the forces of demand and supply, via communication between buyers and sellers, determine market price.

In a market economy, every scarce commodity commands a price, and that price is market determined by the product's demand and supply curves. For example, examine the consequences of an increase in the demand for beef in America. When consumers go to their grocery stores and purchase more beef, they indicate to the grocer that they prefer that product over other goods their money could have bought. These dollar "votes" are cast when consumers purchase the available beef. The grocers must then purchase more beef from the packers. The packers need more beef to supply retailers' increased needs, so they buy more from the feedlots, and the feedlots need more animals, so stock ranchers increase the size of their breeding herd, and so on. This sequence of market relationships, as sketched in Figure 12-3, is much abbreviated as you might easily recognize. A large number of other suppliers also detect and respond to changes in the demand for their products or services. As consumers purchase more meat, the demand curve for meat shifts, which increases the price of meat at the retail level. This higher price is noticed by meat packers as their orders increase. Consequently, packers demand more slaughter cattle and offer higher prices. This signal is passed back through the market to feedlots, feed-grain producers, and cow-calf operators. In a competitive system, producers increase output in response to higher prices because they can improve their earnings by doing so. It is the profit motive that makes the market system work.

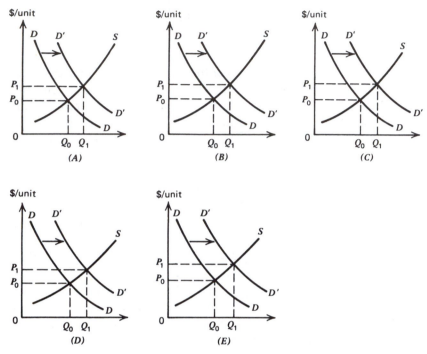

**FIGURE 12-3**   Market interrelationships for beef. (*A*) Retail. (*B*) Packers. (*C*) Feedlots. (*D*) Cow-calf operators. (*E*) Feed grains.

It is easy to see how our economy answers the question of what and how much is produced—it is determined by profitability. Firms will produce those goods from which they can make a profit. They will not produce a commodity if their information indicates they will lose money producing that good.

A competitive market system allocates resources to produce a good at its lowest cost. A producer who does not do so will find earnings declining, even to the point of being forced out of business by competitors. When people sell the services of their labor or other resources, they receive compensation in the form of wages or salaries, rent, interest, or income in general. With this income, they purchase goods and services. In a free–enterprise system the allocation of resources to production determines the distribution of income, which in turn determines how much of the total output each member of society will receive. This system is very complex, but it does work well and without the planning that occurs in such command economies as the former Soviet Union and China.

## The Marketing Process

Success by individuals and firms in achieving their operational goals contributes to success in society's larger goal of economic efficiency in the total system. Marketers must accomplish many specific tasks, all of which are contained in two general aims. One requirement they must fulfill is to determine demand and changes in demand for products in order that the movement of commodities through the marketing system may be expedited. The second is to achieve efficiency in the marketing process by improving pricing and operational efficiency.

If consumers' want change, the resulting changes in demand require that management make the necessary decisions regarding business investments, operational procedures, and products handled so that future consumer desires will be met. Fluctuations in demand can drastically affect the profitability of an operation, making it necessary to anticipate those changes.

A first step in the marketing process is to ascertain what consumers want. To estimate sales, a firm must consider such sales-influencing factors as general economic conditions, population in the market area, incomes, the price of the product, and the prices of substitute or complementary products. Another factor affecting sales is the amount of advertising and promotional effort used in persuading consumers to want the product. Estimates of probable sales may then be compared with the firm's anticipated costs of marketing the product to derive an estimate of profitability, the "bottom line" in deciding for or against the project.

## The Efficiency of Markets

Although we moved toward answering questions of economic efficiency by applying concepts of production and cost economics to commodity-producing firms, economic analysis is not so severely limited. Identical product and resource relationships also exist in firms involved in the marketing process, and the methods of analysis are equally appropriate here.

*Marketing efficiency* is measured by comparing output and input values. Output values are based on consumer valuation of a good, and input values (costs) are determined by the values of alternative production capabilities. Therefore, markets are efficient when the ratio of the value of output to the value of input throughout the marketing system is maximized.

Any change that reduces marketing costs per unit of output is desirable, but if it also reduces consumer satisfaction it may not be an improvement in marketing efficiency. Conversely, a cost increase need not cause efficiency to fall if the value of output (consumer benefits) has increased as much or more than costs.

Because consumer satisfaction cannot be measured directly, changes are analyzed in terms of "technical" efficiency and "pricing" efficiency.

**Technical efficiency** is concerned with the manner in which physical marketing functions are performed to achieve maximum output per unit of input. Technological changes can be evaluated to determine whether they will reduce marketing costs per unit of output. For example, it is relatively simple to investigate the costs of a new apple-packing machine that may handle more apples per day while also reducing labor costs.

**Pricing efficiency** is concerned with the accuracy, precision, and speed with which prices reflect consumer demands and are passed back through the market channels to producers. Pricing efficiency is thus affected by rigidity of marketing costs and the nature and degree of competition in the industry. Activities that may improve pricing efficiency are improvements in market news and information and competition.

The primary reason for firms to increase their marketing efficiency is the expected income improvement; for society, the basic goal of economic efficiency requires marketing efficiency.

# APPROACHES TO THE STUDY OF MARKETING _____

The study of marketing involves various approaches. The most common are the functional, the institutional, and the market structure approaches.

## Functional Approach

The functional approach studies marketing in terms of the many activities that are performed in getting farm products from the producer to the consumer. These activities are called *functions*. They are performed by co-operative and private marketing firms.

Using the functional approach, it is feasible to "cost" these functions and to compare them against others doing the same job or against standards of performance. The following functions are widely accepted for classification purposes: (1) exchange—buying and selling; (2) physical—processing, storage, and transportation; and (3) facilitating—standardization, financing, risk bearing, and market information. Most of these functions are performed in the marketing of nearly all commodities.

*Exchange functions* take place throughout the market channel and include buyers bidding for the supplies of commodities and sellers offering commodities at the best price they think they can attain. The buying function also includes locating supplies of the commodity and assembling them for shipment. The selling function can vary, depending on what stage of the market channel the product is in. It involves packaging, labeling, advertising, promotion and all other merchandising activities.

*Physical functions* add form, time, and place utility (value) to commodities. Processing adds *form utility* to a product by taking, for instance, live beef and transforming it into T-bone steaks, liver, chuck roasts, and so on. In agricultural marketing the processing function encompasses all those manufacturing activities that require agricultural products as raw materials.

Storage adds *time utility* to a product by holding it from harvest or production and distributing it on the market over time as it is needed. For example, flour millers maintain large stocks of grain and producers carry inventories of spare parts for machines and of gasoline and oil for their equipment. This storage function occurs at all levels in the marketing channel.

Transportation adds *place utility* to a commodity. The fact that oranges are grown in Florida and California is of little value to consumers in New York or Wyoming if the oranges are not shipped from producing to consuming areas. Transportation includes moving commodities from the farm to processing or wholesaling facilities and from these facilities to their final destination. Costs incurred in the preparation of goods for shipment and in the loading of goods can be a significant portion of total transportation costs.

*Facilitating functions* improve the performance of the marketing system by increasing operational and pricing efficiency.

Standardization is the establishing of grades and of quality criteria for a commodity. This function makes it possible for buyers to know exactly what they are buying without personal inspection of the goods. Through standardization, buyers know, for instance, what is dark northern spring wheat of ordinary protein. Therefore, the costs of exchange activities are greatly reduced because buyers do not have to travel to look at a specific bin of wheat, or to telephone to receive a verbal description of it. When standardization is employed, the accepted quality characteristics must be enforced. This enforcement is usually the job of an agency of the state or federal government, so that the standards can be applied impartially.

Financing is necessary throughout the marketing process because someone must own a commodity as it moves through the marketing stages. And there is a lag between the time someone buys and sells a commodity. Money tied up in commodities is money that could be in other investments, therefore, interest foregone is a real cost of financing the purchasing and storing of a commodity. If the marketer borrowed the funds to purchase and hold commodities, that person's financing cost is the interest paid on the borrowed funds.

The risk-bearing function falls on the commodity owner who is faced with possible losses due to physical or market risks. Physical risk is the risk of loss due to quality deterioration or destruction. Physical losses can be caused by moisture, heat, wind, fire, hail, and so on, and can be reduced or eliminated through the use of insurance (such as the federal disaster payments or coverage by insurance companies). Whether or not one buys insurance depends on the probability of physical loss and the cost of insurance. Market risk is the risk, borne by commodity owners, of a possible adverse price movement.

Farmers' and merchandisers' losses due to price movements, which occur while holding inventories, can be reduced by using the futures market (discussed later in this chapter). The futures market allows them to hedge their cash position, reducing price risk.

Market information involves collecting, analyzing, and disseminating information. In the USDA most data are collected by the Economic Research Service, the Statistical Reporting Service, and the Consumer and Marketing Service. Market information is necessary for the smooth operation of the price system. If buyers and sellers are well informed about the factors that affect supply and demand, prices will be established that more nearly clear the market.

## Institutional Approach

The institutional approach examines the activities of business organizations or people involved in marketing. These middlemen can be classified as follows: (1) merchant middleman (retailers and wholesalers), (2) agent middlemen (brokers and commissionmen), (3) speculative middlemen, (4) facilitative organizations, and (5) food processors.

Middlemen perform the operations necessary to transfer goods from the producer to the consumer, because of the benefit of specialization and scale that exist in marketing as well as in production. Rather than producers conducting all the marketing functions, middlemen, through specialization and division of labor, reduce total distribution costs.

Retail organizations usually purchase products for resale to consumers. In the food business, grocery stores are an example of a retail operation. These retail *merchant middlemen* purchase from wholesalers, taking title to the products they handle. Wholesalers buy commodities from processors and sell to industrial users or to retailers. There are different types of wholesalers, but their primary function is to hold inventories, package products in lots that meet consumer needs, prepare lots for shipment, and make arrangements for transportation. Wholesalers also provide credit to retailers, offer merchandising assistance, and assume some of the risk the retailer would have to take if the retailer purchased directly from manufacturers.

*Agent middlemen* can be distinguished from merchant middlemen in that they do not take title to goods. These agents engage in negotiations that transfer the title of products from seller to buyer. To accomplish this task successfully, they must have a special knowledge of the product and the markets they serve.

Brokers perform the duty of bringing buyers and sellers together. The broker may represent either side of a sale, but usually represents the seller. The broker's fee ordinarily is based on the amount of the sale. Brokers represent grain or livestock firms, food processors, fruit and vegetable shippers, and other producers and handlers of agricultural commodities.

Commissionmen for agricultural producers are primarily interested in grain, livestock, and fresh fruits and vegetables. They have more authority than brokers in the selling of products. Any product consigned to them, is sold at the best price available. They collect on the sale of the product, deduct their expenses, and remit the balance to the seller. With more decentralization in the marketing of agricultural commodities (less going through terminal or central markets) the use of brokers and commissionmen has declined. Much of the decentralization has been a result of direct buying by retail chains and processors from large country buyer-suppliers.

*Speculative middlemen* take ownership of and hold commodities, thus assuming the risk of loss due to unfavorable price fluctuations. They take ownership of the commodity and attempt to make a profit from uncertain price movements. As an example, some speculative middle.-men are called scalpers, daytraders, or floor brokers on commodity exchanges.

*Facilitative organizations* such as trade associations, are institutions that provide general industry data, or that provide physical facilities for marketing such as grain exchanges or auction yards. These organizations may also guide the rules of trading so that competitive pricing results.

*Processors* transform raw agricultural products into different final products. Almost all agricultural products are processed to some degree between production and consumption.

## Market Structure Approach

Market structure analysis emphasizes the nature of market competition and attempts to relate the variables of market performance to types of market structure and conduct. Market structure is a description of the number and nature of participants in a market. Examples of such dimensions include the number and size distribution of buyers and sellers in the market, the degree of product differentiation, and the barriers to potential entrants.

Market conduct deals with the behavior of firms. Firms that are price searchers are expected to act differently from those in a price-taker type

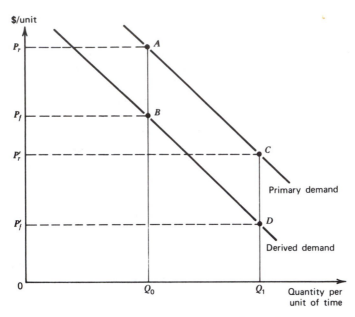

**FIGURE 12-4** Retail and farm demand relationships.

of industry. Price searchers can determine their selling prices or the quantity of output they will sell. In addition, they can use their market power to weaken or eliminate competitors. Market performance is a reflection of the impact of structure and conduct on product prices, costs, and the volume and quality of output. If the market structure in an industry resembles monopoly (one seller, few substitute products, barriers to entry) rather than pure competition, then one can expect poor market performance.

## MARKETING MARGINS _____

The difference between the price that consumers pay for the final good and the price received by producers for the raw product represents marketing costs, or the *marketing margin*. Consumer demand for a product is called *primary demand*. The demand for a product at the farm level is a *derived demand*, meaning it is derived from consumer demand. In Figure 12-4, the vertical distance A-B is the marketing margin, the difference between the retail price ($P_r$) and the farm price ($P_f$) for a given quantity ($Q_0$) marketed. If output increases to $Q_1$ the marketing margin is C-D, which is the difference between the retail price ($P_r$) and the farm price ($P'_f$).

Note that with the marketing margin constant, the vertical distance between the two demand curves remains unchanged. If the marketing mar-

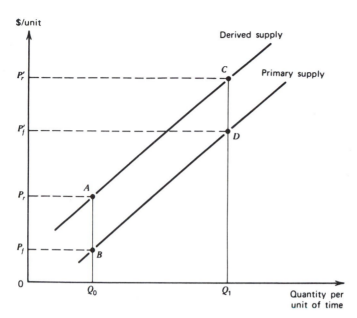

$/unit

**Derived supply**

C

**Primary supply**

$P'_r$

$P'_f$

D

A

$P_r$

$P_f$

B

0

$Q_0$

$Q_1$

Quantity per
unit of time

**FIGURE 12-5**   Retail and farm supply relationships.

gin were to increase with greater output and volume handled, C-D would
be greater than A-B; if the marketing margin declines with increased volume, C-D would be less than A-B.

On the supply side, the supply curve at the farm level is *primary,* since
all food production is based on farm raw products. The *derived supply*
curve is at other market levels and is primary supply plus the marketing
margin, as shown in Figure 12-5. If marketing costs are a constant amount,
then the marketing margins are shown at output level $Q_0$ as A-B and at
output level $Q_1$ as C-D, and A-B equals C-D.

Figures 12-4 and 12-5 have been combined to produce Figure 12-6. The
derived supply curve and primary demand curve determine the *retail*
price. The primary supply curve and the derived demand curve determine the *farm* price. The difference between the retail price and the farm
price is the marketing margin. Under competitive conditions the effect of
changes in marketing costs on retail prices and farm prices can be determined.

Suppose, for instance, an improvement in transportation technology
causes transportation costs to decline. The effect of this decrease will be
a reduction in the marketing margin. On the graph, this reduction trans-
lates into a downward shift in the derived supply curve and an upward
shift in the derived demand curve. Thus, retail price falls and farm price
increases result in a smaller marketing margin.

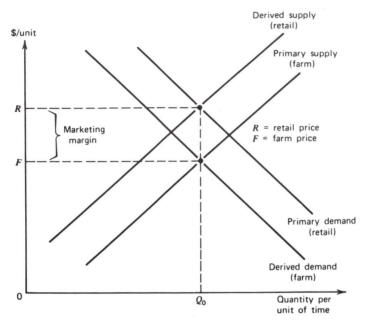

**FIGURE 12-6**   Market determination of farm and retail prices.

Statistics showing the farmers' share of the consumers' food dollar are based on an aggregate market basket of food. The farmers' share of the cost of that basket has varied over time, amounting to 47 percent in 1950, 39 percent in both 1960 and 1970, and 24 percent in 1995. Another way to state these statistics is to say that in 1995 farmers received 24 cents out of every dollar the consumer spent for food and the marketers received 76 cents.

The share of the consumers food dollar varies widely by food group. For every dollar spent by consumers in 1995 for poultry, the farmer received 42 cents; for meat products, 35 cents; for dairy products, 33 cents; for fresh fruits, 19 cents; fresh vegetables, 23 cents; and for bakery and cereal products, only 7 cents. There is a tendency for the farmers' share of the consumer's dollar to move up and down with commodity prices because marketing margins are relatively inflexible.

Many people look at the farmers' share data and think the marketers are taking financial advantage of producers. Before such a statement can be made, one must analyze the marketing activities performed and their costs. The farmers' share of the food dollar will normally reflect the amount of processing necessary, the perishability of the product, the bulkiness of the product relative to its value, and the seasonal nature of production.

# THE FUTURES MARKET _____

*Futures markets* such as the Minneapolis Grain Exchange, the Chicago Board of Trade, and the Chicago Mercantile Exchange operate to facilitate the marketing of many agricultural products. These exchanges have been in existence since the mid-1800s, and their purpose is to provide a place in which the activities of buyers and sellers determine the prices of commodities. At these exchanges traders buy and sell *futures contracts*. Futures contracts are written documents calling for the future delivery of a commodity of a specific grade at a particular time and place. All terms and conditions are specified in the contract and are standardized by the futures exchange. Commodity futures contracts are bought and sold as if they were physical commodities. These futures contracts are dated for the month of expiration, or when delivery must take place.

## Beef Futures: An Example

In 1964, the Chicago Mercantile Exchange established live beef futures. The beef cattle futures expires around the twentieth day of the following months: February, April, June, August, October, and December.

Trading in a contract begins about a year in advance of the delivery (or expiration) date. For example, the buying and selling of the December 1995 contract started in November 1994. It terminated about the end of the third week in December 1995. Trading contracts with different delivery months can be compared to buying and selling in different markets. If a person buys a June contract, that buyer must meet the requirements of the contract by either accepting delivery immediately after June 20 or selling a June contract sometime before the June contract expires. Buyers and sellers need not be concerned about the other party to the contract because the particular exchange is responsible for clearing the transaction through its clearinghouse.

Chicago Merchantile Exchange rules call for a minimum delivery unit of 40,000 pounds of yield grade 1, 2, 3, or 4 choice live steers, averaging between 1050 and 1200 pounds, with no individual steer weighing more than 100 pounds above or below the average weight for the unit. An animal weighing less than 950 pounds or more than 1300 pounds cannot be delivered. Delivery units must contain average weights between 1050 and 1125.5 pounds and have an estimated average hot yield of 62 percent, or between 1125.6 and 1200 pounds with an estimated hot yield of 63 percent.[1] Other yields and weights are acceptable at standard discounts. This *contract unit* represents approximately 35 to 40 head of fat cattle. Delivery on a live beef futures contract is possible at Omaha, Nebraska; Peoria

---

[1] "Hot yield" is the carcass weight remaining after slaughter.

**TABLE 12-1**   June 22 Futures Price Quotations, Live Steers

| Month | Open | High | Low | Settle | Change |
|---|---|---|---|---|---|
| August 1990 | 73.87 | 73.97 | 73.25 | 73.27 | –.78 |
| October 1990 | 75.42 | 75.45 | 74.75 | 74.87 | –.73 |
| December 1990 | 75.50 | 75.50 | 74.80 | 74.85 | –.62 |
| February 1991 | 75.00 | 75.05 | 74.60 | 74.65 | –.45 |
| April 1991 | 76.00 | 76.10 | 75.60 | 75.70 | –.50 |
| June 1991 | 73.00 | 73.00 | 72.70 | 72.77 | –.30 |

and Joliet, Illinois; and Sioux City, Iowa. Deliveries may be made from Guymon, Oklahoma, at a discount of $0.50 per cwt.

When persons who are not members of an exchange buy or sell futures contracts they pay a broker who acts as their agent. A "round turn" is one purchase and one sale (or vice versa) of a specific contract, such as the purchase and later sale of a June 1995 contract.

### Futures Price Quotations

A futures market generates prices. If competitive conditions exist, the futures market generates prices reflecting today's best available estimates of the prices to be received during the respective delivery months. However, as new information arrives minute to minute, this estimate will change too, as you notice every day in the futures price quotations in your local newspaper. The open, high, low, settle, and change are given for each contract. An example of the price information that is released after each market day is illustrated in Table 12-1 for June 22.

In these price quotations, the first trade of the day (for June 22) on an August contract opened at 73.87 cents per pound. The highest trading price for an August contract during the day was 73.97 cents, and the lowest price was 73.25 cents per lb. The last trade of the day (the settle or closing price) for an August contract was 73.27, and the change from the previous close (the last trade made on June 21) for an August 1990 contract was down 0.78 cents per pound.

## THE HEDGING PROCESS _____

*Hedging* is one way an individual or firm could use the futures market.[2] In the standard example of "insurance hedging," the hedger establishes

---

[2] Robert F. Bucher and Gail L. Cramer, "Beef Cattle Futures: A Marketing Management Tool," Montana Agricultural Experiment Station Bulletin 663, Montana State University, Bozeman, August 1972.

a position in the futures market that is opposite from the position held in the cash (product) market. To illustrate hedging, assume that a cattle feeder with 40 head of steers determines in October that a selling price of $67 per cwt in the following June is needed to cover all costs, including a normal rate of return on the investment. If the feeder is fearful that the market price will be less than $67 at the time at which the steers are ready to be sold, that person may wish to sell a futures contract provided it can be sold at or above $67 per cwt. If the feeder sold steers in June for $63 per cwt, and if the futures and cash market prices have moved up or down together, that person would also be able to buy back the futures contract at the same time at $63 per cwt. Thus, the feeder would make $4.00 per cwt on the futures transaction which would balance the $4.00 per cwt loss suffered on feeding operations. This situation is illustrated in the following example.

|  | Cash Market | Futures Market (June Contract) |
|---|---|---|
| October | Desired selling value | Sell |
|  | 1000 # @ $67/cwt = $670 | 1000 # @ $67/cwt = $670 |
| June | Sell | Buy |
|  | 1000 # @ $63/cwt = $630 | 1000 # @ $63/cwt = $630 |
|  | –$ 40 | +$ 40 |

The cattle feeder ended up with $670 for each steer, $630 for the sale of the steer and $40 earned from the futures transaction. This example is what is called a perfect hedge and is a simple view of hedging.

Usually a feeder selling a futures contract as a hedge against a price decline on cattle in the feedlot will sell cattle on the local market and buy back futures at the same time. If the futures price is above the local cash price by more than the cost of making delivery, the feeder can deliver cattle by shipping them to Omaha, or sell the cattle at home and buy cattle in Omaha for delivery on the futures contract.

In actual practice, feeders far from the delivery point do not deliver their own cattle on the futures contract. They buy back the futures contract and sell their cattle at home, or wherever the return is highest.

The price of cattle in the cash market will not always differ from the price of the futures contract in the futures market by the same amount. Economic conditions affecting the cash market do not exactly equal those influencing the futures market. However, because of speculative activity, the cash market price of fat steers at the time when a futures contract matures or expires will be very close to the price of that futures contract. Although it is impossible to eliminate price risk completely, it is possible to reduce that risk through the use of the futures market.

## Hedging Cattle Fattening Operations

At the time of purchasing feeder cattle, the feeder hedges those animals by selling a beef futures contract that expires at the time that fattened cattle will be ready for market. This hedge only establishes an approximate selling price. The feeder must first estimate whether the approximate selling price will be profitable.

A feeder should know (or determine) certain facts to decide whether or not to hedge. These facts include the following:

1. The initial cost of feeder cattle at the feedlot (per cwt).
2. The best estimate of total cost of weight gain (per cwt).
3. The current price of the futures contract for the month in which cattle will be sold (per cwt).
4. The estimated "basis," which is the difference between prices of live cattle at the local selling point and the futures contracts at the time fat cattle will be sold, or the cost of delivering cattle to Omaha (per cwt).
5. The cost of hedging (per head).

The cost of feeder cattle includes the purchase price, the buyer's commission, transportation costs, and any other expenses incurred in securing cattle for the feedlot. If feeders raise their own calves or yearlings the cost is the price for which the calves or yearlings could have been sold at the time they are put in the feedlot. For ease in estimating the effect of hedging, feeder cattle costs are used in terms of *cost per cwt.*

A feeder's best estimate of *cost of gain* can be calculated using the previous year's feeding records as a benchmark. Cost of gain includes the costs of feed fed, interest on the value of feed, interest on the value of the calf, death loss, bedding, hired labor, fuel, marketing costs, and any other variable costs incurred in the operation.

The current price of the futures contract for the month in which cattle will be sold can be found on the market page of some newspapers or by calling a broker handling futures contracts.

The cost of shipping fat cattle from Billings, Montana, and selling them in the Omaha market is approximately $3.20 per cwt. (Freight of about $2.50, and yardage commission and inspection of about $0.70.) The cost of shrink must be added to this cost. If shrink is 3 percent and the fat cattle price is $55, the shrink cost is $1.65 per hundredweight.

A feeder who plans to sell cattle at Billings in June will be concerned about the basis, or the June difference between fat cattle price in Billings and the price of the June futures contract. Past records show that this varies from almost nothing to – $2.73 (i.e., Billings fat cattle price is $2.73 under futures). Therefore, some feeders use the cost of transporting and selling cattle in Omaha, about $3.20 per cwt, as an estimate of the basis. Then if the June difference between Billings fat cattle and futures is less than the cost of shipping to Omaha, they will sell their cattle in Billings. If the difference is larger, they will ship the cattle to Omaha and sell there.

The cost of hedging includes the $60 broker's commission per contract, plus interest on $1200 margin over the length of the contract.[3] Assuming interest at 12 percent and 8 months as the life of the contract, hedging cost per head can be estimated as follows:

| | |
|---|---|
| Commission (round turn) | $ 60.00 |
| Interest on margins | |
| $1200 x 0.12 x 8/12 of year | $ 96.00 |
| Total | $156.00 |
| Hedging cost per head | $  3.90 |
| ($156 ÷ 40 head/contract) | |

The following is a description of estimating possible profits and of executing a hedge. It is based on the assumption that the feeder is planning to sell cattle at Billings, with costs estimated for an October to June feeding period.

The following assumptions will be used in the example (dollars are rounded to the nearest whole in order to keep the arithmetic simple)

| | |
|---|---|
| 1. Cost of calves (400 lbs) at the feedlot in October | $70.00/cwt |
| 2. Cost of gain (600 lbs) | 60.00/cwt |
| 3. Current (October) price of June futures | 70.00/cwt |
| 4. Expected basis at Billings, Montana | 3.00/cwt |
| 5. Cost of hedging | 4.00/head |

## The Six Steps of Hedging

The process of hedging a cattle-fattening operation breaks down into six steps as follows:

1. Estimate the potential feeding profit.
2. Determine the number of contracts to sell.
3. Arrange financing.
4. Place the order to sell futures ("place the hedge").
5. Record the order.
6. Place the order to buy back the futures ("lift the hedge").

---

[3] This margin is the security deposit that must be deposited with the broker when the futures is sold.

**TABLE 12-2**   Profit Estimate Made in October

| | |
|---|---:|
| 1. Cost of calf 400 lbs × $70/cwt | $280.00 |
| 2. Cost of gain 600 lbs × $60/cwt | 360.00 |
| 3. Hedging cost per head | 4.00 |
| 4. Total cost of 1000 lb fat steer | 644.00 |
| 5. Sale of 1000 lb fat steer × $67/cwt | 670.00 |
| 6. Estimated profit per head | 26.00 |

With the required knowledge, the feeder can estimate the total cost of a 1000 pound fat steer by computing the necessary cost items (1 through 4) as shown in Table 12-2. The sale value of that 1000-pound steer may be calculated by subtracting the estimated basis ($3.00 per cwt) from the current price ($70.00 per cwt) of the June futures contract and multiplying the resulting $67.00 per cwt by 10 cwt (the weight of the finished steer), as shown in item 5. By subtracting item 4 from item 5, the estimate of profit is $26.00 per head (item 6).

The feeder can now decide whether to run the risk of a price decline or to hedge. The risk can be shifted by selling the June futures contract at the time the calves are put in the feedlot. If a price decline occurs, the feeder will sell the cattle for less than the amount estimated, but will be able to buy back June futures for less than the initial sales price. The loss on the cattle on the cash market will be offset by the gain on the futures transaction so that the $26.00 feeding profit will be "locked in."

Related to the chance of gain is the chance of loss. In certain situations, a feeder may want to minimize a loss. Suppose a feeder has cattle on feed and estimates that a $5.00 per head loss would result if hedged and further is convinced that beef prices are going lower and a loss greater than $5.00 per head could result. The $5.00 loss can be locked in by hedging, thus avoiding a greater loss.

To determine the number of contracts to sell, the feeder must first determine how many fat cattle will equal the contract weight of 40,000 pounds (for the Chicago Mercantile Exchange Contract). The feeder can determine the number of contracts required by dividing the total number of cattle to hedge by the number of cattle per contract. For example, 100 head of cattle divided by 40 head per contract equals 2.5 contracts. Because one-half of a contract cannot be traded, this feeder would sell two contracts and leave 20 head of cattle unhedged. A rule suggested for hedging is "never hedge cattle you do not have" because that would simply be speculation.

Memory often is short, so the feeder should write down the details of the futures transaction. This memo should include:

1. Broker's name, address, and telephone number.
2. Number and name of contracts sold (e.g., "2 June—live beef.").
3. Price limit of the order (if the order sells at or above the desired price, the broker will confirm the execution of the sale).
4. Date of sale, price of sale, and margin paid to broker.

The hedge may be lifted by buying back the contracts any time up to contract maturity. It usually is done before the tenth day of the month of maturity. The hedge must be lifted by the nineteenth of the contract month unless the feeder wants to deliver cattle in Omaha to satisfy the contract. Feeders avoid making delivery for several reasons. One is that many of the delivered fat cattle will not meet contract requirements. For example, cattle not acceptable in the contracts include heifers, dairy steers, steers of other than British breeds, steers of less than 950 pounds weight, packages of steers that grade "good," and packages of steers that vary too much in individual weights. So even though a feeder may have cattle in Omaha to deliver, one still may have to sell cattle in the Omaha market and get a buyer to purchase other steers to make up the delivery package. Therefore, feeders ordinarily sell their cattle wherever they can get the best net return over freight and selling costs, and make the necessary futures market transactions so as to avoid making a contract delivery.

If convinced that beef prices were going to continue rising, a feeder probably would not hedge, but if the cattle were hedged the hedge might be lifted before the cattle were fat. Of course, when the hedge is lifted under these conditions the feeder would suffer a loss in the futures market that would be balanced by a gain in the cash market. However, if prices continued to rise, the feeder would get the benefit of the rise that occurred between the time the hedge was lifted and the fat cattle were sold. On the other hand, if prices fell after the hedge was lifted, the feeder would lose money on the cattle with no balancing gain on futures contracts.

Some hedging guidelines:

1. Calculate an acceptable profit.
2. Know basis of market.
3. Keep banker informed of hedging plans.
4. Do not hedge cattle not owned (this is speculation, not hedging).
5. Lift hedge by buying contracts before tenth day of contract month.

Pitfalls to avoid:

1. Locking in losses instead of profits because of unknown or undetermined costs or because a basis discount that is too small is used.
2. Being unable to meet margin calls because banker did not know of hedging plans.

3. Getting trapped into selling in an erratic futures market just before delivery date.

## Speculation

Cash or current prices reflect current supply and demand conditions. The futures market reflects current best estimates of demand and supply in the specified futures month. Typically, agricultural producers use the futures market to hedge against price movements by selling futures contracts. This selling of contracts exerts a downward pressure on futures prices. Balancing producers' sales are purchases of contracts by people on the long (buying) side of the market. Some of these people may be exporters, processors, or speculators.

Exporters may hedge their future deliveries too. If they buy futures contracts to hedge, then someone must have sold a contract. It may be a speculator who thinks prices are going to fall, or a producer who is hedging a crop. Regardless of the amount of hedging that is occurring, adequate speculative activity is required. *Speculation* is a transaction in goods or financial obligations that assumes a risk in the hope of a favorable price movement. Speculation helps create the volume of trading to allow producers to get into and out of contracts, to keep the market competitive, and to keep the average cost of futures transactions as low as possible.

To economists, speculation is not gambling. A gambler takes unnecessary risks, risks deliberately created. Therefore, the gambler serves no useful purpose to society. The speculator, however, assumes a risk that is already present. For example, when cattle feeders buy cattle to feed, they face the risk that prices of fat cattle may drop to a point at which they will suffer large losses. Unavoidable price risks exist. Someone must assume these risks because they are part of our economic environment. The only question is whether the cattle feeder will assume the risk or transfer it to a speculator through hedging operations.

# TRADING IN AGRICULTURAL OPTIONS _____

Differing from trading in futures contracts, **agricultural options** are a type of financial instrument that gives the holder the right to buy or sell *futures contracts* at the price specified in the option within a given period of time. A one time payment (premium) is made to purchase an option.

A producer or other trader reduces the risk of adverse price movements by taking opposite positions in the futures and cash markets. Such position hedging in the market sets up offsetting gains and losses from price changes during the period of the futures contract. By linking options trading to one's own hedging transactions, a person may benefit from price changes at an already known cost of that expected benefit. The reason for this is that options contracts do not require the holder to exercise the buying or selling right specified in the option. One may expect the price of a commodity to change in a particular direction and trade in options accordingly. An opposite price movement would make unprofitable the exercise of the right granted by the option, so the only sacrifice to the trader is limited to the premium the trader has paid for that option.

The price of an option is determined by the actions of buyers and sellers making their bids and offers on the exchange. The price specified in the option is called the *striking price*. Since offers and bids for options contain terms and prices that differ from buyer to buyer and seller to seller, a number of different striking prices are available to traders at any one time for these options. The price that a buyer pays for an option is called the *premium*, with the size of the premium depending on that trader's price expectations for the commodity and the time remaining until the option expires. Thus, the holder of an option has paid a *premium* for a fixed *striking price* at which that person would be willing to exercise the rights conferred by that option.

Two types of options contracts are traded—*put* options and *call* options. If you *buy* a put option, you have bought the right but not the obligation to sell a futures contract at the price specified (the strike price) at any time up to the expiration date of that option. Buying a call option gives you the right but not the obligation to buy a futures contract on or before the option's expiration date at the price specified in that option.

Selling (writing) puts or calls confers obligations to the original option seller that are opposite to the above rights of option buyers. The sale of a put is the same market position as someone who has a long futures position. Also, the sale of a call is the same position as a short futures position (if the option is exercised), and the risk to the seller is the same as someone who has a short futures position. The seller of the option, however, receives the option premium for the risk of writing the option which has the same risk as participants in futures markets. On the other hand, the financial exposure for an option buyer is the option premium that is paid at the time of purchase.

Let's suppose, for example, that you are a corn producer. You have incurred variable costs in all your tillage and planting operations with the expectation that, by your mid-December marketing time, the price of corn will at least be high enough to cover all these costs. Because the price of corn may fall by that time, sometime between planting and harvesting you hedged yourself against the possibility of a price decline by selling December futures. By this hedge, however, you have also foregone a possible gain from a rising price of corn because you will have to buy back that December futures at a correspondingly higher price.

Now, suppose it is early October. The current price of corn is $3.50 per bushel and you have reason to expect it to increase to $4.00 or more by the time your December futures expire. Assume that you now buy a call option on December futures at a striking price of $3.50 per bushel, and that you pay a $0.50 premium for that option. The call option grants you the right to buy December corn futures at $3.50 per bushel. No matter how high the price of corn might rise, you may exercise your buying rights at any time (at $3.50 per bushel, to the limit specified in your call option) up to the expiration of the option.

At any December futures price of $3.50 or less you would not exercise your option but simply let it expire, and your loss is limited to the $0.50 premium that you paid for your call option. At any December futures price greater than $3.50, the option would be exercised because you can buy December corn futures for $3.50 and immediately sell them at their currently higher price. You gain by the difference between the $3.50 striking price and the December futures price minus the $0.50 premium.

If, on the other hand, you expect the price of corn to fall rather than rise, you would buy a put option. Suppose that you can buy that option also at $3.50 per bushel with a premium of $0.50. You now have the right to sell December corn futures at the striking price of $3.50 per bushel. If your expectations materialize, you will exercise the right, selling December futures at $3.50, and gain by the difference between the current December futures price and your $3.50 striking price minus the $0.50 premium.

In either of these examples, if the price of corn moves adversely to your market position, you simply do not exercise your option right. So you have insured your participation in a favorable market price change at a maximum cost of the premium that you had to pay for that option.

The several different striking prices and their different premiums offered by the exchanges are affected by the trading activity of buyers and sellers. Thus, the choice problem faced by a producer in deciding whether to buy or sell puts or calls is made more complex by the array of striking prices and premium alternatives available on the exchange.

# DECENTRALIZED MARKETS _____

Few agricultural products are produced in a single location; rather they are produced in many different areas, each of which, frequently, also has a local market for the product. Producers usually sell their products in the local market where they can receive the highest net price, after taking into account transportation and handling costs (**transfer costs**).

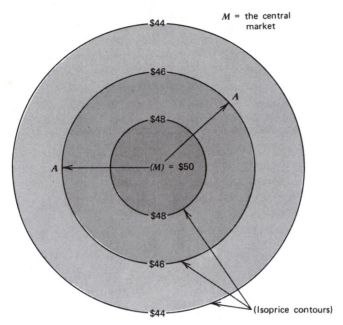

**FIGURE 12-7** A geographic price surface.

If the product is produced and sold in a competitive economic environment, a producer can be aware of what products will bring at many different markets, because a competitive market provides adequate market news information. The producer can determine the farm price by subtracting transfer costs from the quoted market price.

Using the ideas of market price and transfer costs, a *price surface* can be constructed, as in Figure 12-7. Suppose the figure represents a local area market for hogs, with the price of hogs at $50.00 per cwt in the central market (*M*). The farm price at any location is equal to the central market price less the appropriate transfer costs, and is shown by a specific isoprice contour. The isoprice contours (the concentric circles) show that the net farm prices are equal at equal distances from the central market for a given quality and quantity of the product marketed. Thus, hogs produced at a distance of *A-M* miles in any direction from the central market (*M*) bring a net farm price of $46 per cwt. Similar calculations can be made for other distances from the central market.

With two or more markets for a single product, the producer will ship to that market at which one can receive the highest market price less *transfer costs*—all those costs arising from moving commodities from one market to another. Therefore, one can see why market news information is so important to producers. It would be impossible for a producer to make rational marketing decisions if that person were not aware of different

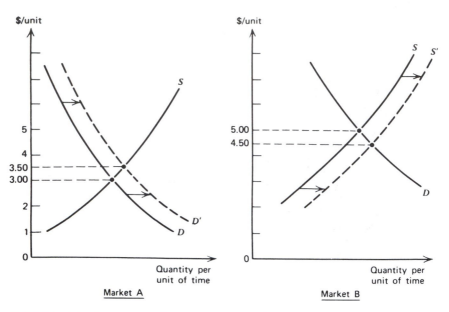

**FIGURE 12-8**  Spatial equilibrium in two markets.

geographic market prices. As long as goods can flow freely between competitive geographic points, markets will be related to each other by transfer costs. However, lags do exist in commodity flows and producers can take advantage of these lags.

## How Markets Are Related

Intermarket competition forces the market prices at two distant markets to differ most of the time by transfer costs for a given quality product. Assume transportation costs between markets A and B are $1.00. If, as in Figure 12-8, their prices differ by more than $1.00, buyers will purchase the product in market A at $3.00, spend $1.00 in getting the product to market B, then sell it in market B at $5.00, leaving a profit of $1.00 per unit. How many times have you seen a farmer market hay in another community because that price is greater than the local price by more than the cost of delivery? This buying and selling action will increase the demand in market A and increase its price, and selling it in market B will shift the supply curve and lower its price there. These transactions (called "arbitrage") will continue until the possibility for such profit is eliminated and the price difference between markets again equals the transfer costs. This action shifts the quantities of goods from one market where the good is less highly valued to another where it is more highly valued.

Thus, the two markets now do a better job of allocating goods between consumers than could be done if it weren't for people who are willing to undertake whatever risks may be involved in these transactions.

## Marketing Orders and Agreements

Marketing orders and agreements were first established to make more orderly the marketing of milk and milk products, fruits, vegetables, and nuts. Through orderly (controlled) marketing it was thought that agricultural producers could even out the flows of sales of commodities on the market through the marketing year, increasing producers' net prices and therefore their incomes. These programs originated in the Agricultural Adjustment Act of 1933 and were clarified and modified under the Agricultural Marketing Agreement Act of 1937.

A *marketing agreement* is voluntary whereas a *marketing order* is mandatory. An agreement is a contract that is endorsed by the Secretary of Agriculture and the handlers of the particular commodity. This contract is binding only on those who sign it. These marketing agreements are exempt from the antitrust laws. A marketing order, on the other hand, is binding on all handlers of a commodity in a production or marketing area if two-thirds of the producers, with at least one-half of the volume of the commodity, approve the order through a referendum vote.

Most milk marketing orders attempt to protect farm income through classified pricing plans and by fixing minimum prices handlers must pay producers for various classes of milk. For example, prices of milk for fluid uses are set at a higher level than those for manufacturing uses.

Commodity marketing orders attempt to protect farm income by specifying the quality and quantity of a commodity that may be shipped, by prohibiting unfair methods of competition and unfair trade practices, and by establishing marketing research and development programs to promote the products.

Marketing orders are usually more effective than marketing agreements because agreements are only binding on those who sign them, whereas marketing orders, once approved, are binding on all handlers. This arrangement prevents the individual from undermining the agreement by acting in a personal rather than group interest.

# SUMMARY

Marketing is concerned with those productive activities that add time, place, and form utility to agricultural commodities. The existence of marketing is a result of specialization and trade in the economic system.

A market need not be a place. A market consists of buyers and sellers with facilities to communicate with each other. A well-organized market enhances the quality of communication between buyers and sellers so maket prices can be determined. These prices, in turn, guide producers' production decisions and consumers' purchasing decisions.

Efficiency is highly regarded in our society. Market efficiency can be measured by comparing the value of output to the value of inputs.

Three approaches to the study of marketing are the functional, institutional, and market structure. The functional approach analyzes marketing activities such as the exchange, physical, and facilitating functions that must be performed. The institutional approach examines the organizations or people involved in marketing. The market structure approach stresses the nature of competition in markets.

A marketing margin is the difference between the price consumers pay for the final product and the price received by producers for the raw product. Margins vary widely among agricultural commodities because of such factors as the amount of processing that is necessary, perishability of the product, bulkiness of the product, and the seasonal nature of production. Large marketing margins may not mean that marketers are taking financial advantage of producers and consumers.

Futures markets determine prices for many agricultural commodities. They can be used to forward price commodities through hedging operations. Other marketing alternatives are available through the use of put and call options.

Marketing orders and agreements were established in the 1930s to even out the flow of products on the market. The purpose of orderly marketing is to increase producers' net prices and, therefore, their incomes.

## CHAPTER HIGHLIGHTS

1. Marketing costs are those costs incurred in assembling, transporting, processing, and distributing farm food to consumers.
2. The major marketing costs are for labor, packaging, and transportation.
3. Agricultural marketing starts with production and continues until a product is sold to the final consumer. Thus, marketing also includes the input-supply firms that serve farmers and ranchers.
4. Marketing developed as a result of specialization and trade.

5. All economic systems must answer the questions of (1) what and how much to produce, (2) how to allocate resources or inputs, (3) what production methods to use, and (4) how the product is to be rationed.

6. A market consists of buyers and sellers with facilities to communicate with each other. Markets may be local, regional, national, or international in scope.

7. In a market economy, market prices guide most production and consumption decisions.

8. There are three general approaches to the study of marketing: (1) functional approach, (2) institutional approach, and (3) market structure approach.

9. Marketing is productive because it adds form, time, and place utility (or satisfaction) to agricultural commodities.

10. The marketing margin for a commodity is the difference between the retail price and the farm price.

11. The farmers' share of the consumers' food dollar reflects the costs of providing marketing services. Marketing services vary from one commodity to another.

12. Futures markets are important forward price discovery mechanisms that are used to price many farm products.

13. Futures contracts are standardized.

14. Hedging is establishing a position in the futures market that is opposite from the position one has in the cash (or product) market.

15. Speculators who take the opposite side of futures contracts are imperative in the futures market. In addition, they increase the volume of contracts traded to allow producers easy entry and exit from the market and keep the cost of hedging low.

16. Speculation in futures markets is not gambling.

17. A basis is the difference between cash and futures prices.

18. The basis is affected by economic influences; it is not fixed, but is continually changing.

19. In a perfectly competitive market, the farm price of a commodity at any location is equal to the market price minus transfer costs.

20. Marketing orders and agreements are used to even out the seasonal variations in marketing patterns. Their purpose is to increase producers' prices and incomes.

# KEY TERMS AND CONCEPTS TO REMEMBER

| | |
|---|---|
| Agent middlemen | Facilitative organizations |
| Agricultural options | Form utility |
| Exchange function | Futures contracts |
| Facilitating function | Futures markets |
| Hedging | Place utility |
| Law of comparative advantage | Price surface |
| Market | Pricing efficiency |
| Marketing | Processors |
| Marketing agreement | Speculation |
| Marketing efficiency | Speculative middlemen |
| Marketing margin | Technical efficiency |
| Marketing order | Time utility |
| Merchant middleman | Transfer costs |
| Physical function | |

# REVIEW QUESTIONS

1. Why have marketing costs been increasing? What factors are causing marketing costs to rise?
2. How did markets develop? What role did specialized production and trade have in this development process?
3. Define a market. Why can markets be local, regional, national, or international in scope?
4. What is marketing efficiency? Are all changes that reduce marketing costs per unit of output an improvement in marketing efficiency? Explain.
5. Explain the three approaches to the study of marketing problems presented in this chapter. Is one approach preferred to the others?
6. Does a decrease in the farmers' share of the consumer dollar spent for farm food mean that marketers are taking financial advantage of producers? Discuss.
7. On January 28, the local cash price for winter wheat is $3.50 per bushel and the March future at Kansas City is $3.70 per bushel. On March 4 you deliver your winter wheat to your local elevator and the cash price is $3.45. The March future on March 4 is $3.80. What price would you obtain for your wheat if you hedged your wheat on the March basis as of January 28?

8. What is the "basis"? Calculate the basis for a commodity produced in your area. You can find the futures price information in the *Wall Street Journal* and obtain the local cash price from your local auction, elevator, or first handler.
9. Why would you expect intermarket competition to force market prices at two distant markets to differ only by transfer costs? Do you assume competitive markets?
10. How would a farmer use options to establish a minimum price for selling his crop at harvest? Would this farmer use a put or call option? Explain.

# SUGGESTED READINGS

Chicago Board of Trade. *Options on Agricultural Futures: A Home Study Course*, Chicago: Board of Trade of the City of Chicago, 19??.

Dahl, Dale C., and Jerome W. Hammond. *Market and Price Analysis: The Agricultural Industries*, New York: McGraw-Hill, 1977, Chapters 7–10, 12, and 13.

Heyne, Paul T. *The Economic Way of Thinking*, 7th ed., New York: Macmillan Publishing Company, 1993, Chapter 7.

Hieronymus, T. A. *Economics of Futures Trading for Commercial and Personal Profit*, New York: Commodity Research Bureau, 1971.

Kohls, Richard L., and Joseph N. Uhl. *Marketing of Agricultural Products*, 7th ed. New York: Macmillan, 1990, Chapters 1, 5, 6, 7, 9, 16, and 20.

Moore, John R., and Richard G. Walsh, eds. *Market Structure of the Agricultural Industries*, Ames, Iowa: The Iowa State University Press, 1966. Chapter 15.

Shepherd, Geoffrey S., and Gene A. Futrell. *Marketing Farm Products*, 5th ed., Ames, Iowa: The Iowa State University Press, 1969, Chapters 1 and 2.

Tomek, William G., and Kenneth L. Robinson. *Agricultural Product Prices*, 3rd ed., Ithaca, N.Y.: Cornell University Press, 1990, Chapters 6 and 12.

## AN OUTSTANDING CONTRIBUTOR

**Lowell D. Hill**   Lowell Hill is L. J. Norton Professor in the Department of Agricultural Economics at the University of Illinois.

Hill grew up on a farm in Iowa and, after earning his B.S. degree in vocational agriculture from Iowa State University in 1951, he spent 2 years teaching vo-ag in Iowa (1951–52, and 1954–55) and the following 4 years operating a grain/livestock farm in northcentral Iowa.

In 1959 Hill began his graduate work at Michigan State University, where he earned both the M.S. and Ph.D. degrees in agricultural economics—his M.S. in 1961, and the Ph.D. in 1963. He was then appointed assistant professor of agricultural economics at the University of Illinois in 1963, promoted to associate professor in 1969, then to full professor in 1972.

Hill's contributions have been many and varied. He has supervised the research and graduate work of 37 Master's and 16 Ph.D. students. He has the ability to conduct interdisciplinary research and to integrate their results into understandable language that has had its impact in the agricultural industry.

To date, he has written 38 professional journal articles, more than 100 experiment station bulletins and reports, and more than 100 other articles in trade and popular journals. He also has made hundreds of presentations to various groups, including four to the U.S. Congress.

As a strong proponent of the land grant system, Hill's research program and public service activities have had a strong and direct influence on legislation and regulations affecting grain quality and the nation's export markets. He has been presented the Grain Age Award for service to agriculture in both 1977 and 1979; the Paul A. Funk Award for outstanding service to agriculture in 1979; the Certificate of Superior Performance from the Agricultural Communicators in Education in 1980; the Corn Utilization Award in 1982; a Blue Ribbon Award from the American Society of Agricultural Engineers in 1987; and the USDAs Distinguished Service Award in 1989.

His scholarship on grain quality is internationally recognized, and he has also been recognized by his colleagues. He was presented the American Agricultural Economics Association's Quality of Communication Award in 1980 and 1988. In addition, he received the AAEA Distinguished Policy Contribution Award in 1988 and their Distinguished Extension Programs Award in 1989.

He is married to the former Betty Carpenter, and they have two children, Rebecca and Brent.

# CHAPTER 13

# Natural Resources

$\mathbf{T}$he word *resource* was used earlier (Chapter 4) as meaning any good or service that can be used to produce a good capable of satisfying a human want. A *natural resource* is simply a resource provided by nature, a good existing in the universe, yet not limited solely to physical things.[1]

Such phenomena in nature have no value until they have been discovered and an economic use has been found for them. Natural resources, then, are a function not only of their physical existence, but also depend on our knowledge of them and the technology of use plus our ability to make economic use of them as sources of want-satisfying goods and services. It is quite possible that we have discovered economic uses for only a minute fraction of resources in and on the earth and its surroundings. How much remains to be discovered? How many new uses can be discovered?

Of the wide array of resources around us, the one that most typically comes to mind when the term natural resource is used is *land*—land from which we derive our food and fiber, space over which to transport people and goods, building materials, space for homesites, for recreation and aesthetic purposes.

Other natural resources are the *water* that exists in lakes, streams, rivers, or in underground channels and reservoirs; *minerals* in their many types, locations, and varying concentrations; and self-propagating *plant and animal life,* all of which are of immense usefulness to humankind. And if plant and animal life are properly included, how can we ignore *humans*, who, with their labor and other productive abilities, are probably the most important of all natural resources?

## A Natural Resource Classification

Rational answers to questions of which resources to use, when to use them, and how rapidly to use them are made more difficult because of the characteristics exhibited by these resources.

We may pump the oil from an underground basin, at a rate determined by the costs of getting the oil out of the ground and the value of the pumped oil, but what do we do when the well runs dry? We may have used this resource wastefully by a too-rapid pumping rate. This problem is quite different from capturing a benefit from the flow of the wind. If its velocity has not been diminished, nor its quality fouled in the

---

[1] Were we to be too restrictive in this we would have to find a special definitional slot for such phenomena as wind, tides, and the energy of the sun.

process of its utilization, there has been no reduction or depletion of that resource. Not using this energy source may constitute "waste," since failing to use the wind's power when it is available is a resource-use forever lost.

These are examples of the special characteristics of certain natural resources. Some are fixed in quantity, and using them depletes the amounts remaining. We call these *fund* (or *stock*) *resources* to reflect the fact that their quantities are fixed in their natural state.[2] Many of our natural resources are of this type: coal, oil, natural gas, sand and stone, iron ore and other minerals, and similar natural deposits that are non-renewable resources whose use forever reduces their remaining quantities.[3]

Other natural resources such as sunlight, wind, rain, tides and flowing water are called *flow resources.* Their present use does not prevent possible future use because the available quantity is constantly being replenished. If, in using the wind's power, heat energy from the sun, or the flow of water in a stream does not disturb their continued flow, the amounts available for other uses are left undiminished.

A simple fund or flow classification does not adequately serve to describe all natural resources, however. Many other important resources exhibit some characteristics of both these groupings. Growing, maturing plants (the flow) may be harvested without damaging the productivity of the parent stock (the fund). The product flow can be maintained indefinitely, increased or decreased, depending on the harvesting rate and the manner in which both the fund and flow are managed. This special characteristic has led some authorities to identify these resources as either a subcategory of fund and flow resources or separately as *biological resources.*

Our forests, ranges, livestock, fish, and wildlife yield an annual product that may be taken without harm to the productive source of that output. Yet each can also be used in ways that may either enhance or diminish the quantity (and productivity) of that fund resource, with production increases or decreases following changes in uses or use rates.

Forest cutting can be so intensive in one period of time that future yields are reduced, even eliminated entirely. Rangeland can be over-grazed to the extent that grass regrowth and production declines in later years. Ranchers may sell off brood stock, in addition to young stock, and find

---

[2] Natural processes may very well be continuing the formation of additional amounts of certain fund resources, but at a rate insignificant when compared with existing stocks or present rates of use.

[3] Even though depletable in its natural form, exhausting a fund resource need not mean it is gone forever; many natural resources may be reused. Steel processed from iron ore that is fabricated into automobiles, then junked, may be reclaimed for further use. Disposing of solid wastes in landfills may even provide a future bonanza by having concentrated many scarce materials that may later be extracted and reprocessed to be used again.

total output reduced until the breeding herd is again restored. Fish and wildlife harvesting may be carried on at a rate which also causes a reduction in the fund part of those resources, from which the flow is derived, even to the point of extinction, as in the case of the dodo bird, the California grizzly bear, or the carrier pigeon.

Even the soil itself, although viewed most frequently as a fund resource, also has some of the characteristics of a flow resource. Management practices may have the deliberate objective of reducing, maintaining, or improving the level of plant nutrients held in the soil, depending on whether current practices use those nutrients at a greater, equal, or lesser rate than their inflow to the soil. Much like a savings account in the bank, the present balance of those nutrients is the result of both additions to and subtractions from that account, as well as any original amount.

The importance of economic criteria in these examples, and in similar decision problems with many other natural resources, cannot be ignored. It should be clear that the physical facts of resource existence, and their capabilities to satisfy human wants, do not establish the criteria by which those natural resources are used; physical conditions can only set the limits within which correct answers are determined. In the utilization of any natural (or other) resource we cannot escape making choices as to whether or not to use a resource, the rate at which to use it, and the purpose of its use, causing these choices to be economic questions requiring economic criteria for their solution.

## Resource Inventories

Of all the natural resources that exist, what do we know about them? How much of each is there? How much do we use? How much can we use? Are there better uses for them? Will there be adequate supplies in the future? What will be the effects of increasing or decreasing our rates of use? And of what importance are the numerous "if" conditions that must be dealt with in projections into the future?

Attempts at identifying the amounts of specific natural resources that we now have, and may or may not have in some future period, are not, in themselves, necessarily confusing; interpretational liberties taken with technically developed terms frequently make them so, however. Such **resource reserve** labels as "proved," "known," and "potential" carry special meanings that require a clear understanding of their basic intent and assumptions on which they are based.

Take, for instance, the abstraction shown in Figure 13-1, a conceptual framework that can be applied to any natural (or other) resource. The area of the box itself demonstrates the total physical quantity of a specific re-

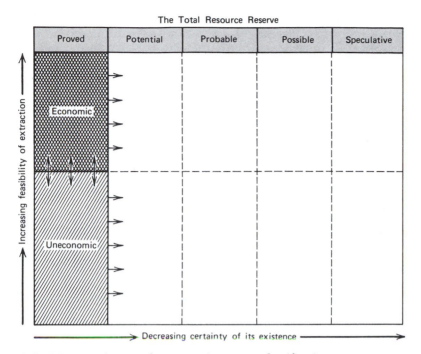

FIGURE 13-1   A natural resource inventory classification.

source as it exists in the natural state.[4] The horizontal axis measures the certainty with which the existence of the total amount of the resource is known. Those concerned with geologic structures and formations and the quantity of the resource contained at various locations and depths of the earth's surface have, in some cases, identified and determined specific quantities of the deposits; in others, sheer speculation may suggest that the resource might be there, and our degree of certainty is so indicated as we move farther to the right in the diagram.

The mere fact of physical presence, however, is insufficient. We must be able to take the resource from its present location, with whatever refining and processing is required, and get it marketed. If this development can be done without sacrificing other, more valuable, goods and services, it is economically feasible; if not, its high costs would cause the

[4] The following categorization of resource reserves is based on Speir Collins, "Reserves, Reserves, Reserves," *Petroleum Today*, 1976/two, Washington, D.C.: American Petroleum Institute, Vol. 2, 1976, pp. 26–29. This classification, and the diagram in Figure 13-1, are similar, but with reserves labels similar to Vincent E. McKelvey, "Mineral Potential of the United States," in *The Mineral Position of the United States, 1975–2000*, Eugene N. Cameron, ed., Madison, Wisc.: The University of Wisconsin Press, 1973, pp. 67–82.

resource to be unutilized. We indicate this along the vertical axis, with the degree of feasibility increasing as we move upward along the axis.

Extraction and related technology make up an important contribution in this distinction. A primitive technology, based primarily on hand labor, for instance, may prevent any use of this resource. As capital investment in improved technology makes possible the movement of sufficient quantities of the resource-bearing material, costs per unit of the extracted product fall low enough that it becomes feasible to utilize the resource, as indicated by the line running horizontally through the diagram. The more technology is improved, the farther downward the line will be pushed, reflecting the fact that it has become possible to reach to deeper levels or to lower concentrations for the resource, as indicated by the downward-pointing arrows.[5]

The other important element in determining economic feasibility is price. As the price of the extracted product increases (or decreases), the horizontal line is shifted downward (or upward), indicated by the upward- and downward-pointing arrows. The higher the price the deeper we can go in extracting the resource, and the lower the concentrations that can be utilized. A falling price, on the other hand, will make uneconomic the harvesting of this resource at the depths and concentrations that formerly were feasible.

That part of the total resource stock that is called the _proved reserve_ depends on past production and exploratory work that has been done to determine the size and content of each deposit. How much of the resource there is in such deposits (the area in the diagram labeled "Proved") is known with considerable certainty. But price, costs, and technological factors, as just discussed, limit the usable portion of the proved reserve to the shaded part labeled "Economic." In spite of knowing with certainty how much (physically) exists, the "Uneconomic" portion[6] is presently unavailable for use because taking the resource from its deposit and getting it to market costs more than it is worth.

With population increases, and possible new uses for the resource, the demand curve for the resource will shift to the right, causing its price to rise, _ceteris paribus_. Not only does this cause additional quantities of the proved reserve to become economic, the higher price will encourage more intensive searching for new sources and further effort at improving the

---

[5] This effect is demonstrated in the petroleum industry. Early technology that depended on (natural) well-pressure could recover approximately 20 percent of the crude oil in place; the technology of water injection increases that recovery to about 30 percent; tertiary methods using chemicals add more to the amount recovered, with higher-cost special polymers (still in the research stage) expected to increase the amount recovered even more. (Reported in _Business Week_, June 14, 1976, pp. 31–32.)

[6] Also referred to as "paramarginal" or "submarginal," depending on how much price would have to increase to make extraction profitable.

technology of detection, as well as its extraction, causing a shift of the vertical boundary to the right. Proved reserves are thus expanded despite the fact that past and current use have constantly been withdrawing from the physical quantity originally available.

Somewhat less certain are *potential reserves*. Depending on the type of resource, estimates are developed from outcroppings, test drillings or other samplings, production data, and projections from geologic evidence.

Because an unexplored area may closely adjoin a producing area, inferences may be drawn as to the quantity that might be found. Such reserves are called *probable*. And at an even greater degree of uncertainty are the *possible* reserves, with their estimated quantities being based on underground structural similarities with a producing area nearby.

The most uncertain of all are the *speculative resources*. The lack of geologic testing in an area forces this speculation, as with the question of finding oil under the U.S. East Coast Continental Shelf. Seismic studies reveal different formations from those in producing areas, but oil is still believed to be trapped in certain locations under that shelf.

## Resource Supplies

Having some knowledge of the physical amounts of various resources that are present is of value only as a limit. The ***physical supply*** of any resource may be defined as the total quantity provided by nature.[7] But that would include the total area of the diagram in Figure 13-1, which, for most natural resources, has little relationship to the amounts used. The concept of greatest relevance is that of ***economic supply***, the part of the physical supply that is used for want satisfaction. We could attempt to equate this physical quantity with the economically feasible portion of the proved reserve (the amount economically available for human use), but this is a total concept—the sum of both the amount actually used and the amount unused. If we limit ourselves to meaning that amount of the resource actually used within some time period (annually, for instance), we have been pushed to considering the equilibrium quantity, a single point on the supply curve of that resource.

With this view of economic supply, we can more readily recognize the importance of both supply and demand on how intensively we use our natural resources, and at what rates they are being consumed in the production process. These two concepts are fundamental to understanding changes that occur in our economic reserves. They make clear that expansions and contractions of the economic supplies of natural resources

---

[7] Raleigh Barlowe, *Land Resource Economics*, 2nd ed., Englewood Cliffs, N.J.: Prentice-Hall, 1958, pp. 19–20.

are caused by resource price changes, which are the result of changes in either or both supply and demand functions.

Natural (or other) resources in themselves have no value; they are valued, and command a price, only because they are capable of producing goods and services people want and are willing to pay for. The demand for a resource is thus a *derived demand*—it is derived from the fact that the resource can produce something else, the demand for which gets reflected back to the resource itself. Hence, the demand for land depends on (is derived from) the intensity of our demand for the products of land—food and fiber, aesthetics, or other valued qualities, and not just because it is "land." We must now consider how this (or any other) resource's value-creation ability is reflected by the market in such a way that its economic value productivity is correctly demonstrated when a price has been paid by someone to gain control over its use.

## Resource Values

In order finally to arrive at a dollar value of a resource (its per unit "price," "market price," "market value," "selling price," "productive value," all of which mean the same thing),[8] we must determine the costs and returns arising with the use of that resource and the net returns resulting from its use.

Resource productivity surely must affect the price one is willing to pay. If the best use of a tract of land could return no more than a zero net income in the future, how much would you be willing to pay for it? Would you be willing to pay more if the net return were $10,000 per year? How much more? Or, if the net return were to be a negative amount, would you pay anything at all for that resource (i.e., pay good and valuable dollars for the privilege of losing money!)?

Note the emphasis on *future*, rather than past or current, returns. Something is worth money only because it can produce a net return in the future, and the higher that future net return the greater its market value. You may own a $50,000 herd bull, prized for the value of the offspring, and you may have "made a bundle" from him over the past few years. Now suppose that he became totally impotent yesterday, and that fact is known to all possible buyers; today that bull is worth only what the slaughter market says he's worth, no matter what his net value productivity has been in the past, because of his inability to produce future income.

---

[8] Although these words *mean* the same thing, the specific dollar amounts can be (and are) very different for different people. For some people, the value to them exceeds the market-determined price for a good and it is a bargain for them, while for others the market price is too high for them to buy it.

There should be (and there is) a systematic method with which to derive answers to such questions of "worth," rather than having to rely only on outright guesses.[9] Recall the discussion in Chapter 4 on the *MVP* of a resource where one more unit of a resource is worth only what it can produce, giving rise to the demand curve that was shown in Figure 4-5. That is the value of a unit of the resource per unit of time (e.g., annually), an easily understood phenomenon for anything that is used up in a one-time use or application. But what if the resource, such as with most natural resources, does not just disappear with one use but can be used again and again, even indefinitely? The net returns earned by that resource occur as an *annual flow* throughout its productive life, and the valuation process appears to be more complex. The theoretical basis (from Chapter 4) still is applicable, however, but we can simplify it here.

Suppose that you presently operate a farm or ranch, and that a nearby tract of land is offered for sale. You might carefully estimate the amount by which your annual receipts would be increased by buying the land, and the amount by which your costs also would be increased.[10] Suppose, further, that after properly deducting all the increased costs associated with using this tract of land, the annual net income to land alone is $20 per acre and that this will continue indefinitely.[11]

Twenty dollars received today is worth just $20 because it will buy $20 worth of goods and services right now. But what about next year's $20? And each additional $20 further and further into the future? What are all those $20 amounts worth right now? We must determine what each of those $20 is worth now to get at one lump sum figure that says "this is what it's worth now." Pay less than this amount and you got more than you have bargained for; pay more and you will have made some unnecessary sacrifices to get the resource.

Our opportunity cost concept is useful here: Suppose that the next best alternative use of your money to pay for the land is to deposit it in a savings account in your local bank that will pay interest at 5 percent. A deposit of $400 at 5 percent will yield that same $20 per year for as long as you wish (assuming the 5 percent opportunity continues), therefore, net

---

[9] Successful decision making ought to have a better foundation than sheer guesswork, since guesses have a nasty habit of being wrong about half the time.

[10] The costs included would be all those costs associated with using this resource except a return on the investment. You would thus have tallied up all the tillage and other production costs, insurance, taxes, and the costs of all the other resources that are combined with this resource to produce a salable product, including the opportunity costs of your owned and unpaid labor, management, and capital.

[11] Predicting a specific number into the distant future may not seem very realistic, but a world of certainty is the only starting base we can use. We can make any later adjustments we wish for uncertain yields, prices, and costs over time. This will do nothing, however, to help us understand the meaning of resource "value" and the process by which this is accomplished.

earnings limit its present value at $400 per acre. You couldn't pay more than $400 per acre or you would be sacrificing greater alternative earnings elsewhere.

We use a formalized expression to adjust ("discount") future sums of money back to their present values, a process referred to as the *capitalization of earnings*, which determines the **capitalized** (or **present**) *value* of future earnings. This capitalization approach may seem complex, but this is exactly what the market has done when it has determined the market price of any good.

The comparison just given said, in effect, "I have a sum of money ($400) which, if invested at 5 percent, will yield $20 per year." The resource valuation approach is just the opposite, saying, "This resource will earn a net return of $20 per year, what is it worth today?"

We will use $P$ as a symbol for present value, $A$ for the amount to be received in the future (with subscripts 1, 2, 3, etc., to signify the year in which the amounts occur), and $r$ as the opportunity rate used for discounting. So, we have $P + Pr = A$, where $P$ is the unknown (the present value); $Pr$, which determines the first year's earnings on that amount; and $A_1$, the amount that can be withdrawn after 1 year (which includes the original amount).

We know $A_1$ (the $20), so we must solve for $P$. Factoring out the common term $P$ from the equation we have $P(1 + r) = A_1$. Transposing, we have $P = A_1/(1 + r) = \$20/1.05 = \$19.05$, the value *today* of *next year's* $20. What about the second year's $20? The present value of that is the amount that would grow to $20 if deposited today and left for 2 years. $P$ deposited today is increased by 1.05 after 1 year, and after 2 years is increased again by 1.05, or, $P(1 + r)(1 + r) = A_2$. The formula for the second year becomes $P = A_2/(1 + r)^2 = \$20/(1.05)^2 = \$20/1.1025 = \$18.14$, which says that at a 5 percent discount rate, $20 to be received 2 years hence is worth $18.14 today.

As we consider more and more years into the future, the process becomes more and more difficult to handle. For $n$ years the equation becomes

$$P = \frac{A_1}{(1 + r)} + \frac{A_2}{(1 + r)^2} + \frac{A_3}{(1 + r)^3} + \frac{A_4}{(1 + r)^4} + \cdots + \frac{A_n}{(1 + r)^n}$$

Carrying out these computations for as many years as our tract of land will last would be extremely tedious. For any resource that will last for a long, long time into the future ("in perpetuity"), the formula sums algebraically to $P = A/r$. Thus, $P = \$20/.05 = \$400$, the present value of the resource able to yield a perpetual net return of $20 per year.[12]

---

[12] This method works well for our perfectly certain world, but in real life we cannot be that exact. Predicting yields, prices, costs, and interest rates 10, 50, or 100 years into the future becomes highly uncertain. On the other hand, how accurate is the actual market price of a

## Resource Conservation

To some, the *conservation* of natural resources is accomplished by a willful reduction in the rate at which these resources are used, leading to the conclusion that conservation and saving (nonuse) are one and the same. And to this is frequently added the admonition that we must "save for future generations." Doing this leads, however, to an untenable contradiction of terms. Faithful adherence to this dictum must result in perpetual nonuse because we are unable to specify which generation in the near or far-off future may have the privilege of use.

Resource saving may be cloaked in economic sounding phrases such as "efficient use," "wise use," "use without waste," but these confuse rather than clarify the meaning of conservation.

To save resources so as to have a larger quantity available in the future is of questionable value because new discoveries (or new technology yet unavailable) may result in very large increases of usable reserves, or new alternative sources of the same service. And who can foretell with any certainty what shifts might occur in the uses to which presently known reserves might be put? The Northern Great Plains (especially North Dakota, Montana, and Wyoming) are endowed with many billions of tons of low sulfur coal. Coal is a natural resource supply that is becoming more valuable as other traditional sources of heat energy are being depleted.[13] To save this coal for some vague future period would be a gamble which predicts that a much better or cheaper energy source will not be discovered— one that could even make these coal deposits worthless—and thus a sheer economic waste of a (presently) valuable natural resource.

To obtain compliance with noneconomic criteria is difficult even with government action "in the name of society"—a government decision process that appears much less bound by the economic costs of its actions. This feeling appears to be widespread, partly because a government's

---

long-lived asset? It can be correct only if a perfectly competitive market (which doesn't exist) has determined that price. The degree of inaccuracy depends on imperfections in the market, which results from all the frailties of human beings and their institutions.

In using the capitalization approach, the nearer to the present the greater the importance of $A$, and the less important is $r$; and the further into the future the less important $A$ becomes, while the relative importance of $r$ increases. Compare the 100th-year $20 with that of the first year, and you'll notice that estimating errors out there are quite unimportant. Twenty dollars to be received 100 years from now is worth only $0.15 today as compared with next year's $20 worth of $19.05 today. The present value of a perpetual annual stream of $20 is $400, and the 100th-year's contribution is only $0.15 out of that total. Thus, the further into the future a return occurs the smaller becomes its contribution to present value, and the less important are our estimating errors.

[13] Strippable coal (with less than 150 feet of overburden) in this general area is estimated to be in excess of 50 billion tons; in *North Central Power Study*, Vol. 1, Washington, D.C.: U.S. Department of the Interior, Bureau of Reclamation, October 1971, p. 9.

planning horizon is so much longer than for individuals or firms, and it can therefore "afford it" while individuals cannot. But this rationalization is erroneous thinking. Given whatever the values and costs to society might be, the fact that these might be hidden (for a time) doesn't mean a wrong decision wasn't made.

The conservation of natural resources is an important part of the basic economic problem of dealing with scarcity by making the proper choices, accomplished only by weighing economic alternatives. As consumers economize in their production choices, the basic requirements of efficiency in resource conservation have been met.

Economic efficiency in resource use dictates that we use our resources at the time, and at the rate, and in those uses where their contribution to consumer satisfaction is the greatest. The present value of their net returns will then be maximized. And their opportunity costs for other less profitable uses will also be maximized, thereby preventing wasteful use. We are then using no more of those resources than is necessary to produce the mix and quantities of all those products that consumers will buy (which minimizes the real costs of those goods). This efficiency achieves part of the saver's basic objective on its only supportable basis— economics.

Recognizing conservation as an economic problem directs our attention to the act of responding, consciously or unconsciously, to prices and costs in the use (or nonuse) of those resources. The problem with natural resources is not simply that a fund resource is depleted with use and that we later will regret its absence. The problem is that the value of services will be lost when the resource is gone *unless* another source of that service has been discovered in the meantime.

We have optimized natural resource conservation when we have so distributed the rates at which those resources are used that we have maximized the present value of the future stream of their net social benefits. The economic meaning of conservation takes the future into account by answering the question: What is the best rate at which to utilize any resource so as to maximize net social benefits over time? Answering this question forces a comparison of values between different time periods, the only basis for deciding what to do with any natural resource. So what we are saying is that the question of economic efficiency encompasses the question of conservation; that *resource conservation cannot be viewed as something separable from the economics of resource use.*

When we compare the values of two different periods we find ourselves making use of the type of information discussed above in the valuation section. Because a dollar in the future is not worth the same as a dollar in hand right now, we are forced to ask: Why not? And we get around, finally, to an admission that human nature is the cause: *We prefer goods now rather than in some future period.*

Given a choice of having a dollar now versus getting that dollar a year (or more) from now, we will choose possession today rather than later. We would choose to have that dollar now because we could either use it to gain whatever present satisfactions can be obtained by buying a desired good now, or invest that dollar in an interest-earning opportunity (a savings deposit, for instance) and have it grow to some larger amount and therefore able to yield more satisfaction later. This preference for goods now rather than later is called our *time preference*, or time impatience.

Individual time preferences can be (and are) widely different, ranging, at any given instant, from those who will lend (invest) money now so as to enhance their future consumption, to those who borrow against the future so as to be able to consume now in preference to waiting until later. Our time preference can thus be expressed as a *rate* at which future values are discounted back to the present.

Exchanging present and future goods is facilitated by the existence of money and financial markets; exchanges being the *reason for* the existence of those markets, rather than the other way around. The market offers an interest rate for savers, or charges borrowers, permitting us to conduct transactions that demonstrate our preferences. The borrower, by the act of borrowing, has said, "I want the goods this money can buy *now* strongly enough that I'll pay the interest premium to escape having to wait until I have the cash in hand." Whatever the interest rate paid, the borrower has exhibited a sufficiently high time preference that made consumption now "worth more" than the value of that satisfaction later, with a difference in present and future values being equal to the amount of the interest paid.

The only difference between individuals and society in this attribute is the magnitude of the time preference and the length of time that is relevant. One cannot easily make (and carry out) plans for, say, 100 years from now; that's too far into the future and would thus be severely discounted, encouraging earlier use and consumption rather than later. Society, however, can expect (plan for) a much longer life span, which results in a lower discount rate and a postponement of use to a more distant future.

## Economics of Intertemporal Resource Use

The discussion of intertemporal allocation that follows focuses exclusively on stock resources. In particular, the fundamental logic of the efficient timing of stock resource extraction is clarified by examining how the owners of petroleum deposits choose between two alternatives. The first is to pump oil out of the ground now and to invest the proceeds (or *rents*), which amount to the difference between revenues gained from oil sales and expenditures on extraction. The second is to delay extraction until some time in the future, say next year. The extraction of the latter alternative has to do with expected increases in rents, resulting from price increases, declining costs, or both.

As more and more resource owners opt for a deferral of extraction, oil prices will rise in the present, due to a reduction in supply, and fall in the future, because of a supply increase. As a result, the *per unit rent*, defined as the difference between price and per unit extraction cost, will rise more slowly.

Obviously, there is a limit to how much petroleum will be conserved. at some point, at least some owners will find **depletion** (defined as bringing forward the extraction schedule, which is precisely the opposite of how the word *conservation* is being used here) to be more profitable. Depletion, of course, accelerates the growth over time in per unit rents, by lowering current prices relative to future prices.

Let us consider a simple numerical example of these sorts of calculations. Suppose that extracting petroleum costs $5.00/barrel and that financial assets yield a 10 percent annual return. Let us also say that the current price of oil is $15.00/barrel, and that next year's price is expected to be $16.00/barrel. Within the context of this simple case, this year's per unit rent is $10.00 ($15.00 less $5.00) and next year's will be $11.00 ($16.00 less $5.00).

This combination of current and future per unit rents happens to be entirely consistent with intertemporal equilibrium. That is, resource owners have no reason to alter production schedules because the gain associated with conserving one more barrel (until next year) is exactly comparable to what would be earned by pumping out one more barrel and investing the per unit rent (again, until next year). To be specific, change over time in per unit rents,

$$\frac{[(\$16.00 - \$5.00) - (\$15.00 - \$5.00)]}{(\$15.00 - \$5.00)} = \frac{[\$11.00 - \$10.00]}{\$10.00} = 10 \text{ percent,}$$

is neither greater nor less than the annual return on financial assets. In the context of our simple example, the latter is equivalent to the *opportunity cost of capital*.

To appreciate that a resource commodity market would not be in intertemporal equilibrium if there were a discrepancy between per unit rent growth and financial returns, consider the adjustments that would take place if the former were larger than the latter. Let us say, for instance, that the future market value of a barrel of oil is $17.00 instead of $16.00. If this were the case, then per unit rent would be increasing by 20 percent per annum ($12.00 less $10.00 divided by $10.00). Responding to this disequilibrium, petroleum companies would hold on to their environmental wealth. Conservation would, in turn, cause current prices to rise, due to a supply reduction, and future prices to fall, because of a supply increase. Eventually, an intertemporal market equilibrium would be reached, which would feature a smaller difference between current and future per unit rents.

Exactly the opposite sequence of changes would be observed if growth in the per unit rent were less than the opportunity cost of capital. Say that resource owners expected the value of a barrel of oil to be $15.50 a year from now. Under these circumstances, per unit rent growth would be 5 percent ($10.50 less $10.00 divided by $10.00). Intertemporal equilibrium established as accelerated depletion caused the current price, and therefore the current per unit rent, to fall relative to the future price and per unit rent.

As can be gathered from a demand shift in any given period, which would alter relative prices and per unit rents, the petroleum market would shift to a new intertemporal equilibrium. Change would also be observed if extraction costs rose or fell or opportunity costs of capital changed.

# COMMON PROPERTY RESOURCE PROBLEMS _____

For many of the nation's natural resources, property rights are withheld from private ownership. They are owned by the government "in the name of the public," an ownership referred to as *common property.* Examples of such resources are the federal forests and grazing lands in the West, ocean and lake fisheries, a variety of offshore mineral and petroleum deposits, and the air that we breathe.

With common ownership in these resources, individuals use them at rates optimized by their own private costs and returns. We each disregard any other costs that may arise as a consequence of that use because they are external to that decision maker. Factories will pollute the commonly owned air because it is not in their private economic interests to take that cost into account. In spite of the technological devices attached to today's automobiles, we still pollute the air when we drive. We each ignore that cost because it is not in our economic interests to incur further pollution control costs for the benefit of others.

Common property resources do not readily lend themselves to the valuation process inherent in the institutions of private property and the marketplace. And, because we usually are free to use these resources without regard for the costs that are external to our private cost and returns, these resources are used at more intensive rates than is economically justified from the standpoint of society as a whole.

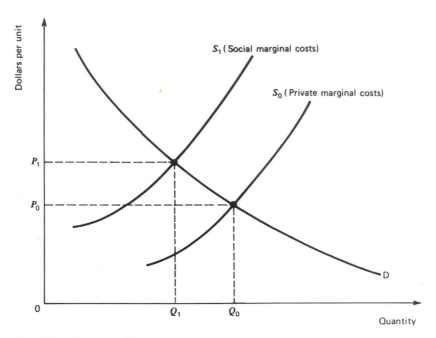

**FIGURE 13-2**   Equilibrium in common property usage.

We illustrate this problem with an example diagrammed in Figure 13-2. Assume a fishery under two different institutional arrangements. The first is one of open, uncontrolled fishing, and the second, a method of licensing that controls the number of fishers and the fish harvest. With an open fishing arrangement, only the private costs and returns of fishing are considered by those engaged in that activity. The marginal cost curves of all fishers in this market are summed horizontally to derive the supply curve $S_0$ in the diagram. Given the demand for fish—the demand curve $D$—the optimal price of fish is $P_0$ and the total amount of fish harvested is the quantity $Q_0$. As a result of open fishing, the fishery stock would be depleted over time because each will harvest at a rate that ignores the impact on the present stock of fish. This arrangement will cause the cost of fishing to increase, as the people fishing will have to make longer trips and spend more time catching a given number of fish from the dwindling stock of fish. These external costs, imposed on others, plus the private costs of fishing are termed *social costs*. The sum of the social costs of fishing is shown in the supply curve $S_1$. The optimal quantity of fish harvested, taking all costs into account, is the quantity $Q_1$ at price $P_1$, rather than the quantity $Q_0$ at price $P_0$. This overharvesting problem is similar to many other common property resource problems wherever external costs are not included in the individual's decision-making process.

External costs of fishing are internalized for all who want to fish by assigning specific property rights to particular individuals, at a cost. Most states and the federal government prevent the imposition of external costs on others by licensing and setting individual catch limits, thereby preventing the stock depletion that would occur under open fishing. Adding the costs of property rights to all individuals' fishing decisions increases their marginal costs and causes them to reduce their fishing activities and fish harvesting rate to the more optimal $Q_1$, the rate at which the stock of fish can be maintained.

# AGRICULTURE'S NATURAL RESOURCES _____

Those who concern themselves with the productive uses of land sooner or later raise the question of whether we may run out of agricultural land in some future period. Despite the serious concern over an impending food crisis, U.S. agriculture's problem has been just the opposite for more than 60 years. Rather than finding food costs increasing as a result of agricultural inadequacies, the problem has been one of being able to produce more than can economically be marketed.

An overview of trends in the nation's developing agricultural industry helps focus on the changes that have occurred over a long period of years (Table 13-1). Sharp increases in output have occurred in a manner that could appear to have resulted from fortunate accidents rather than as conscious responses to changing economic conditions.

The opening of the continental United States to westward settlement added many millions of acres of productive land. As population grew from about 5 million to more than 75 million during the 1800s, probably three-quarters of a billion acres of agricultural land were also added. By 1920, the effects of land-settlement programs and World War I food needs had increased the amount of land in farms to more than 950 million acres.

Two important changes in agriculture contributed to the large growth in output. First, the development of mechanical power made possible a conversion from animal power, which became especially rapid during the pre-World War II period. Then, from about the 1930s, scientific discoveries in improving soil fertility, controlling weeds and plant diseases, genetic improvement of plants, and better management practices have all combined to create a large reservoir of productive capacity. The volume of total output has quadrupled since 1900, with only about a 17 percent increase of land in farms.

Except for short periods of international crisis, returns in agriculture have been relatively low. This, coupled with improved off-farm opportunities, both pushed and pulled farmers to other pursuits. Farm numbers

**TABLE 13-1** A Growing U.S. Agriculture

| Year | U.S. Population | Number of Farms | Farmland (Acres) | Index of Output |
|---|---|---|---|---|
| | | Millions | | (1982 = 100) |
| 1800 | 5 | — | — | — |
| 1850 | 23 | 1.5 | — | — |
| 1900 | 76 | 5.7 | 839 | 26 |
| 1920 | 106 | 6.5 | 956 | 37 |
| 1940 | 132 | 6.4 | 1061 | 43 |
| 1950 | 152 | 5.6 | 1162 | 51 |
| 1960 | 181 | 4.0 | 1176 | 63 |
| 1970 | 205 | 2.9 | 1107 | 73 |
| 1980 | 228 | 2.4 | 1039 | 91 |
| 1985 | 238 | 2.3 | 1016 | 103 |
| 1990 | 250 | 2.14 | 987 | 108 |
| 1991 | 253 | 2.11 | 982 | 108 |
| 1992 | 255 | 2.10 | 979 | 116 |
| 1993 | 258 | 2.08 | 976 | 108 |
| 1994 | 261 | 2.06 | 973 | NA* |
| 1995 | 263 | 2.07 | 972 | NA* |

SOURCES: Compiled from annual issues of *Agricultural Statistics* and *Statistical Abstract of the United States*, both U.S. Government Printing Office, Washington, D.C.; and published data of the Economic Research Service, USDA, Washington, D.C.

* NA means not available.

declined nearly 65 percent from their 5.7 million in 1900, and by more than almost 70 percent from the high reached in the mid-1930s.

Food production capability is influenced by many factors. Land, people, water, energy, minerals, capital, management practices, and climatic conditions are integral parts of the production process. Each has contributed to an output increase of about 70 percent since 1960, even though 17 percent less land is now being used to produce agricultural products.

The total set of resources used in agriculture is far from constant. Quantities and proportions of resources are frequently adjusted in response to economic and other conditions. Technological improvements in the use of one type of resource, with their resulting cost effects, cause increased usage of some resources while the use of others is reduced—the type of resource substitution discussed in Chapter 5.

Other cost effects come from competitive forces outside of agriculture, such as increased urban or recreational demand for land. With the increased price of land from such demand shifts, producers are forced to substitute other resources for land in order to remain competitive in their product markets. Higher wages elsewhere draw labor out of agriculture and become a cause of increased agricultural labor cost, resulting in the substitution of capital and other inputs for labor. Industrial competition for water is viewed as a threat to agricultural use of that resource. And given the rate of output of energy and mineral resources that are of use in agriculture, the greater their use elsewhere in the economy the higher become their costs to agriculture. The increased competitive demand for resources, both within and outside of agriculture shows up in higher resource costs, shifts in resource use, and changes in agriculture's relative income position. The remaining sections in this chapter note some of the broader changes that have occurred for major resource categories.

## Land

The total land area of the United States presently amounts to nearly 2.3 billion acres (Table 13-2). Of this total, agriculture used more than one-half in 1992 (including forest land grazed). Occasional remeasurement and increases in reservoirs constructed account for the 7-million acre decline in total land area since 1959.

Through the decades shown, total cropland held at slightly less than one-fifth of all land uses, decreasing by 5 million acres in the last 5-year period but still 2 million acres greater than in 1959.

The 658 million acres of grassland pasture and range on which livestock grazed in 1992 made up almost one-third of all land uses for that year. A total of 41 million acres has been shifted out of pasture and range since 1959. These lands provided about 80 percent of all grazing, with the balance being from forestland grazed and cropland used only for pasture.

Within the "special uses" grouping are acres dedicated to uses that, by their nature, prevent agricultural use. Although their total acreage changed by 158 million acres during the 33 years shown, not all that land was removed from agriculture. The largest change within this category has been in the land devoted to recreation and wildlife. A sizable portion of the change resulted from 10 million acres of Alaskan public land being reclassified as a wildlife preserve.[14]

During the 33 years shown by these data, population grew by about 85 million, increasing the nonagricultural demand for land. Not only were greater acreages used for residences and industrial and commercial establishments, but additional acreages were also required for highways, roads, and airports.

---

[14] *Our Land and Water Resources*, Washington, D.C.: Economic Research Service, U.S. Department of Agriculture, Misc. Pub. No. 1290, May 1974, p. 10.

**TABLE 13-2**  Trends in Land Use for Major Uses, Selected Years, United States

| Type of Use | 1959 | 1969 | 1974 | 1978 | 1982 | 1987 | 1992 |
|---|---|---|---|---|---|---|---|
| | | | | Million acres | | | |
| Cropland[a] | 392 | 384 | 382 | 395 | 404 | 399 | 394 |
| Pasture and range[b] | 699 | 692 | 681 | 663 | 662 | 656 | 658 |
| Forested land[c] | 728 | 723 | 718 | 703 | 665 | 648 | 648 |
| Special uses[d] | 123 | 142 | 147 | 158 | 270 | 279 | 281 |
| Other[e] | 329 | 323 | 336 | 345 | 274 | 283 | 283 |
| Total | 2271 | 2264 | 2264 | 2264 | 2265 | 2265 | 2264 |

SOURCE: Arthur B. Daugherty, *Major Use of Land in the United States*, 1992, Economics and Statistics Service, USDA, Washington, D.C., September 1995.

[a] Includes cropland harvested, idle, fallow, and crop failure.

[b] Includes an estimated 76 million acres of cropland used only for pasture in 1978.

[c] Both public and private, exclusive of forest land devoted to parks, wildlife preserves, etc., but including forest acres grazed.

[d] Urban uses, farmsteads, highways and roads, parks, and preserves, military installations, airports, and railroad rights-of-way.

[e] Desert, swamp, and other areas of little (present) agricultural use.

Economic efficiency of land use means that higher-valued uses (those with the greatest present value of discounted future net returns) will displace lower-valued uses. As urban and urban-related uses take cropland from farms, pastureland, for instance, will be pushed out by cropland, and pastureland will, in turn, "bump" other lower-valued uses. This process is not especially visible in our data, but it has been occurring through time.

The listing of total cropland disguises land losses to other uses. Total acres show only what is available for a variety of crop-related uses, not acreages used. Changes in cropland harvested more directly reflect operator responses to market and other conditions. Between 1959 and 1964, acres from which crops were harvested declined by 26 million acres. Productivity gains caused output to grow more rapidly than did the markets for many crops and required increased acres to be idled under the federal farm programs. By 1969, program-idled land totaled 51 million acres, so that cropland neither harvested nor pastured increased by 20 million acres.

During the 9-year interval from 1969 to 1978, the change in total cropland was a sharp response to the release of 33 million acres of land from program restrictions and the large increase in grain sales abroad.[15] The acreage of crops harvested in 1992 was 2 percent less than for 1959, even though total cropland acres increased by about 1 percent.

_____
[15] Ibid., p. 3.

**TABLE 13-3** U.S. Water Availability and Use, Selected Years[a]

| | | | | Consumption | | | |
|---|---|---|---|---|---|---|---|
| Year | Average Annual Runoff[b] | Total Withdrawn | Industrial | Public Supply[c] | Rural, Domestic and Lstk. | Irrigation | Acres Irrigated (million) |
| | ———— Thousands of million gallons per day ———— | | | | | | |
| 1950 | 1271 | 184 | 77 | 14 | 4 | 89 | 26 |
| 1960 | 1271 | 273 | 138 | 21 | 4 | 110 | 34 |
| 1970 | 1271 | 373 | 211 | 27 | 5 | 130 | 39 |
| 1975 | 1271 | 419 | 245 | 29 | 5 | 140 | 41 |
| 1980 | 1271 | 445 | 255 | 34 | 6 | 150 | 50 |
| 1985 | 1271 | 392 | 211 | 36 | 8 | 137 | 50 |
| 1990 | 1271 | 408 | 225 | 38 | 8 | 137 | 51 |

SOURCE: Wayne B. Solley, Robert R. Pierce and Howard A. Perlman, "Estimated Use of Water in the United States in 1990," Geological Survey Circular 1081, U.S. Department of Interior, Alexandria, Va., 1993; *Our Land and Water Resources,* Economic Research Service Misc. Pub. No. 1290, May 1974; *Our Nation's Land and Water Resources,* ERS-530, August 1973; and *Agricultural Statistics* (1982), U.S. Government Printing Office, Washington, D.C.

[a] Forty-eight contiguous states and District of Columbia.

[b] Compare Alaska's average runoff of about 650 million acre-feet per year.

[c] Urban use.

## Water

Given present technology, only a very small part of the world's physical supply of water is usable. It is from the less than 1 percent not in the oceans or in frozen polar ice caps that we derive the water used for life support.

Within the continental United States, about 4,193,000 million gallons per day of water falls as precipitation annually, providing water for growing plants, for streams and lakes, and underground aquifers. Of that amount, less than one-third is available for use by human direction, the other 70 percent escaping naturally by evaporation or transpiration from plants. The balance, about 1,271,000 million gallons per day, is the natural run-off from which our economic supply of water is derived (Table 13-3).

As our population has grown and agriculture has increased its output to meet the expanding demand for food, the amount of water withdrawn for use has increased greatly. Total withdrawals have more than doubled over the 184 million gallons per day taken in 1950. But withdrawal does not mean that much water has disappeared, however. In more recent years, consumptive use (i.e., disappearance by evaporation, transpiration, human and animal consumption, or incorporation in products) has averaged less than 25 percent of the total amount of water withdrawn. The difference between withdrawals and consumption is the amount returned to surface or ground water sources for subsequent use.

Although urban uses of water have increased, households use only a very small proportion of total urban use. The largest nonagricultural users of water are industrial and steam-electric generating plants. Because much water is used for cooling, rather than becoming a part of their products, consumptive use by such plants amounts to less than 5 percent of the water withdrawn.

More than 35 percent of all the water withdrawn in the United States is for agricultural uses, with more than 94 percent of agricultural withdrawals being used for irrigation. Because of its high consumption rate, agricultural uses of water amount to almost 85 percent of total American consumption.[16]

In the 40 years from 1950, irrigated land has almost doubled to 51 million acres, making up about 17 percent of all harvested cropland by 1990. In the face of restricted or depleting supplies, such practices as canal and ditch lining, and waste water control, and the increasing use of sprinkler irrigation systems have reduced per acre water use by more than 20 percent over the last 40-year period.

An average of 625,000 acres of irrigated land has been developed annually since 1950. For many of those years the primary growth was provided by government-sponsored irrigation projects. Irrigation is a large water user in the West. The western states accounted for 90 percent of the total water used for irrigation. The major irrigation water users by state are California, Idaho, Colorado, Montana, and Texas.

Urban-related water problems are quite dissimilar in different parts of the country. In the higher rainfall areas of the East and Southeast, water-supply problems stem more from inadequate storage capacity than from insufficient precipitation. In the Southwest and certain parts of the Great Plains the demand for water exceeds local supplies, requiring the transfer of large amounts of water to population centers.

Except for water-supply problems in the arid West, the most serious urban problem has been one of water pollution from disposal of wastes. Stream and lake pollution, both from household and industrial waste loads, has intensified the need for large investments in sewage treatment plants and industrial water recycling facilities. This is a problem common to all urban areas of the nation. Agricultural pollution of waterways has also become serious in many areas. The long life of residual herbicides and pesticides causes them to be present both in the soil and in wa-

---

[16] Wayne B. Solley, Robert R. Pierce and Howard A. Perlman, "Estimated Use of Water in the United States in 1990," Geological Survey Circular 1081, U.S. Department of Interior, Alexandria, Va., 1993.

ter supplies, with uncertainty over their full ecological effects. Declines in water quality also result from siltation, increased salinity, and the eutrophication caused by leaching inorganic nitrate and phosphate fertilizers.

## Energy Use in Agriculture

Energy of one form or another has played an increasingly important role through the long history of coaxing greater quantities of food from the earth. The succession of energy types from human power only, to animals, then inanimate sources, has served to reduce the energy cost component of the food we eat.

Fossil fuels have made possible the mechanical power improvements that have become so important a part of modern agriculture. Our heavy dependence on finite energy has been the cause of growing criticism. And the Arab oil embargo in 1973 made us even more aware of how much we rely on fossil energy. The sharp rise in energy costs, not only fuel energy, but fertilizers, pesticides, and other petroleum-based products as well, brought demands for energy conservation. Research has since attempted to discover and identify uses and rates of use throughout the economy.

Table 13-4 summarizes estimates of use rates within the food system. Data that are both current and accurate are not yet as abundant as for

**TABLE 13-4** Energy Use Estimates for Major Components of the U.S. Food and Fiber System[a]

| Component of System | Petroleum Fuels | Natural and Propane Gas | Electricity | Other | Percentage Share of System Total |
|---|---|---|---|---|---|
| | -------- Percentage of own energy used -------- | | | | |
| Farm production | 38 | 48 | 14 | — | 15 |
| Food processing industries | 25 | 42 | 18 | 15 | 42 |
| Marketing and distribution | 43 | 35 | 21 | 1 | 15 |
| Restaurants and cafes | 54 | 28 | 17 | 1 | 28 |
| Percentage shares of system total | 38 | 38 | 18 | 6 | 100 |

SOURCE: Economic Research Service, "Energy Requirements in the U.S. Food System," *Agricultural Outlook,* Washington, D.C.: U.S. Department of Agriculture, AO-8, March 1976, pp. 18–21, and R. Thomas Van Arsdall and Patricia J. Devlin, "Energy Policies: Price Impacts on the U.S. Food System," Washington, D.C.: U.S. Department of Agriculture, Agricultural Economic Report No. 407, July 1979.

[a] Forecast for 1991.

many other segments of the economy. The studies from which this information was derived were, in themselves, incomplete estimates of separate components of the system.

Increased energy prices, plus conscious efforts of firms and households to reduce their energy consumption, have undoubtedly had some effect. Gasoline provided farms with 23 percent of their total energy, as compared to 51 percent from diesel fuel. Most new farm tractors bought since the mid-1970s are diesel fueled, which has altered the relative energy contributions of these two fuels. Likewise, home weatherstripping and insulation, the manufacturing of more energy-efficient home appliances, and the conversion of many industrial buildings and electrical power plants to coal heat all contribute to changing fuel-source proportions.

The food system components described in Table 13-4 include four selected major activities involved in making food available to the consumer. The major divisions of the system included in the study are farms; all food processing; wholesale, and retail firms and the transportation network; and public food-service establishments.

The food system is estimated to use 14 percent of total American energy use. Within this system, agricultural production uses about 15 percent, which amounts to less than 1.8 percent of the United States' total.

Of the energy sources, petroleum fuels (gasoline, diesel, and fuel oils) provide 38 percent of total energy used by the food and fiber system. The same share, 38 percent, is derived from natural and propane gases, with the remaining 24 percent obtained from electricity, coal, and other minor fuels. Because of its heavy emphasis on mobile power, the farm production sector derives more than 85 percent of all its energy from petroleum fuels and natural and propane gas.

Food system energy use in 1991, both in British thermal units (Btu) and dollar amounts,[17] are shown in Table 13-5. Identified components of the food system account for about 14 percent of total energy consumption. The largest user of energy in the system is food processing, using 36 percent of the total energy used by the six identified sectors. It also takes about 38 percent as much energy to store, refrigerate, and prepare food for consumption—restaurants and cafes and in-home uses—as it does to get our food produced and delivered to these users.

Food system energy costs amounted to about 20 percent of the $486 billion U.S. consumers spent for food in 1991. Given assumed changes in the food system, higher energy use is projected.

_____

[17] Data compiled from R. Thomas Van Arsdall and Patricia J. Devlin, "Energy Policies: Price Impacts on the U.S. Food System," Washington, D.C.: U.S. Department of Agriculture, Agricultural Economic Report No. 407, July 1979.

**TABLE 13-5**   Energy Use and Costs in the Food System With Projections For 1991[a]

| System Component | Energy Use 1991 | Energy Costs 1991 |
|---|---|---|
| | (Quadrillion Btu) | Millions |
| Farm production | 824 | $ 8,030 |
| Agricultural chemicals | 676 | 3,674 |
| Food processing | 4,176 | 30,163 |
| Marketing and distribution | 1,491 | 14,164 |
| Restaurants and cafes | 2,806 | 23,735 |
| In-home preparation | 1,150 | 18,834 |
| Total | 11,523 | $98,600 |

SOURCE: R. Thomas Van Arsdall and Patricia J. Devlin, "Energy Policies: Price Impacts on the U.S. Food System," Washington, D.C.: U.S. Department of Agriculture, Agricultural Economic Report No. 407, July 1979.

[a] Forecast for 1991.

The longer term investments in energy-using equipment, in all areas from energy production through industrial, commercial, and home use, causes our response to the energy problem to be slow. But opportunities for energy conservation exist and are being adopted. Responses to energy price changes take the form of adjustments both in rates of use and energy-form substitution. Use-rate changes depend on the user's price elasticity of demand for energy; factor-substitution is a longer term response, often requiring new machinery, equipment, and appliances able to utilize different energy sources than the items being displaced.

Because of our reliance on imported oil for liquid fuels, and the instability of much of that supply, there has been a major effort to convert grain crops into ethanol to use as an extender for gasoline. A mixture of 10 percent ethanol and 90 percent gasoline is called *gasohol*. Gasohol may be used in vehicles without engine modifications. U.S. ethanol production in 1995 was 1.4 billion gallons. This amount would yield 14 billion gallons of gasohol, roughly 9 percent of U.S. consumption of automotive fuel. Currently, ethanol production is being subsidized by a federal tax exemption of $0.54 per gallon, exemption from state taxes in some states, and federal loans and loan guarantees for those investing in alcohol distilleries to produce ethanol. Whether or not gasohol from grain crops becomes an economically viable source of fuel depends to a considerable extent on future price relationships between grains and gasoline.

## Human and Other Resources

The human input in producing our supply of food and fiber is derived from farm families and other workers who provide the labor services to agriculture. Through most of the long history of mankind, increased food output has been closely linked to the number of people engaged in agricultural production. During much of this period, technology changed very little so that the labor/land/output proportions remained quite stable.

In the more primitive societies, most if not all of agriculture's output must be used simply to support the human and animal power used to produce food. With no surplus output beyond resource support needs, economic progress leading to other consumer goods is inhibited. Only with the ability to produce in excess of one's own food needs can workers be released to produce better housing, medical services, education, and other beneficial services. Over the past century, the technology of production improved so as to increase the amount of food a worker could produce.

The combination of labor and mechanical devices in agricultural production began with the invention and widespread adoption of such basic capital items as the moldboard plow, mechanical reaper, and the internal combustion engine during the 1800s. As such capital inputs became more generally available, the substitution of capital for labor intensified and the productivity of labor was greatly enhanced in the process.

Purchased inputs have become increasingly important in agriculture production (Table 13-6). Feed, seed, mechanical power and other equipment, fertilizers and agricultural chemicals represent capital inputs, all of which contribute to the increasing productivity of agriculture, and a lower real cost of food and fiber for consumers.

Feed, seed, and livestock purchases increased 44 percent between 1960 and 1980. Also durable equipment increased 48 percent over the same period. With the large acreage reduction programs of the 1980s, less capital was used in agriculture and, as a result, major purchased inputs have declined. Durable equipment use has dropped 41 percent since 1980.

The use of fertilizers (nitrogen, phosphate, and potash) and agricultural chemicals have served to help increase yields on both a per acre and a per worker basis, further reducing the labor cost per unit of output. The combined effects of these and other such inputs have reduced the farm labor requirements in production by 56 percent since 1960.

Table 13-7 indicates how the use of labor has declined in recent years. Just since 1950, we have seen the U.S. population increase by more than 73 percent, and that has been accompanied by higher per capita incomes and increased effective demand for agricultural products. At the same time, however, total labor use in agriculture declined by 70 percent.

**TABLE 13-6**  Major Purchased Items Contributing to Productivity

| Year | Feed, Seed, and Livestock Purchases (1982 = 100) | Durable Equipment (1982 = 100) | Energy (1982 = 100) | Agricultural Chemicals (1982 = 100) | Farm Labor (1982 = 100) |
|------|------|------|------|------|------|
| 1960 | 71 | 69 | 82 | 59 | 186 |
| 1965 | 70 | 69 | 89 | 72 | 160 |
| 1970 | 83 | 78 | 92 | 76 | 133 |
| 1975 | 78 | 89 | 101 | 92 | 120 |
| 1980 | 102 | 102 | 110 | 133 | 102 |
| 1985 | 93 | 86 | 90 | 100 | 91 |
| 1990 | 93 | 65 | 92 | 98 | 85 |
| 1993 | 96 | 60 | 92 | 106 | 81 |

SOURCES: *Changes in Farm Production and Efficiency,* Washington, D.C.: Economic Research Service, U.S. Department of Agriculture; and *Agricultural Statistics,* 1995–96. Washington, D.C.: U.S. Government Printing Office.

**TABLE 13-7**  The U.S. Farm Labor Resource

| Year | Total Agricultural Labor | Family Labor | Hired Labor | Number of People Supported by One Farm Worker |
|------|------|------|------|------|
| | -------------------- Thousands -------------------- | | | |
| 1950 | 9342 | 7252 | 2090 | 16 |
| 1955 | 8237 | 6341 | 1897 | 20 |
| 1960 | 7057 | 5172 | 1885 | 26 |
| 1965 | 5610 | 4128 | 1482 | 37 |
| 1970 | 4523 | 3348 | 1175 | 48 |
| 1975 | 4357 | 2501 | 1273 | 58 |
| 1980 | 3705 | 2402 | 1303 | 76 |
| 1985 | 2941 | 1904 | 1037 | 85 |
| 1990 | 2869 | 1965 | 904 | 96 |
| 1995 | 2835 | 1967 | 868 | NA[a] |

SOURCE: *Economic Report of the President,* 1993, U.S. Government Printing Office, Washington, D.C. and *Agricultural Statistics,* 1995–96, U.S. Government Printing Office, Washington, D.C.

[a] NA means not available. This number is no longer published, but the author's estimate is about 101 in 1993.

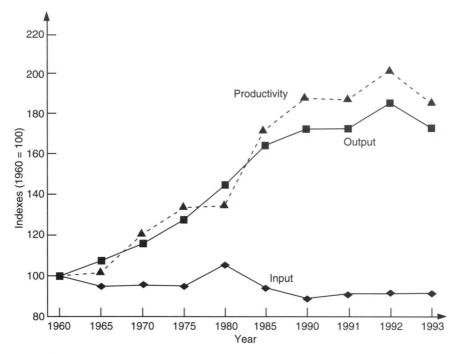

**FIGURE 13-3**   Index measures of agricultural use and production.

Until the 1960s, more than three-fourths of all agricultural labor was provided by the farm family. By 1995, hired labor amounted to 31 percent of all labor used on U.S. farms and ranches.

Because of the seasonality of production, many farmers find it impossible to utilize their full labor force on a year-round basis. Migrant foreign nationals and migratory workers have been used to meet such special needs.

The effect of the overall resource mix on the productivity of labor is demonstrated by the final series in the table, which shows the number of people supported by each agricultural worker. In 1900, one farm worker was able to produce enough food and fiber for about six people. As farms have utilized more effective combinations of capital and labor, the number of people supported by an agricultural worker grew to about 101 by 1993.

This huge growth in production per farm worker—increasing the number of people supported by six times in 43 years—is not due only to enhanced farm productivity. As the nation's food and fiber system has developed, a number of early-stage activities have been shifted from farm to nonfarm workers, and this series does not reflect those changes.

Figure 13-3 summarizes with indexes the changes that have occurred in the productivity of the agricultural industry. We use 1960 as the base 100 for this diagram only for its value as a visual benchmark.

The overall resource commitment for producing our food and fiber has changed relatively little through the years shown in the diagram. From an input index of 106 in 1952, input use declined steadily and remained below 100 for the 15-year period 1961 through 1975. The combination of farm program restrictions and reduced land, labor, and machinery inputs since the early 1980s has dropped the index well below its base 100. That, combined with unusually severe flooding in the midwest in 1993, caused the output of all farm products to fall by almost 7 percent from their 1992 level.

Continually increasing output, however, has been a characteristic of the industry and agricultural output rebounded in the mid-1990s due to better growing conditions. Agricultural output in 1993 was over 70 percent higher than what was produced in 1960. And the further back in time we might wish to go in making comparisons, the greater becomes the percentage of increase. For instance, 1985 output was more than 90 percent greater than in 1950, more than double that of 1940, and triple that of 1936.

The line in the graph labeled Productivity is a measure of unit resource productivity, obtained by dividing the index of output by the corresponding input index. The result gives an estimate of output per unit of input for all agricultural resources together.

# SUMMARY

Natural resources are those resources provided by nature. They become valuable when the technology of their utilization and the demand for their products makes use feasible.

Some natural resources such as wind, the tides, and sunlight are available for use on a continuing or flow basis. Use of these resources need not diminish their future availability. Other resources, however, are present in a fixed (fund) quantity. Present use of such resources diminishes the physical amounts remaining for use in the future.

Whether land, petroleum, water, or coal, the depleting supply of any resource will cause its price to increase, *ceteris paribus*, provided the market is permitted to function. With price increases that truly reflect economic conditions surrounding the use of a resource, a number of forces are set in motion: (1) the rate of depletion is reduced as marginal users are denied economic use of the resource; (2) the search for alternative sources of the same or similar service is intensified; (3) the economic supply expands as the increased price makes possible the use of lower quality sources of that resource; and (4) the higher resource price is reflected in the prices of certain consumer goods, causing consumers to restrict their purchases by shifting to alternative goods.

The largest use of land by far is agricultural, amounting to about 50 percent of total United States land area. Is this a sufficient acreage to meet our future requirements? Land use trends through history would suggest the problem is much less serious than many would have us believe. We need as many acres as it takes to produce the food and fiber that we are willing and able to buy. As the supply of land for food and fiber declines (either relative to demand, or absolutely), food and fiber prices will rise. Increasing prices for food and fiber will be reflected back to land with the price of land also increasing. As the price of land increases, land conversions to other uses will be retarded or even reversed.

The relative prices of all resources used in agriculture are ever changing. Responses to these changes are implied by relative changes in the rates of use of land acres, water for irrigation, family and hired labor, energy in all its forms, and capital in the form of power, machines of all kinds, fertilizers and agricultural chemicals, and so forth. Resource substitution and the adoption of new technology have permitted a near tripling (in effect) of the economic supply of land since the mid-1930s. And this increase occurred despite a decline in physical acres used for food and fiber production.

## CHAPTER HIGHLIGHTS

1. Natural resources are all those resources provided by nature, including land, water, minerals, plants, animals, and humans themselves.
2. A natural resource has value only when an economic use has been discovered for it.
3. Fund resources are those whose quantities are finite, such as petroleum and minerals whose present use reduces the physical amount available in the future.
4. Flow resources are renewable, such as sunlight, wind, and flowing water, whose present use need not reduce their future supply.
5. Biological resources such as growing plants and animals exhibit characteristics of both fund and flow resources.
6. Proved reserves depend on what we know about the existence of a resource, but the economically usable part of proved reserves depends on the demand and supply of the resource.
7. The demand for a resource is derived from the demand for that resource's product.
8. The value of a resource is determined by the present value of its future net earnings.
9. The difference in value between present and future goods is determined by the rate at which we (each) discount those future values.

10. Individual rates of discount vary by the individual.

11. Resource conservation is an economic problem.

12. Common property resources are frequently used at uneconomic rates because all costs of using them are not considered.

13. Social costs include all the costs (private plus external) of an economic activity.

14. The total United States land area amounts to about 2.3 billion acres, with agriculture using approximately one-half the total.

15. Cropland amounts to less than one-fourth of the total land area.

16. The consumption of water amounts to less than 25 percent of the total water supply.

17. Water pollution problems result from both urban and agricultural uses of water and other resources.

18. Direct farm production uses less than 2 percent of all the energy used in America.

19. The agricultural marketing system uses more than three times as much energy as do the raw food and fiber producers.

20. More energy is used to prepare food for consumption than farmers and ranchers use to produce that food.

21. Energy costs make up about 20 percent of total food costs.

22. The increased use of capital inputs increases the productivity of the other resources.

23. The use of labor in agriculture has declined dramatically, yet each laborer is able to support many more people than could be done earlier.

24. Output has increased greatly, even though there has been no significant increase in overall inputs since the mid-1950s.

## KEY TERMS AND CONCEPTS TO REMEMBER

Biological resources

Capitalized value

Common property

Conservation

Economic supply

Flow resources

Fund or stock resources

Natural resource

Physical supply

Present value

Resource reserve

Social costs

Time preference

# REVIEW QUESTIONS

1. What are "natural" resources, and why do we value them?
2. What do we mean by the economic supply of land (or any natural resource) as compared with the physical supply?
3. Is the physical supply of a natural resource of any meaning to society?
4. Is economic growth incompatible with preserving the environment? Is it incompatible with conservation?
5. Why not just pass laws preventing the use of disappearing natural resources?
6. How much land should we reserve exclusively for agricultural use? Surely this is an economic question, so can that question be answered with anything but economic criteria?
7. "Because the supply of water is so limited, especially in the arid West, economic wisdom dictates that we reserve most of this resource for agricultural use." Analyze this problem carefully, taking into account the meaning of economic efficiency both for this resource and the economic well-being of all the members of society.
8. If banks (in a position of charging interest on borrowed money) aren't the true cause of interest, why should we have to use an interest rate in determining the present value of any resource?
9. What is $1000, to be received 3 years from today, worth to you? What could you sell that obligation for? Might these two values (the value to you versus the value to a buyer) be different?
10. American agriculture is frequently criticized as being "energy inefficient" because we use far more fossil energy to produce each unit of food and fiber than do foreign producers. Is this a valid efficiency criterion? If it isn't, what criteria should be used in judging the energy efficiency of U.S. agriculture?

# SUGGESTED READINGS

Barlowe, Raleigh. *Land Resource Economics,* 4th ed., Englewood Cliffs, N.J.: Prentice Hall, 1986.

Brehm, Carl. *Introduction to Economics.* New York: Random House, Chapter 7.

Collins, Speir. "Reserves, Reserves, Reserves," *Petroleum Today,* Washington, D.C.: AmericanPetroleum Institute, 1976/two.

Cotner, M. L. "Land Use Policy and Agriculture: A National Perspective," Washington, D.C.: Economic Research Service, USDA ERS-630, July 1976.

Cotner, M. L., M. D. Skold, and O. Krause. "Farmland: Will There be Enough?" Washington, D.C.: Economic Research Service, USDA ERS-584, May 1975.

"Farmland Resources for the Future," Washington, D.C.: Economic Research Service, USDA, Agricultural Information Bulletin No. 385, April 1975.

Gwartney, James D., and Richard Stroup. *Economics, Private and Public Choice,* 6th ed., Fort Worth: Dryden Press, 1992, Chapters 28–30.

"Our Land and Water Resources," Washington, D.C.: Economic Research Service, USDA, Misc. Pub. No. 1290, May 1974.

Pavelis, George A. "Natural Resource Capital in American Agriculture," Washington, D.C.: Economic Research Service, USDA, Working Paper No. 37, September 1977.

Taylor, Robert C. "Natural Economic Impact of Using Crop Residues and Grain to Produce Alcohol," Report Prepared for the U.S. Congress, Office of Technology Assessment, Washington, D.C., December 1978.

U.S. Department of Agriculture, "Cutting Energy Costs," *The 1980 Yearbook of Agriculture,* U.S. Government Printing Office, Washington, D.C.

# AN OUTSTANDING CONTRIBUTOR

**Daniel W. Bromley**   Daniel W. Bromley is currently Anderson-Bascom Professor, a distinguished chair position, and is also Chair of the Department of Agricultural and Applied Economics at the University of Wisconsin—Madison. He was reappointed to the Department Chair in 1995 after formerly serving from 1981 to 1986.

Bromley received a B.S. degree in range ecology from Utah State University in 1963 and developed an interest in agricultural economics applied to resource problems. He was Outstanding Senior in the College of Natural Resources at Utah State University in 1963. He received an M.S. degree and Ph.D. in natural resource economics in 1967 and 1969, respectively, at Oregon State University. He started his agricultural economics career at Madison in 1969 and has worked most of his career on natural resource economics issues. He is a leading authority on natural resource economics and has concentrated his work on the welfare-theoretic foundations of public policy.

Bromley has established many professional service affiliations outside the University of Wisconsin-Madison, such as (1) Associate Fellow, London Environmental Economics Center; (2) Honorary Fellow, Evaluation and Territorial Economy Scientific Study Center, Italy; (3) Advisory Board Member, Economics Institute, Boulder, Colorado; (4) Consultant to the World Bank on Environmental Policy and Land Reform in South Africa; and (5) Consultant to the Ministry for the Environment in New Zealand. He has also served on numerous University of Wisconsin committees and state organizations to advise on natural resource issues. He has evaluated agricultural economics programs as part of a site-visit team at Montana State University, University of Massachusetts, and the University of Nebraska.

Dr. Bromley's teaching duties at Madison have included natural resource economics, economics of public decision making, water resource economics, and institutional economics. He has been recognized as a major contributor in making the University of Wisconsin's graduate program in natural resources world-renowned. He has also presented invited lectures and seminars at Cambridge and Oxford in England, and at universities in Germany. He has traveled extensively to serve on various university committee appointments and as a consultant, and is currently External Examiner for Universiti Pertanian Malaysia and Sri Venkataswara University in India. He has an impressive record as a highly effective thesis advisor, having served as major advisor for over 30 Ph.D.'s. Bromley's students currently serve in leading universities in the United States, Australia, and New Zealand, as well as in important international organizations such as the World Bank.

Bromley has published extensively on public decision making, natural resource economics, and economic development. He wrote *Economic Interests and Institutions: The Conceptual Foundations of Public Policy* in 1986–87 when he was a Visiting Fellow at Wolfson College, Cambridge University. Since that time he has written *Environment and Economy: Property rights and Public Policy* and has edited three recent books on public policy and environmental economics. These books have drawn wide attention as important contributions to the development of new methodology in the profession. Also, he has over 40 refereed journal publications, including 13 in *AJAE*.

Bromley has received several professional honors including Fellow of the American Agricultural Economics Association in 1992. He has served as editor of *Land Economics* since 1974, a leading professional journal dealing with environmental, natural resources, urban, and public utility issues. He was appointed as the first Anderson-Bascom Professor of Agricultural Economics in 1986 in recognition of his distinguished research and teaching record.

CHAPTER **14**

# Rural Development

**R**ural development may be defined as making rural America a better place in which to live and work.[1] Emphasis is directed to the well-being of people rather than on economic growth itself. Development of rural areas may or may not be associated with an increase in real per capita incomes. Many studies of rural development concentrate on how to attract industry to an area to improve living standards. But this is only one part of rural development. Other concerns of rural development are poverty problems, population distribution, rural housing, public services, and employment opportunities. All these are included in the meaning of the term *quality of life*.

The quality of life in rural communities is directly influenced by local agricultural producers; and the well-being of agricultural producers is affected by their rural communities. Rural communities are the service and the shopping centers for agricultural producers. This is where producers buy most of their farm supplies, market their products, handle their finances, and purchase their food, clothing, cars, education, services, recreation, and entertainment. Rural communities depend on agricultural producers to provide them with employment and a desirable place to live. Most farm people believe that the quality of life in rural America should be comparable to life in the rest of the country. Farm families are concerned, as they should be, about the quantity, quality, and costs of the public and private goods they purchase.

## RENEWED INTEREST IN RURAL DEVELOPMENT

Historically, rural development was restricted to improving conditions on farms and in economically depressed areas. Today the emphasis on rural development is to provide greater equity for all rural people in incomes, housing, health care, and other goods and services. Public policy is being used to disperse population and alter economic growth patterns. Many believe that by decentralizing people and industry it will increase economic efficiency, improve their social well-being, and improve communications in the political process.

---

[1] "Community Improvement—The Rural Component," Report of the Young Executives Committee, USDA, June 1973.

The crisis of the large cities in handling their social problems has had a large impact on renewing interest in rural development.[2] Major U.S. cities are plagued with clogged transportation facilities, air, water, and noise pollution, housing and building decay, and high crime rates. At present, about three-fourths of the population lives on about 2 percent of the land area.[3] When people are packed into such environments, it brings out the worst in human behavior. This type of dehumanization is unnatural, unhealthy, and is the cause of many social problems. It has been estimated that it would take $400 billion over a 10-year period to rebuild and remodel our major cities.[4]

In the past, the migration patterns of people have been from farm and rural areas to cities, putting pressure on already heavily populated centers. City problems, and the imbalance in population distribution, provide support for the belief that many of the social and economic difficulties were caused by "big cities being too darn big."[5] Thus, current policy is attempting to reduce the migration from the countryside, and possibly even to reverse the trend.

# NATIONAL POPULATION DISTRIBUTION _____

The population of the United States has increased from near 4 million in 1790 to about 263 million in 1995. Before World War I, immigration was a significant factor in population growth. Other important factors in recent population growth have been a reduction in the death rate and an increase in the number of births.

Regional population patterns have been influenced by early settlements, westward migration, and urbanization. In 1790, most of the population lived along the Atlantic coast, with just 3 percent of the population living in the interior part of the country. By 1995, the East Coast population comprised about 37 percent of the total. Between 1800 and 1900, there was a large movement of people westward to settle farmland.

About this time the pace of industrial development began to increase. Industries located along the Atlantic and Pacific coasts, the Gulf of Mexico, and the Great Lakes, where water transportation was available. This shift

---

[2] "Potential for Economic Growth in Montana—The Pros and Cons," Bozeman: Montana State University, p. 119.
[3] "A New Life for the Country," The Report of the President's Task Force on Rural Development, March 1970, p. 2.
[4] Ibid., p. 2.
[5] Op. cit., "Potential for Economic Growth in Montana—The Pros and Cons," p. 119.

began the rural to urban migration pattern that was to dominate regional population trends from about 1900 to 1950.

In 1790, only 5 percent of the population lived in communities of more than 2500 people. The United States was considered primarily rural until 1920 when the census showed just under 50 percent of the population in rural communities. By 1980, about 75 percent of the U.S. population was living in cities of more than 2500 people, and 34 percent of the population lived in urban areas with a population of more than 50,000. The declining sectors of agriculture, mining, railroading, fishing, and forestry were major reasons for the rural to urban migration. The substitution of capital for labor in these industries has induced outmigration. Also, people were moving to population centers, seeking higher incomes, and hoping to escape from poverty.

The South is expected to continue to be the most populated region in the United States. Population projections also show that the increase in U.S. population will continue to be concentrated in the South and West.

# DECLINE AND GROWTH OF RURAL AREAS AND COMMUNITIES _____

Increases in population and changes in its distribution have provided substantial benefits to our level of living, but many undesirable secondary effects of migration patterns have emerged. We find rural areas having large amounts of poverty, a smaller percentage of the rural population graduating from high school and college, fewer medical and dental personnel, evidence of substandard housing and lower quality public services such as police and fire protection and sewage disposal, less air transportation, and fewer recreational and cultural opportunities.

Even with lower quality services, many people desire a place to live with a lower density of population. A Gallup poll showed that 47 percent of the residents living on farms and ranches preferred to remain on farms, 29 percent preferred a small town of less than 2500, 18 percent preferred the suburbs, and only 6 percent preferred the city. For persons residing in cities of less than 2500, 23 percent indicated farm, 69 percent small towns, 8 percent suburbs, and none indicated a preference for the city. For persons in cities with a population between 2500 and 50,000 the survey showed 10 percent preferred farms, 62 percent small towns, 23 percent suburbs, and 5 percent preferred city life. Residents of the largest cities, those over 50,000, preferred the larger cities and suburbs. Six percent of these people preferred the farm or ranch, 23 percent small towns, 35 percent suburbs and 36 percent liked cities of more than 50,000 popula-

**TABLE 14-1**   Characteristics of the U.S. Population

| Year or Period | Total | Metropolitan | Nonmetropolitan |
|---|---|---|---|
| | | Millions | |
| 1960 | 179,311 | 133,072 | 46,239 |
| 1970 | 203,302 | 155,939 | 47,363 |
| 1980 | 226,543 | 172,455 | 54,087 |
| 1990 | 248,700 | 197,800 | 50,900 |
| 1994 | 260,400 | 207,500 | 52,900 |
| | | Percent change | |
| 1960–70 | 13.4 | 17.2 | 2.4 |
| 1970–80 | 11.4 | 10.6 | 14.2 |
| 1980–90 | 9.8 | 11.9 | 2.8 |
| 1990–94 | 4.8 | 4.9 | 3.9 |

SOURCE: Bureau of the Census, U.S. Department of Commerce, Statistical Abstract of the United States, 1992; Economic Report of the President, 1996; and Economic Research Service, U.S. Department of Agriculture, Rural Conditions and Trends, 1995.

tions.[6] Other more recent opinion polls show similar results. There are many people in concentrated populated areas who would prefer rural living.

As economic development expanded in the United States following the end of the Civil War, urban migration increased rapidly. The rural to urban movement continued until the 1970s. Currently, there are about 53 million rural people living outside Metropolitan Statistical Areas (MSAs). The rural population includes those people, plus about 6 million others who live in "rural settings" inside the MSAs. Thus, the total rural population is about 59 million people.

Between 1960 and 1970, metropolitan areas grew by 17.2 percent compared to 2.4 percent for nonmetropolitan areas (Table 14-1). This changed during the 1970s, when nonmetropolitan areas grew 14.2 percent as metropolitan areas grew by only 10.6 percent. Most of the growth in nonmetropolitan areas has occurred in counties adjacent to MSAs. In those counties that are not close to an MSA, most of the population growth has been in areas with a city of 10,000 or more population. These growth rates were reversed between 1980 and 1990, with the population in metropolitan areas increasing by 11.9 percent and nonmetropolitan population increasing by 2.8 percent. From 1990 to 1994, the metropolitan population

---

[6] R. P. Devine, "Citizens' Attitudes Toward Their Cities," in Edward Henry, ed., *MicroCity*, Collegeville, Minn.: St. John's University Center for the Study of Local Government, 1970, p. 42.

increased 4.9 percent while the nonmetropolitan population increased 3.9 percent. Nonmetropolitan retirement–destination areas and recreation areas have had the most rapid growth.

Employment growth in the rural areas was greater than in urban areas during the 1970s, but it was less during most of the 1980s. Total employment in the United States increased by 18.2 percent from 1982 through 1992, with a growth rate of 33.8 percent in metropolitan areas and a negative 17.6 percent in rural areas. Total rural employment decreased by 5.3 million in those years. However, from 1990 to 1994 rural employment has increased at nearly double the growth rate of metropolitan areas. The growth in employment is in the service, transportation and construction industries.[7]

# PROBLEMS IN RURAL COMMUNITIES _____

## Income and Employment

The median income of all households in the United States was $32,264 in 1994.[8] Average family farm income was $30,270. Total farm household income was higher than nonfarm households for the first time in 1990. An income differential also exists between average family incomes in metropolitan and non-metropolitan areas. Median metropolitan income per family was $34,251, and average nonmetropolitan income was $26,249 in 1994. The regional variation in family income is large. In the South median income per household in 1994 was $2,243 less than in other regions of the United States.

In 1994, the Bureau of the Census reported over 45 million Americans with incomes below the poverty line of $15,570 for a nonfarm family of four (Table 14-2). These persons accounted for about 17 percent of the U.S. population. The number of low-income persons is more than it was in 1959. Since the 1969 low of 24 million, however, the number of people below the poverty line has been rising because of high rates of inflation and higher unemployment rates.

Nonmetropolitan areas had about 20 percent of the population in the United States, but 24 percent of the poverty. Farms had 2 percent of the U.S. population and 2 percent of the poverty. About one in every nine people in rural America had incomes below the poverty level. This condition affected almost 2 million rural families.

---

[7] Economic Research Service, U.S. Department of Agriculture, "Rural Conditions and Trends," Spring 1995.

[8] Bureau of Census, U.S. Department of Commerce, Unpublished data, March 1995.

**TABLE 14-2**   Residence of Persons of Low Income Status, U.S.

| Persons | Number Below Low-Income Level | | | | | | | |
|---|---|---|---|---|---|---|---|---|
| | 1959 | 1969 | 1975 | 1981 | 1985 | 1988 | 1991 | 1994 |
| | Million people | | | | | | | |
| Total | 38.8 | 24.1 | 25.9 | 31.8 | 33.1 | 31.9 | 35.7 | 45.2 |
| Nonfarm | NA | 22.1 | 24.6 | 30.6 | 32.0 | 31.3 | 35.1 | NA |
| Farm | NA | 2.0 | 1.3 | 1.3 | 1.1 | 0.6 | .6 | NA |
| Metropolitan areas | 17.0 | 13.1 | 15.3 | 19.3 | 23.3 | 23.2 | 26.8 | 38.1 |
| Nonmetropolitan areas | 21.8 | 11.0 | 10.5 | 12.5 | 9.8 | 8.7 | 8.9 | 7.1 |

SOURCE: Bureau of Census, U.S. Department of Commerce, Series P-60, No. 181, 1995.
NA means not available.

Regionally the South has the highest proportion of low-income residents, but poverty is widespread and is evident nationwide. The rural areas have fewer absolute numbers of poor, but the poor represent a higher proportion of the total rural population than the poor in the cities represent of the total urban population. A major trend in the 1970s was the increasing proportion of poor in the central cities. The central cities have 30 percent of the urban population, but 43 percent of the nation's poor.

## Housing

The number and quality of housing units have improved since 1950. In 1990, the Census of Housing showed 92 million occupied housing units in the United States. This is a 7 percent increase over the 86 million units available in 1980, and nearly 35 percent over 1970. Much of the large increase is attributable to various federal and state programs to help people obtain adequate housing.

Twenty-four percent or about 22 million occupied housing units are in rural areas. Only 1.2 million of these units are substandard. A unit is classified as a substandard unit if it is overcrowded or lacks complete plumbing. The inside plumbing must include one or more of these facilities: hot water, a private flush toilet, and a private bathtub or shower.[9] The percentage of substandard housing in rural areas dropped from 59 percent in 1950 to 5.2 percent in 1990. This decline shows how effective insured or guaranteed loans have been through Farmers Home Administration and Federal Housing Administration programs. During the

---

[9] Allan Bird, and Ronald Kampe, "Twenty-Five Years of Housing Progress in Rural America," Agricultural Economic Report No. 373, Economic Research Service, USDA, June 1977.

1950s and 1960s, federal housing assistance to rural residents was provided by the Federal Housing Administration and the Veterans Administration. Currently, the Rural Housing and Community Development Service (RHCDS) is assisting with home financing in towns of less than 20,000 population located outside metropolitan areas.

## Government Expenditures

Rural Americans do not share proportionately in federal government programs according to recent studies of the geographic distribution of federal spending.[10] The federal government funded programs having an impact on development, with funding obligations totaling $1,328 billion in fiscal year 1994. These programs can be categorized into programs for human resource development, community and industrial development, housing, agriculture and natural resources, income security, and defense and space (Table 14-3). Nonmetropolitan counties received less than one-fifth of the federal outlays. On a per capita basis, rural areas received $4472 compared with $5261 in urban counties. Sparsely populated rural counties, however, received the largest amount on a per capita basis in 1994.

Federal expenditures per capita for human resource development and income security programs such as social security, health care, education, and employment training were less in metropolitan as compared with nonmetropolitan counties. Urban areas were granted $2879 per capita compared with about $3100 per capita in rural areas. Totally rural counties received $3304 per capita.

Per capita federal outlays for community and industrial development were much lower in nonmetropolitan areas in 1994, being $674 per capita in metropolitan areas and $364 per capita in nonmetropolitan counties. Federal outlays were highest in urbanized rural areas at $403 per capita.

National defense and space outlays are weighted heavily toward metropolitan areas. Per capita defense expenditures were $786 in metropolitan areas as compared to $385 in rural America.

Over the last decade there has been a substantial increase in federal expenditures in rural areas. The pressure for more evenly balanced urban–rural growth has caused government decision makers to direct greater per capita allocations to rural areas.

---

[10] J. Norman Reed, Charles I. Hendler, and Eleanor Whitehead, "Federal Funds in 1979—Geographic Distribution and Recent Trends," Economic Research Service, USDA, Economic Development Division, Washington, D.C., April 1982.

**TABLE 14-3**   Distribution of Federal Outlays in Specified Program Areas, Metropolitan and Nonmetropolitan Counties, Fiscal Year 1994

| Program Type and Allocation | Total | Metropolitan Counties | Nonmetropolitan Counties | | |
|---|---|---|---|---|---|
| | | | Urbanized | Less Urbanized | Totally Rural |
| Agriculture and natural | | | | | |
| resources (billion $) | $20.1 | $6.3 | $1.7 | $8.3 | $3.7 |
| Percent of total | 100 | 31.3 | 8.6 | 41.5 | 18.6 |
| Dollars per capita | $77 | $30 | $103 | $279 | $601 |
| Community resources | | | | | |
| (billion $) | $175.4 | $156.1 | $6.7 | $10.4 | $2.1 |
| Percent of total | 100 | 89.0 | 3.8 | 5.9 | 1.2 |
| Dollars per capita | $674 | $753 | $403 | $346 | $340 |
| Human resources | | | | | |
| (billion $) | $18.6 | $14.3 | $1.4 | $2.4 | $0.6 |
| Percent of total | 100 | 76.6 | 7.3 | 13.0 | 3.1 |
| Dollars per capita | $72 | $69 | $81 | $81 | $92 |
| Income security | | | | | |
| (billion $) | $746.7 | $584.1 | $48.1 | $93.8 | $20.6 |
| Percent of total | 100 | 78.2 | 6.4 | 12.6 | 2.8 |
| Dollars per capita | $2879 | $2816 | $2876 | $3133 | $3304 |
| National defense | | | | | |
| (billion $) | $183.5 | $163.1 | $7.9 | $10.5 | $1.9 |
| Percent of total | 100 | 88.9 | 4.3 | 5.7 | 1.1 |
| Dollars per capita | $741 | $786 | $473 | $352 | $311 |
| Total outlay (billion $) | $1,327.8 | $1091.3 | $75.3 | $131.3 | $29.8 |
| Percent of total | 100 | 82.2 | 5.7 | 9.9 | 2.2 |
| Dollars per capita | $5100 | $5261 | $4504 | $4386 | $4774 |

SOURCE: Elliot J. Dubin, "Geographic Distribution of Federal Funds in 1985," Agriculture and Rural Economy Division, Economic Research Service Staff Report AGES 89–7, USDA, Washington, D.C.: March 1989 and Rural Economy Division, Economic Research Service, U.S. Department of Agriculture, 1996.

# Education

The United States devotes huge amounts of resources toward educating its citizens, and a great many employable people forego substantial earnings while taking advantage of the educational opportunities available to them. Both society and the individuals incurring those costs must have decided that an education is worth its cost.

In looking at the value of an education, we will separate the noneconomic from the economic rewards of an education. Directing our attention to the latter does not imply that noneconomic rewards are unimportant. Both are somehow considered when a person weighs the pros and cons of an education.

Two elements in the economic returns from an education can be identified: the *investment component* and the *consumption component*.[11] The investment component regards the process of education as an investment in an individual's increased productivity that yields increased future earnings. Educational investments in improving the knowledge and productivity of people is an investment in **human capital.** It involves committing present costs for expected future returns, and can thus be analyzed in the same manner as investments in physical capital. The consumption component includes both the immediate and longer run satisfactions (utility) that the individual derives from an education. Consumption benefits accrue primarily to the student, but there frequently are satisfactions to his or her family as well.

The net return on a student's education is determined by the estimated benefits associated with that student's increased productivity and satisfactions and the cost requirement of those gains. In economic terms, the student will spend an amount on education that equates private marginal benefits to private marginal costs, both in terms of their present values.

When goods are bought and sold in a perfectly competitive market, an equilibrium is determined by the intersection of the demand and supply curves. At that intersection point the last unit consumers buy yields a benefit equal to the cost of producing that unit. Its benefits and costs at the margin are equal, and net benefits to society are maximized.

But there are special requirements that must be met in this equilibrium. If there are other benefits or costs beyond those involved in the transaction, the equilibrium cannot be a social optimum.

You will buy a steak dinner by equating your own private marginal benefits and marginal costs; no other costs or benefits are imposed on others by your decision. Suppose, however, that you are buying plants and flowers to beautify your home's appearance. If others also derive pleasure from this, total benefits exceed your private marginal costs. Or, you may optimize the use of your stereo by listening to a rock concert at 1:00 A.M. (at high volume, of course). If this prevents people in the apartment next door from sleeping, a disutility (cost) has been incurred. Such benefits or costs to others, being external to your cost and returns equation, mean that a social optimum is not achieved with your private optimizing. In the landscaping example, too little of the product is bought; in the latter, too much has been purchased.

---

[11] T. W. Schultz, *The Economic Value of an Education,* New York: Columbia University Press, 1963, p. 5.

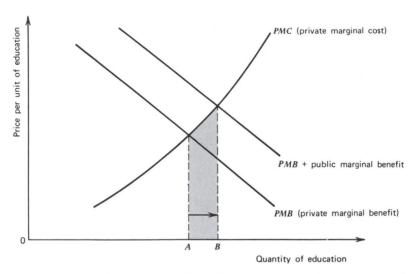

**FIGURE 14-1**  The economic effect of externalities in education.

An interesting aspect of education is that society derives benefits in addition to the individual's benefits. These added benefits to the larger society (they could also be costs) are called *externalities* or *spillover effects.* They have important repercussions on economic decisions. If education is to be socially efficient and equitable, it requires that the cost burden of an education also be shared. This is one of the major arguments for public funding of education.

Educational expenditures must be extended beyond the point where *private marginal costs (PMC)* equal *private marginal benefits (PMB).* In this case the socially efficient expenditure on education is where the private marginal costs plus external marginal costs equal private marginal benefits plus external marginal benefits. To illustrate, let's assume a student is considering the amount of education to purchase as shown in Figure 14-1. The student would equate private marginal benefits and private marginal costs and purchase quantity *A.* This student could not take into account the added benefits to society of his or her education. Hence, usually by public funding of education, and regulations or graduation requirements, the student equates private marginal costs (in this example we assume no external marginal costs) with private marginal benefits plus the public marginal benefits. A larger quantity of education is therefore consumed. Quantity *B* is purchased rather than *A*; from society's viewpoint, quantity *B* is optimal.

Educational achievement increased rapidly between 1950 and 1991. In 1950, the median years of schooling (50 percent above and 50 percent below this number) completed by persons 25 years old and over was 9.3 years. Recently, the median education level had increased to 12.7 years.

The percentage of metropolitan residents 18 and over completing 4 years of high school was 82.2 percent and nonmetropolitan 76.5 percent. Metropolitan students who completed 4 or more years of college was 22.8 percent and nonmetropolitan 13.6 percent.

These figures tend to mask some of the differences between urban and rural residents. Many researchers have found that regardless of the measurement system used, rural students "consistently rank lower" than their urban counterpart.[12] A Missouri study measuring educational quality found that the average score on the Ohio State University psychological test for the smallest rural schools (0-199 students) was 43.6 compared to large urban schools (1000 or more students) with a score of 56.5. The average for all urban schools was 15 percent higher than for all rural schools.[13]

The president's National Advisory Commission on rural poverty contends that the extent to which rural people have been denied equality of educational opportunity is evident not only in the output of the educational system, but also in the resources that go into the system.[14] The commission concluded that upgrading rural schools would increase educational opportunities for rural poor, but would not solve the problems of economically deprived students.

Although rural youth are getting a better education than their parents had, they continue to lag behind urban students, but the gap may be narrowing. A Montana study shows that on basic skills (as measured by the Stanford achievement test) at the senior level in high school, no significant differences exist in mean test scores based on size of the school.[15] At the sophomore level, however, the larger urban high-school students scored better on these achievement tests than did students from the smaller rural schools. It is necessary to realize that these achievement tests evaluate basic skills only and do not include the enrichment programs of the larger high schools.

In Chapter 13 we discussed how the present value of any resource is determined. That same concept is appropriate in appraising the economic value of an education, and the rate of return on that investment.

The rate of return on investment or "payoff from schooling" has been relatively high. Although the studies were completed some years ago,

---

[12] "Rural Development Progress," Fourth Annual Report of the Secretary of Agriculture to the Congress, USDA, January 1977, p. 32.

[13] Jerry G. West, and Donald D. Osburn, "Quality of Schooling in Rural Areas," *Southern Journal of Agricultural Economics,* July 1972, p. 86.

[14] Report by the President's National Advisory Commission on Rural Poverty, *The People Left Behind,* Washington, D.C.: U.S. Government Printing Office, September 1967, p. 41.

[15] John W. Kimble, Gail L. Cramer, and Verne W. House, "Basic Quality of Secondary Education in Rural Montana," Agricultural Experiment Station Bulletin 685, Bozeman: Montana State University, April 1976.

their conclusions still should be valid. These studies show a rate of return on primary education of 10 to 20 percent, on secondary education of 15 to 30 percent, and on postsecondary education of 10 to 20 percent.

A comparison by areas (rural versus urban) shows that the educational payoff is higher for rural white males than urban males. What this comparison says is that public educational efforts have been less intensive in rural than in urban areas. All is not well in rural education. There were 500,000 rural adults in the early 1970s who had not been to school at all and 2 million rural adults who were functionally illiterate.[16] Also, there were at least 5 percent of all rural school-aged children who were not enrolled in school.[17] In recent years, the nation's functionally illiterate population has been increasing at about 2.5 percent per year, with a total of 27 million adults who have difficulty performing in modern society.

# COMMUNITY SERVICES _____

Wherever one might look in rural America, change has been the order of the day, much of it with undesirable results. Agricultural firms have been forced to adjust the kinds and quantities of resources used, and to adopt new technologies, in order to remain competitive. In the process, farm numbers have fallen from a high of near 6.8 million in 1935 to 2.1 million by 1995. During the same period, farm population declined to 4.6 million from its high of 32.3 million people.

Declines in farm numbers and farm population became most rapid in the 1945–65 period. Over those 20 years the number of farms declined by 130,000 per year. Farm population dropped in that same period by an average of 600,000 each year. The 15 years from 1980 to 1995 saw a considerable slowing in their rates of decline, however. Farm numbers declined by about 24,000 per year, about one-fifth the previous rate, while the 95,000 per year decline in the number of farm people was about 80 percent less than the previous rate.

For rural communities, once organized to serve the needs of more than 32 million farm people, a loss of more than 85 percent of their potential customers has had a serious impact. The most severely affected communities are to be found in the Great Plains and the Mountain states. Communities in other areas of the United States are much less concentrated, with scattered pockets of hard-hit areas. Other regions, such as Appalachia, face problems that stem more from urban origins than from the loss of farm people. Yet other central communities have been more fortunate.

---

[16] Op. cit., "Rural Development Progress," p. 32.

[17] Ibid., cited from Marion W. Edelman, Marylee Allen, Cindy Brown, Ann Rosewater, et al. *Children Out of School in America*, Cambridge, Mass.: Children's Defense Fund, 1974, p. 37.

Located near growing urban centers, farm losses have been more than off-set by people establishing homes farther and farther out in the rural area. This movement has caused other urban-related problems, but the effect of depopulation is not one of them.

Farm mechanization and the corresponding need to increase acreages is only one important factor in changing rural communities. With improved transportation and related conveniences, it became easier for rural community residents to obtain necessary supplies and services in the larger towns and cities, to the detriment of their own community's service center.

In choosing between daily travel to elementary or high school for their children, some opted for a second residence in the town of the preferred school. Local health-care facilities have suffered both patronage and personnel losses in the same way and for the same reasons.

Most if not all public and private services are greatly influenced by population density of the serviced area. Not only do local stores of all kinds suffer from a declining population, but problems of inventory maintenance and upkeep of facilities cause the cost of goods sold to increase sharply.

On a per capita basis, roads and transportation, schools and libraries, hospital and health-care facilities, farm electrification and telephone services, newspapers, radio and television, and police and fire protection among other public and private services all are provided at much higher cost in sparsely populated areas. Either the scale of each of these is reduced to the size that can serve the local market's needs or the geographic area served must be expanded, both of which have a direct bearing on the quantity, quality, and delivery costs of these services. Thus, there can be no parity of costs for the more isolated rural areas in comparison to their urban counterpart.

Concern over rural–urban disparities in obtaining services begs the question of what can be done to avoid or alleviate these problems. Questions need to be directed at what forms of community reorganization might lead to improved quality of services and reduce their costs. Rural areas have an institutional structure that is more appropriate to heavily populated areas. But their problem is one of trying to revise institutions that might overcome distance-related problems and escape the high "social cost of space."[18]

The village, town, or city holds a position of dispenser of goods and services to its outlying trade area. The size of that area depends on how

[18] Social cost of space is a term first used by A. H. Anderson, "Social Cost of Space," *Journal of Farm Economics,* Vol. 32, August 1950, pp. 411–430. The costs of public and private goods and services were related to greater delivery distances in sparsely settled rural areas.

**TABLE 14-4**  Scale Economies in Public Services

| Service | Result |
|---|---|
| High schools | ATC is U-shaped, with a minimum at about 1700 pupils |
| School administration | ATC is U-shaped, with a minimum at about 44,000 pupils |
| Fire protection | ATC is U-shaped, with a minimum at about 110,000 population |
| Police protection | ATC is about horizontal |
| Hospitals | ATC is declining to about 800 beds |

SOURCE: Condensed from W. Hirsch, *The Economics of State and Local Government*, New York: McGraw-Hill, 1970, p. 183.

NOTE: ATC = average total cost.

well those functions are performed, and on local-patron loyalty to and identification with that service center.

How large must a town be to offer the variety of services a community requires at reasonable costs? We may conceive of "units" of services in the same way as we recognize units of resources, or units of products produced. We then may consider the average total costs per unit of those services in the same manner as with physically produced goods.

The efficiency of providing public services depends on the economies of scale in production, population density of the area served (thus, the demand for services), and the costs of organization and delivery of those services. Estimates of approximate populations required to provide the most cost-efficient high school, school administration, fire protection, police protection, and hospital services are shown in Table 14-4.

## The Local School Problem

As can be seen, a substantial population base is needed to be able to provide these services at least cost. Because they have a smaller population, rural areas typically must bear higher costs for these services and also receive lower-quality services. A good example of this is the problem of the small rural school.

Most rural schools are too small to exhibit any degree of cost effectiveness in providing building space, educational equipment aids, and faculty. Faced with low enrollments and sharply rising costs, rural school districts early bore the brunt of having to "do something" in both low population density and depopulating areas.

To demonstrate community size requirements, let's use the high school information from Table 14-4. If we estimate the size of an average rural family as four people, only about one-fourth of the children under 18 will be of high school age. Thus it would take a community of more than 25,000 population to minimize the cost of high school educational services. Few of the nation's rural counties have that large a population.

School district consolidation is one approach toward reducing the costs of educating rural youth. Less than 30 percent of the more than 67,000 school districts that existed in 1952 remain today. At the small end of the scale, the push to consolidate has been of both administrative and citizen reaction in origin. The state threatens or denies accreditation because of unacceptably low program quality, and taxpayers find that minimum quality entails unbearably high costs. Transportation costs (bus ownership, operation costs, and driver wages) become an important component of school costs, a cost per student that varies inversely with population density. Another cost not included in cost tallies, but a real cost nevertheless, is the time spent riding to and from school. For many students, this travel time amounts to more than 2 hours a day, and they find it difficult to use this time productively.

Although average total cost for high schools is minimized at about 1700 students (Table 14-4), the more dramatic cost reductions are to be had in combining the smaller schools.[19] But an inadequate school is often the rural area's sole community "identity," with its athletic programs (especially football and basketball) being the focus of that attachment. Important as such a unifying institution might be, the cost of maintaining a small school in the local community for this purpose is high. And the young bear this cost in their restricted educational opportunities.

## Hospitals and Health Care[20]

Because of their low population density, rural residents also face cost and quality problems in obtaining health-care services locally. Increasingly specialized professional skills, new medical knowledge and technology, and greater consumer expectations all have served to raise both the capacity and fixed costs of health-care facilities. The costs of providing health care may be indicated by the following equation:

---

[19] The degree of cost saving through consolidation is indicated by some Montana data relating costs per student and size of school. Small rural high schools (those with less than 50 students) had an ATC per student that was more than three times greater than larger urban schools with more than 1000 enrolled: From the Biennial Report of the Montana State Department of Public Instruction, Helena, Mont.

[20] More completely discussed in R. J. McConnen, "Communities and Their Impact on Rural Life in the Years Ahead," Great Plains Agricultural Publication No. 44, Bozeman, Mont., 1972.

$$\begin{array}{l}\text{Average total} \\ \text{patient cost}\end{array} = \dfrac{\begin{array}{c}\text{Total fixed cost} \\ \text{of the minimally} \\ \text{acceptable facility}\end{array}}{\begin{array}{c}\text{Number of} \\ \text{patients treated}\end{array}} + \dfrac{\begin{array}{c}\text{Total variable cost} \\ \text{of the minimally} \\ \text{acceptable facility}\end{array}}{\begin{array}{c}\text{Number of} \\ \text{patients treated}\end{array}} + \begin{array}{c}\text{Average travel} \\ \text{cost for patient}\end{array}$$

To assign numerical values in this equation, a minimally acceptable facility must first be determined. The minimum variety and quality of services, the minimum amount of equipment, the greatest distance, and the maximum number of patients per physician that are locally acceptable all have a bearing on the facility and the level of its costs at any point in time. These factors define the standard of acceptability—a social norm—which is very much influenced by comparable facilities available to urban residents.

The components of cost for such a facility change through time, as society's expectations change, and as advances are made in the body of medical knowledge and in the sophistication of medical equipment.

Once the minimally acceptable facility has been specified, the amount of fixed (or overhead) costs may be determined. Total fixed costs are all those annual costs incurred in providing the facility itself. These costs remain the same no matter how many patients are treated (including such cost items as property taxes, depreciation and maintenance, utilities, and salaries for the minimal staff). These costs are unaffected by the intensity with which the facility may be used, being the same whether 5 or 500 patients are treated there.

Those costs that we call variable change as patient numbers increase or decrease. Some of these costs, such as medicines and medical supplies used, vary directly with the volume of use, while others such as office salaries do not vary directly with the patient load.

Travel costs are those costs incurred by the patient in getting to and from the health-care facility. These costs include transportation plus other expenditures for meals, lodging, and the opportunity cost of the time that it takes to obtain the needed treatment.

Figure 14-2 compares the average cost per patient in two different periods, 1980 and 1990. The average costs are shown for each year at various rates of use of the facility.

Assume that the facility treated 100 patients in 1980 with an average total cost per patient of $100. Under this assumption, the facility treated only one-half the number of patients that it could have handled. Average total costs would have been somewhat higher than if the facility had been used to its capacity.

Over the 10-year period, certain factors have changed that affect patient costs. Inflation shows up in increased fees and salaries, equipment costs, taxes, medicines and other supplies, insurance, and other operat-

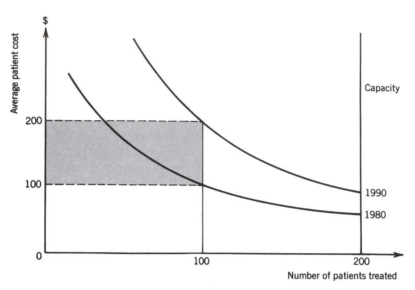

**FIGURE 14-2**   Patient treatment costs and the effects of time and utilization rate.

ing costs. During this period, society's interpretation of a minimally acceptable facility also changed. With new developments in the field of medicine came increased expectations regarding the quality as well as quantity of medical treatment provided by the facility. The adoption of higher-cost diagnostic equipment and other new medical technology have greatly increased the unit's fixed costs. These cost changes are indicated by the shape and location of the average cost curve for 1990.

Historically, these changes have served to increase the ratio of fixed to variable costs. As fixed costs have increased relative to other costs, the average total cost curve will exhibit a steeper slope than it had in the earlier period. What was acceptable in 1980 was below the acceptable level for 1990. As a result, the average cost curve of providing a minimally acceptable health facility increased.

The cost increase is much greater at the less intensive use rates and approximates the experience of sparsely populated rural areas that have attempted to provide modern health-care facilities locally. With 100 patients also treated in 1990, costs were increased to $200 per patient, which is much higher than for the earlier period. It is higher also than the comparable cost per patient had the facility been used to its capacity.

Schools and health care are but two examples indicating how the very attribute of extra space, so strongly desired by some people, also causes large costs to those residents. The social cost of space is further magnified by the fact that rural incomes are lower than urban incomes. Rural residents cannot afford to pay as much for services as the city dweller must pay, yet comparable services frequently cost much more in rural areas. Thus the rural resident often is forced to do without a particular service or to accept less, both in quantity and quality.

## Alternative Solutions[21]

Differences between rural and urban social expectations have been reduced, making rural people less willing to accept second-rate services. Modern communication systems have made Americans much more aware of the factors affecting the quality of life. And transportation improvements have made it easier for rural people to make greater use of services available in the larger communities. It thus becomes a question of a rural community's ability to compete effectively for local residents' service needs by using properly organized services. The alternative is to permit a sometimes distant urbanized area to do the job that might have been done at home, and possibly even more economically.

As mentioned earlier, the increasingly higher costs of presently structured institutional services are part of the problem. Appropriately structured to serve more populous areas, both cost and quality of services suffer in low population density areas. They cannot offer the variety of services expected of them at a cost sufficiently low to attract local users to those services, thus the seeds of community decline have been planted. Because institutions themselves are part of the cause of rural area decline, their reorganization ought therefore to hold promise of favorable results.

One possible approach would make use of electronic technology. Computer terminals could be located in rural stores, which would link them to a central warehouse capable of serving a large number of rural areas. The selection of items available to those rural stores would be much larger and more varied than any of them could provide individually. Such an operation would greatly reduce local storage needs, reduce inventory investment costs, cut insurance costs, and be able to take advantage of bulk buying.

Another approach would establish agricultural shopping centers in rural areas. These centers would provide one-stop shopping for rural residents at lower costs than do the separately housed enterprises ordinarily found in rural towns. Some rural towns have already made changes from separate-building businesses to the shopping center concept.

A number of rural areas have adopted an encouraging approach in the health-care field. They have organized multicounty regional medical centers that link local area service units as outlying branches of the main cen-

---

[21] Ibid.

ter. Scheduled travel to the center by medical professionals provides the needed specialty services, leaving outpatient treatment to local doctors and nurses. Such use of modern communication facilities permits a high degree of coordinated service between an isolated community and the highly trained specialist who must locate in a more densely populated area. More importantly, highly specialized services are thereby made available to rural areas whose costs would be prohibitively high if they were to attempt to provide them locally.

A fourth proposal concerns education facilities for rural youth. Through modern technology, such as two-way television, instruction can be transported long distances. Now there is a choice: transport students to achieve the desired class size or carry televised instruction to the students. Cost savings can thus be achieved even while improving instructional quality.

These examples do not exhaust the possibilities open to rural areas. Many other potential cost savings can be attained in both public and private services. Part of the reason for the declining population in many rural areas is the disparity of services (both cost and quality) between rural and urban areas. Many rural communities must, somehow, overcome local cost and quality problems so as to prevent further deterioration.

The answer may lie in adopting new institutions, or the use of new cost-reducing technologies, or a combination of both. Improvements in rural government services might be achieved in a number of ways, one being the outright consolidation of two or more counties. This solution has been politically unattractive in the past, however, because it would result in one or more of these counties losing their identity. That problem can be avoided by a regional approach in which the participating counties each provide different specialized services for the other counties in the group while continuing to maintain their separate identity.

In other instances some of the more sparsely settled counties have contracted with adjoining counties for snow removal, road repair and maintenance, and similar services requiring investment in heavy equipment. The per capita burden for those services is thereby reduced.

Within county boundaries similar services by county and city offices may be combined to effect savings to local taxpayers. Sheriff and police departments have been consolidated in a number of rural areas with favorable results.

Whether it be the consolidation of schools or other governmental units, the competition between counties, towns, or county and town must first be overcome. Barring that, savings still might be realized through some form of interunit agreement that shares costs and services.

# LEGISLATION FOR RURAL DEVELOPMENT _____

## The Rural Development Act of 1972

Although other rural development legislation passed by Congress has had effects on rural areas, none has had the force or commitment of the Rural Development Act of 1972. This act contains six titles:

*Title one* makes provisions for business and industrial loans, community facility loans, and industrial development grants. These programs were administered by the Farmers Home Administration.

*Title two* of the act provides for cost-sharing in watershed protection and flood prevention programs.

*Title three* authorizes technical assistance to, and cost-sharing arrangements with, public agencies and organizations in order to provide rural community water supplies, control and abatement of agriculturally related pollution, disposal of solid wastes in rural areas, and water storage in rural areas for fire protection.

*Title four* authorizes the secretary of Agriculture is authorized to provide financial and technical assistance to equip local areas to prevent and control wildfires in rural areas.

*Title five,* because of the low-income problem in agriculture, funds research programs that emphasize new approaches to the management of small farms.

*Title six* is a miscellaneous part of the act directing federal organizations to give priority to rural areas in locating new government offices and facilities. This section also authorizes the government to pay part of the cost of agricultural conservation programs and related pollution abatement and control programs on 10-year contracts.

The most important section of the act on rural development is title one, which is designed to attract business to rural areas. It is hoped that this will promote development in rural communities and improve living conditions.

While most rural development programs under the former Farmers Home Administration were small, they may receive increased emphasis in future years. Appropriations for 1996 rural business and industrial loans were only $300 million, or about $6 million per state. These loans were to be used by public, private, or cooperative organizations to improve the economic and environmental atmosphere in rural communities. Business and industry loans are available in communities of up to 50,000 population. At earlier levels of business and industry funding, this program was expected to create about 20,000 jobs per year nationally, less than 7 percent of the number needed to stop rural out-migration and to increase labor participation in rural regions.

Community facility loans are made to communities with populations of 20,000 or less. In 1996, $300 million was available to public bodies and nonprofit associations. Of the $300 million, $225 million was made available in direct loans and $75 million was in guaranteed loans. About $6 million per state is very inadequate to finance such things as fire facilities, ambulance service, or industrial parks.

Many grants are also available for environmental concerns, but these are also small and inadequate. Much more effort must be directed at rural America if it is to provide the quality of life offered in urban areas. Earl Heady of Iowa State University has said that "no land grant university has put rural development as the major item on its agenda of affairs presented to the state."

## Rural Development Efforts by the Eisenhower and Kennedy Administrations

Low-income problems of small farmers were given specific attention in the Rural Development Program of 1955. The major objectives of this program were to increase employment opportunities in rural areas, to raise the income level of rural people, and to use local committees to determine and guide developmental plans. This program was administered by the USDA through the Cooperative Extension Service.

Congress provided $2 million to operate the program. By 1960, it was estimated that 210 counties participated in the program and that 18,000 new jobs had been created. However, it did not touch the "hard core underemployment-poverty problem in rural areas."[22]

During the Kennedy Administration the Rural Development Program was replaced by the Area Redevelopment Act of 1961. The objective of the act was to provide continued employment in areas of unemployment or underemployment. About $100 million was allocated for depressed rural areas.

The Department of Commerce administered the Area Redevelopment Act. An office of rural areas development was established in the U.S. Department of Agriculture to handle the rural portion of the act. This program was larger than the 1955 program, and it worked with 853 rural counties. Between 1961 and 1966 the rural areas development program promoted 20,000 projects ranging from community facilities to industrial parks, but again it was not a large enough program to solve the rural de-

---

[22] Willard W. Cochrane, *City Man's Guide to the Farm Problem*, St. Paul, Minn.: University of Minnesota Press, 1965, p. 202.

velopment problem.[23] The Area Redevelopment Act authorized only $394 million. Only $10 million was authorized to assist unemployed workers or small farmers while they were training to improve their skills. Also, only $4.5 million was available to finance retraining programs. Most of these funds went to industrial and commercial projects and grants and loans to communities for public facilities such as water and sewage systems.

Although city people may glamorize rural America for its nice, easy, clean, quiet life, rural people frequently are unhappy with their situation. America's rural areas require greater funding in the solutions of their problems before satisfactory progress can be made.

## Title VII of the 1996 Federal Agricultural Improvement and Reform (FAIR) Act

The 1990 farm bill created a Rural Development Administration (RDA) to consolidate USDA's rural development efforts. Most rural development programs are coordinated by the Secretary of Agriculture. The RDA took over FMHA divisions that handled water, sewer, other community facilities, and business and industrial loan or grant programs. A new Rural Community Advancement Program (RCAP) was initiated with the 1996 FAIR Act under which the Secretary was authorized to provide grants, direct and guaranteed loans, and other rural development needs across the country. RCAP funding was allocated to three areas: Rural Community Facilities, Rural Utilities, and Rural Business and Cooperative Development. As of 1994, USDA's Rural Economic and Community Development (RECD) consists of three new services including: (1) Rural Housing Service (RHS) to help finance new or improved housing for low to moderate-income rural families, (2) Rural Business and Cooperative Development Service (RBCDS) to help build competitive rural businesses and cooperatives that can prosper in the global marketplace, and (3) the Rural Housing and Community Development Service (RHCDS) to make and guarantee loans to develop essential community services in rural areas and towns of up to 20,000 in population.

## SUMMARY

Rural development is concerned with the quality of life in rural communities. Rural communities affect agricultural producers because they are the service centers for agricultural producers. Rural communities in return depend on farmers and ranchers for employment and a place to live.

---

[23] Ibid.

Rural areas have large amounts of poverty, a smaller percentage of high school and college graduates, fewer medical personnel, lower quality public services, and fewer cultural opportunities. These rural–urban disparities need attention, and ways must be found to help alleviate these problems.

## CHAPTER HIGHLIGHTS

1. Rural development emphasizes the improvement of rural life.
2. Agricultural producers influence, and in turn are influenced by, the quality of rural communities.
3. Recent interest in rural development has been heightened by social and economic problems of the city.
4. Outmigration from farms and ranches became most rapid after 1945.
5. Population redistribution has contributed to rural area problems.
6. Since 1970, there has been a renewed interest in rural living.
7. Average rural family incomes are lower than urban family incomes.
8. Farm-dwellers comprised about 2 percent of the American population and 2 percent of its poverty.
9. About one in six rural people has an income below the poverty line.
10. The quality of rural housing has improved considerably since 1950.
11. Government expenditures per capita are increasing in rural areas, yet are below urban counterparts.
12. Education is important in rural America in order to improve opportunities for rural youth.
13. Education can be viewed as an investment in human capital in the same manner as investments in nonhuman capital.
14. The externalities of an education justify public support of education.
15. Because of the distance factor, low population density in rural areas creates serious cost and quality problems for community services.
16. Some form of institutional reorganization would appear to hold promise for improved rural services at reasonable costs.
17. There are economies of scale in public services, thus a substantial population base is required to provide these services at least cost.
18. Increased government emphasis for rural development projects has been made since 1972, but it remains inadequate.

# KEY TERMS AND CONCEPTS TO REMEMBER

| | |
|---|---|
| Consumption component | Investment component |
| Externalities | Rural development |
| Human capital | Spillover effects |

# REVIEW QUESTIONS

1. What is rural development? Is it primarily concerned with economic growth? Discuss.
2. Why is there a renewed interest in rural development? What impact will this interest have on farmers and ranchers?
3. What factors are important in explaining the recent decrease in population in nonmetropolitan areas?
4. Explain the concept of poverty. Why do nonmetropolitan areas have a larger proportion of people in poverty?
5. What are externalities? Write down some examples, and discuss the problems they cause.
6. Why does it cost more per unit to provide public services in rural areas than those same services cost in urban areas?
7. What alternative forms of community service organizations might lead to improved quality of services and reduce their costs?
8. Do economies of scale exist in public services? Explain with an example.

# SUGGESTED READINGS

Barkley, Paul W., and David W. Seckler. *Economic Growth and Environmental Decay,* New York: Harcourt Brace Jovanovich, 1972, Chapters 4 and 8.

Hirsch, Werner Z. *The Economics of State and Local Government,* New York: McGraw-Hill, 1970, Chapter 7.

Mansfield, Edwin. *Economics: Principles, Problems, Decisions,* 7th ed., New York: Norton, 1974, Chapters 18 and 28.

Reynolds, Lloyd G., George D. Green, and Darrell R. Lewis, eds. *Current Issues of Economic Policy,* Homewood, Ill.: Richard D. Irwin, 1973, Readings 29 and 33.

Summers, Gene F., Sharon Evans, Frank Clemente, E. M. Beck, and Jon Minkoff. *Industrial Invasion of Nonmetropolitan America,* New York: Praeger Publishers, 1976, Chapters 4 and 5.

Tweeten, Luther G. *Farm Policy Analysis,* Boulder, Col: Westview Press, 1989, Chapter 10.

# AN OUTSTANDING CONTRIBUTOR

**Vernon W. Ruttan**   Vernon Ruttan is currently professor in the Department of Agricultural and Applied Economics and in the Department of Economics at the University of Minnesota.

Ruttan was born on a 120-acre farm in Kalkaska County in the cut-over area of northern Michigan. He indicated that "the farm on which I was born, was, in the mid-1920s, a generation behind the farms of the corn belt in their transition to modern technology," and that when his father retired in 1964, "he was the last full-time farmer in his township."

Ruttan attended Michigan State University in 1942–43. He received his B.A. degree from Yale University in 1948 and his M.A. and Ph.D. degrees from the University of Chicago in 1952 and 1954. He has held academic appointments at Purdue University, where he served as assistant professor, associate professor, and professor between 1954 and 1963, and at the University of Minnesota where he served as professor and as head of the Department of Agricultural and Applied Economics from 1965 to 1970, and as director of the Economic Development Center from 1970 to 1973. He also served as visiting professor at the University of California, Berkeley, in 1958–59.

Probably the most significant contribution that Dr. Ruttan has made to the literature of agricultural economics has been on the economics of technical change and agricultural development. He made original contributions to the development and testing of the theory of technical and institutional innovation. His book *Agricultural Development: An International Perspective* (co-authored with Yujiro Hayami and published by Johns Hopkins in 1971) has become a basic reference in the field of agricultural development.

Ruttan has also had substantial nonacademic experience. His first professional position was in the Government Relations and Economics staff at the Tennessee Valley Authority (1951–54). In 1961 and 1962, he was staff economist with the president's Council of Economic Advisors. Between 1963 and 1965 he was agricultural economist with the Rockefeller Foundation at the International Rice Research Institute in the Philippines. From 1973 to 1978 he was president of the Agricultural Development Council. Ruttan has also served on the Research Advisory Committee of the U.S. Agency for International Development (1968–73), the Technical Advisory Committee to the Consultative Group on International Agricultural Research (1973–79), and a number of other advisory committees and boards.

The quality of Ruttan's work has been honored by the American Agricultural Economics Association with six of their published research awards (1956, 1957, 1962, 1966, 1967, and 1979). He was elected Fellow of the AAEA in 1974, to membership in the American Academy of Arts and Sciences, and to the Council on Foreign Relations. He served as president of the American Agricultural Economics Association in 1972.

He is married to the former Mabel Mayene Barone, and they have four children. Ruttan says, "My children sometimes refer to me as a migrant worker."

CHAPTER **15**

# Comparative Agricultural Systems

$\text{T}$he term *economic system* encompasses all methods by which resources are allocated and goods and services are distributed. From country to country the systems differ, primarily on the basis of social and political considerations. In some countries the system is capitalism, in others it is socialism, and in still others it is a mixture of capitalism and socialism.

# MAJOR TYPES OF ECONOMIC SYSTEMS

Types of economic systems in their pure forms do not exist, but it is still worthwhile to study the characteristics of each. Such study makes it easier to compare and analyze existing economic systems.

## Capitalism

The economic system labeled *capitalism* is a self-regulating system with a government that is uninvolved in economic decisions. A capitalistic system depends on market forces to determine prices, allocate resources, and distribute income.

The factors of production are privately owned and controlled, resulting in decentralized decision making. Each owner makes production decisions motivated by a desire to realize a profit. Production is begun or terminated by individual choice. The profits earned or losses incurred are a direct result of right or wrong business decisions.

Individual freedom of choice prevails in making consumption and occupational decisions under a system of capitalism. The consumer attempts to maximize satisfaction, given money income. A resource owner attempts to maximize income from the sale of services or resources controlled by that individual.

Under capitalism, free market prices are the sole guides to individuals and firms in making production, exchange, and consumption decisions; and competition is stressed in all types of economic activities.

## State Socialism

The basis of *socialism* as an economic system is collective (state) ownership of all productive resources with state-directed decision making. Industries are "owned by society" as a whole. Control of all property is held by the state for the mutual benefit of the people. This necessitates centralized decision making by government planners and eliminates individual economic incentives.

State socialism involves the complete planning of all economic activity, with resources allocated according to these plans. The state also establishes and administers all prices. Direct economic competition is therefore eliminated, and the state alone initiates new business activity.

## Mixed Economic Systems

Most economies of the world would be termed mixed, a combination of the characteristics of both capitalism and socialism. For example, the United States is generally considered capitalistic, yet the U.S. government guides production in many industries, and government regulations exist in nearly all sectors of the economy. Government subsidies and grants are used to provide incentives and disincentives to produce specific goods and services. In total, however, the American system is relatively capitalistic and public enterprises are held to a minimum.

Elements of capitalism are also found in socialistic systems. In Sweden, a highly socialistic economy, many large-scale enterprises are nationalized. The state owns much of the coal, public transportation, ship building, steel, and forestry industries. On the other hand, private property does exist and many small businesses have been left to individual operation and control.

Even in centrally planned economies such as the former Soviet Union (FSU) and China, some elements of capitalism have always existed. Before the current transition to a free-enterprise economy, Soviet collective and state farms allowed some families a small private plot on which to raise crops and livestock. These products may be used for private consumption or sold at a local market. In China, much of the land is divided into small farms that are publicly owned but privately cultivated.

The general economic system existing in a country dictates the type of agricultural system to be found. However, agriculture is difficult to plan and control. The agriculture industry does not lend itself to centralized control and planning as well as many manufacturing processes. Those countries with socialistic agricultural systems have come to realize that agricultural campaigns often go awry owing to weather, disease, distance from the planning headquarters, and a lack of incentives to encourage individuals and groups to meet targeted production goals.

# SOVIET AGRICULTURE OVER THE YEARS

No economy or agricultural system in the world is completely planned, but the Soviet economy is one of the oldest economies that is still highly command-oriented and should provide some insight into how agriculture is organized in some other parts of the world. What follows is a general overview of Soviet agriculture and planning organizations. Keep in mind that significant changes are currently in progress, but the movement to a free enterprise economy will be very difficult and will take several decades. It takes years to build new institutions and develop a country.

Agricultural production in Russia is fairly well limited to European Russia, the Volga, and Siberia. Because of the geographical location of the Soviet Union, one would expect problems in agriculture resulting from the severity of climatic conditions and poor agricultural soils. More than two-thirds of the land area sown to grains in the Soviet Union is located in areas with insufficient precipitation. Only 1 percent of the arable land in the FSU lies in areas with annual precipitation of 28 inches or more, and 40 percent receives only 16 inches.[1] Severe droughts occur about every third year. Also, because most of the FSU's agricultural land is north of the 49th parallel, there are short growing seasons and short frost-free periods. Because of these conditions, the FSU has always had difficulty getting any stability in its agricultural output.

Throughout the history of the FSU, there has been relatively little success in agriculture and many remedies and reforms have been implemented. The most dramatic reforms were those based on proposals by Marx and Lenin. They stated that large-scale agriculture was necessary to increase the productivity of the agricultural sector and to provide the food necessary in their plans for urbanization and industrialization. Therefore, based on the Decree on Land of 1917 under which the right of private property was abolished, millions of peasant farm families were combined into various collective organizations to form large-scale operations.[2] In 1928, prior to collectivization, the average sown cropland per household was only 5 hectares (12.4 acres). In 1990, the average collective farm was farming 5873 hectares, and the average state farm had 15,277 hectares.

---

[1] Fletcher Pope, Jr., Valentine Zabijaka, and William Ragsdale, "Agriculture in the United States and the Soviet Union," *Foreign Agriculture Economic Report*, No. 92, Economic Research Service, USDA, October 1973.

[2] V. V. Matskevich, "Socio-Economic Changes and the Socialist System of Agriculture in the USSR," *Agriculture of the Soviet Union*, Moscow: MIR Publishers, 1970, p. 20.

**TABLE 15-1**  Number and Size of Collective Farms, Former Soviet Union.

| Year | Number | Average Size (acres) |
|------|--------|----------------------|
| 1940 | 236,945 | 3,531 |
| 1950 | 123,747 | 7,564 |
| 1960 | 44,944 | 15,928 |
| 1970 | 33,561 | 15,073 |
| 1976 | 28,600 | 16,100 |
| 1979 | 26,500 | 16,556 |
| 1980 | 25,900 | 16,309 |
| 1985 | 26,200 | 15,817 |
| 1988 | 26,900 | 15,543 |
| 1990 | 29,100 | 14,512 |

SOURCE: Economic Research Service, USDA, Washington, D.C.

Collectivization was begun on an entirely voluntary basis. However, when in 1929 the grain shortage was acute and the collective and state farms still represented only a minor part of agricultural production, Stalin instituted a plan of forced collectivization. The forced collectivization of agriculture took a heavy toll in terms of human life and livestock. The mass collectivization was in part a method by which to control the independent nature of the peasants as well as a way to channel agricultural efforts to meet national goals. Stalin's concession to the peasants was to allow each farm family to cultivate a private plot of land, ranging in size from one-half to 1.5 acres, depending on the productivity of the soil.

## Collective Farms

*Collective farms* are large-scale farms organized by the Soviet state to bring households together in an attempt to apply modern agricultural technology and, thereby, increase agricultural output. All land in the FSU was state-owned and managed under a central state plan. The number of collective farms or *kolkhozes* decreased from 236,945 in 1940 to 29,100 in 1990, but the average size of these farms grew from 3531 acres to 14,512 acres over the same period, as can be noted in Table 15-1. A mass merger was instituted by Nikita Krushchev, pressuring the *kolkhozes* to merge into "supersized" farms, which was expected to increase efficiency. In 6 months during 1950, the number of collectives fell from 252,000 to less than 124,000. In 1954, state farms started absorbing collective farms. This policy was especially apparent where agriculture remained weak after World War II, and heavy state investment was needed to increase productivity.

**TABLE 15-2**  Number and Size of State Farms, Former Soviet Union.

| Year | Number | Size (Acres) |
|------|--------|--------------|
| 1940 | 4,159  | 30,146 |
| 1950 | 4,988  | 36,818 |
| 1960 | 7,375  | 64,740 |
| 1970 | 14,994 | 51,397 |
| 1976 | 18,100 | 47,200 |
| 1979 | 20,800 | 43,490 |
| 1980 | 21,057 | 42,501 |
| 1985 | 22,687 | 39,783 |
| 1988 | 23,300 | 38,466 |
| 1990 | 23,500 | 37,749 |

SOURCE: Economic Research Service, USDA, Washington, D.C.

Collective and state farms still occupy over 80 percent of the total crop-land sown in Russia and the Ukraine.

## State Farms

*State farms* are operated like "factory" enterprises, and labor is paid a wage. Collective farms, on the other hand, pay wages for certain jobs, but compensation is usually a share in the residual income after all production costs have been met and expenditures for investment have been made.

State farms were organized just after the Bolshevik Revolution in 1917. More than 573 million acres of land were taken from the large landowners. These large estates were not divided among the peasants, but were converted to nationalized state farms operated by the state. These farms were to demonstrate the advancements in agriculture to collective farms. They test new technologies, and their information is used by collective farms for increasing agricultural output.

Each state farm or *sovkoz* is operated by an appointed director who acts as a one-person management. The director is responsible for organizing production, and hiring and rewarding specialists and foremen who manage the farm's production subdivisions.

State farms are gigantic. The number of these farms tripled between 1960 and 1980. As shown in Table 15-2, the average size increased from about 37,000 acres in 1950 to nearly 65,000 in 1960, largely because of the development of new land in Soviet Asia. With the growth of urban populations in the European portion of the FSU, the average acreage of state farms has been declining as new state farms of smaller size have been established near cities and industrial centers. These new state farms specialize in vegetables, dairy, poultry, and livestock fattening to meet the needs of the nearby large urban populations.

Private plots are also farmed by state farm members in their spare time, with the products being used for personal consumption or sold in the open market.

## Soviet Agriculture in Change

The FSU's problems in agriculture have been related to a host of factors. Recurrent crop failures, while influenced by adverse weather, are intertwined with management dilemmas, poorly germinating seeds, shortages of insecticides and weedkillers, and the lack of repair parts for farm equipment.[3] Also, Soviet leaders have found it difficult to plan and control the 29,100 collective farms that included 12 million households in 1990.

Diversified collective farms are being divided and merged with state farms into specialized agricultural units. *Interfarm units* are also being established. Under the interfarm system, several collective farms are joined together to form large-scale specialized agricultural enterprises.

The interfarm complexes are operated by specialists, but are owned by the participating collective farms. The purpose of these complexes is to increase labor efficiency above the level now existing on collective farms. There are many factory-type poultry and egg farms, livestock units, feedmills, and artificial insemination centers.[4] Also, there are numerous machinery and repair stations to serve both collective and state farms. However, many of these large farms, particularly livestock farms, are not efficient and are unable to compete with imported foodstuffs without government protection. State and collective farms continue to depend on state subsidies and controls on agricultural imports to stay in business in much of the FSU.

## Planning Soviet Agriculture

Agricultural production in the FSU was not as thoroughly planned as was industrial production, and total agricultural production really was not planned at all. What was planned was the amount of agricultural products that the state would purchase from the farms at specified prices. Once delivery quotas were met, farms could receive bonuses for additional out-

---

[3] Alexander M. Derevany, "Soviet Agriculture at the Crossroads," *Ag World*, St. Paul, Minn., December 1975, p. 12.

[4] Alexander M. Derevany, "Soviet Agriculture at the Crossroads," Part II, *Ag World*, St. Paul, Minn., January 1976, p. 12.

put. Collective farms could also sell their excess products in the *kolkhoz* markets. "Private plot" agriculture was not centrally planned even though the private plots produced about one-fourth of the total Soviet agricultural output.

The Council of Ministers and the State Planning Commission (*Gosplan*) coordinated the planning of production, utilization, and export of agricultural commodities. The council of ministers was the cabinet of the FSU headed by the premier. This policy-making body included 100 people who administered the entire economy. It made decisions on the production of all commodities, investments, military production, consumer goods, foreign trade, construction, wages, and prices. *Gosplan* took the general directives from the council of ministers and drafted a detailed economic plan. National acreage and production goals were broken down into goals for each of the 15 republics, and these were further refined by regions and districts. With these objectives in mind, the Soviet government prepared its 5-year plans. The purpose of these plans was to set forth goals and priorities. With respect to agriculture, each 5-year plan sets forth amounts of capital investment, fertilizer use, crop production, and land reclamation. These 5-year plans stipulated the amounts that the state purchasing organization would purchase each year at a guaranteed stable price.

Proceeding from the *State Stable-Order Plan*, each farm drew-up its own plan. This plan took into consideration the farm's own resources, past production records, and national goals. In this process, compromise was necessary to arrive at production goals. These plans were also for 5-year periods, but were divided into 1-year components as well.

A difference between production goals and quotas should be noted. Production goals took resource requirements into consideration and were basic to overall planning in the Soviet system. These goals set production targets for each farm, while farm quotas were the amounts that the farms must sell to the state purchasing organization. Since production goals are idealistic, quotas were always smaller quantities than the production goals.

## Changes in Public Policy

Increasingly it was being recognized that the large grain imports by the FSU in 1972–73 and 1979–85 were due as much to a change in priorities as to poor grain crops.[5] Consumer programs have brought visible im-

---

[5] D. Gale Johnson, "Soviet Agricultural and World Trade in Farm Products," *Prospects for Agricultural Trade with the USSR*, Economic Research Service, USDA, 1974, p. 43.

**TABLE 15-3** FSU Food Program Goals

| Commodity | Actual Output | | Five-Year Plans | |
|---|---|---|---|---|
| | 1985 | 1990 | 1981–85 | 1986–90 |
| | Million metric tons | | | |
| Grain | 191.7 | 235.0 | 238–243 | 250–255 |
| Sugar beets | 82.1 | 81.7 | 100–103 | 102–103 |
| Potatoes | 73.0 | 63.6 | 87–89 | 90–92 |
| Sunflower seeds | 5.3 | 6.6 | 6.7 | 7.2–7.5 |
| Soybeans | 0.6 | 0.9 | 1.4 | 2.2–2.3 |
| Meat | 17.1 | 20.0 | 17.0–17.5 | 20.0–20.5 |
| Milk | 98.6 | 108.4 | 97–99 | 104–106 |

SOURCE: Foreign Agricultural Service, USDA, *Foreign Agricultural Circular; Grain and Live-stock Situation and Outlook* (various issues).

provements in the well-being of Soviet people.[6] During the last few decades prior to the collapse of the FSU, the quality of diets improved considerably as did the quantity of food available. Despite these gains, the level of living still lagged behind that of many countries in Europe, and that of the United States. Soviet leaders were faced with an increasingly affluent and sophisticated population that wanted more and better housing, personal automobiles, and more quality foods, especially meat.[7] In this effort, livestock product consumption was heavily subsidized to keep retail prices below production and marketing costs.

The FSU had been increasing imports of meat and meat products, milk and milk products, wheat flour, rice, fruit, vegetables and sugar. Purchases of slaughter animals were consistent with the ambitious goals the FSU had set for itself in its 1981–85 plan. Also, their system of raising livestock in huge mechanized complexes showed the high priority placed on meat-production goals. In 1986–90, the FSU adopted a policy of moderate feed, oilseed, and meat imported with increased domestic feed production. This program permitted slow increases in per capita consumption of livestock products. Soviet livestock rations were deficient in digestible protein. Hence, imports consisted of oilseeds or oilseed meal to improve the protein balance.

When comparing the 1986–90 plan with what actually occurred during that period, one can see that some goals were met while others were not (Table 15-3). Production of grain, potatoes, sugar beets, sunflowers, and

---

[6] Gertrude E. Schroeder, "Soviet Economic Growth and Consumer Welfare: Retrospect and Prospect," *Prospects for Agricultural Trade with the USSR*, Economic Research Service, USDA, 1974, p. 4.

[7] Ibid., p. 6.

soybeans fell short of the planned amounts, while meat and milk met their projections.

The 5-year plan covering the years 1986–90 showed increases in most areas, over both the 1981–85 plan and the levels actually achieved during the 1981–85 period. However, recent 5-year plans for Soviet agricultural development have been a bit less optimistic than previous plans.

A striking feature of the 1986–90 plan was the high-targeted growth of livestock product output. Prior to 1992 personal incomes and government subsidies have been increasing, resulting in increased consumption of meat and animal products.

The Soviet government's 1986-90 plan anticipated a new agreement to import substantial amounts of grain from the United States. The knowledge that a source of grain was assured, allowed the planners to redirect some emphasis from grain production to other products.

Grain production was not to be ignored in the 1986–90 plan. Under that plan, production was to increase, with much of the increase due to the application of additional amounts of fertilizer. To a lesser degree, a shift to higher-yielding varieties of grains accounted for the proposed increases over past production.

The production increases set forth under the 1986–90 plan in all areas were dependent on continued growth in the number and size of interfarm units. As mentioned earlier, the emphasis on interfarm units was an attempt for better control, both in the setting of more realistic goals and priorities, and in achieving those goals.

After two decades of planned expansion of the livestock sector during the 1970s and 1980s, the region's livestock sector declined in the 1990s as a result of the new economic reforms as well as a major shortage of foreign exchange to continue importing large quantities of agricultural commodities. Feed imports to the FSU in the 1990s have plunged to a fraction of the level reached in the late 1980s. The major economic reforms initiated in both Russia and other FSU countries in 1991 led off with price liberalization combined with widespread reductions in producer and consumer subsidies. The rapid increase in food prices following liberalization caused consumers to substitute less expensive foods, such as bread and potatoes, for meats. Livestock products in particular became much less affordable to consumers after consumer subsidies were removed. Livestock production has remained inefficient and noncompetitive with imports because of the local low feed quality, poor technology, and weak incentives to use inputs efficiently. Production subsidies and import controls have continued to be used to maintain livestock production and have hindered progress in improving production efficiency.

Grain production has also declined in the FSU; for example, 1995/96 grain production was estimated at 126 mil. mt, 11 percent less than for 1994/95, and the lowest in 20 years (Table 15-4). However, price liberal-

**TABLE 15-4** Grain Area, Yield and Production (cleanweight) in FSU, 1990–1995[a]

| Country | Average 1986–90 | 1990 | 1991 | 1995 |
|---|---|---|---|---|
| Area (mil. acres): | | | | |
| Russian Federation | 162.2 | 155.8 | 152.7 | 136.0 |
| Ukraine | 38.4 | 36.0 | 36.3 | 34.4 |
| Other FSU | 80.3 | 78.8 | 77.9 | 69.6 |
| Total FSU | 280.9 | 270.6 | 266.9 | 240.0 |
| Yield (mt/acre): | | | | |
| Russian Federation | 0.64 | 0.75 | 0.58 | 0.47 |
| Ukraine | 1.23 | 1.41 | 1.06 | 0.99 |
| Total FSU | 0.70 | 0.81 | 0.60 | 0.52 |
| Production (mil. mt): | | | | |
| Russian Federation | 104.3 | 116.7 | 89.1 | 63.4 |
| Ukraine | 47.4 | 51.0 | 38.7 | 33.9 |
| Total FSU | 196.5 | 218.0 | 161.0 | 125.7 |

SOURCE: Economic Research Service, USDA.

[a] Grains include wheat, barley, rye, corn, oats, millet, unmilled rice, and pulses.

ization in the FSU has led to higher grain prices and lower grain use, especially for feed. Therefore, only Armenia, Azerbaijan, Georgia, Kyrgyzstan, and Tajikistan continue to be highly dependent on grain imports to maintain adequate levels of consumption. A major factor restricting grain imports is the decrease in food aid and export credits since the early 1990s. After falling for many years, FSU grain production has been forecast to increase in response to higher grain prices in 1995.

As a result of the continued price liberalization since 1991 and declining real per capita income, consumer demand has been severely dampened for livestock products. The average real incomes in most FSU countries fell by more than one third from 1991 and 1995 while average per capita meat consumption fell by more than a quarter. However, as the FSU has a cost disadvantage in production, there has been a major increase in high value food imports, especially livestock products, replacing the larger grain imports in the 1970s and 1980s. The United States supplied 12 percent of the $1 billion in high value imports to the FSU in 1995 including half of all meat imports. Total grain imports in 1995 were less than $0.2 billion to the FSU.

Several FSU countries including Russia, Ukraine, Kyrgyzstan, and the Baltic nations have applied for World Trade Organization (WTO) accession. The reform requirements for WTO membership and the application process itself has helped bolster market reforms in these countries.

# USA–FORMER USSR AGRICULTURAL COMPARISONS _____

Agriculture had received relatively more attention in the FSU. Agriculture represented 18 percent of the Soviet gross national product, and 2 percent of the U.S. gross national product in 1990. The amount of capital invested in Soviet agriculture was less than in the United States, and the output produced was also less.

American agricultural farms are, on the average, much smaller than FSU farms. There are 2.1 million farms in the United States, at an average size of 469 acres. By comparison, Soviet state farms averaged 37,749 acres in 1990; collective farms averaged 14,512 acres.

The administrative set-up for Soviet farms resembled that of American corporations, with either appointed or elected managers. The Soviet state farm directors and collective farm chairmen, along with their cadre of specialists, made production management decisions. These decisions were made in accordance with the 5-year plan.

The Soviet labor force was as large as that of the United States. Agriculture employed 19 percent of the large labor force in the FSU; in the United States, less than 2 percent of the labor force is employed in farming.

The land area of the former Soviet Union is about two and one-half times larger than that of the United States, but much of it is not suitable for agriculture. The Soviets have a cultivated land area that is 45 percent larger than the United States, but there is greater drought susceptibility.[8] Lying 15° to 20° farther north, the FSU has a growing season and frost-free period similar to that of Canada.

Soviet agricultural production has grown less rapidly than U.S. agricultural output in recent years: farm output actually decreased in the FSU while increasing in the United States.

Table 15-5 shows large increases in sown acreage in the FSU between 1950 and 1995. The United States, on the other hand, in some years has reduced total sown acreage in an attempt to cut its surpluses. These data also show that most of the FSU's grain acreage is sown to food grains (wheat and rye) whereas in the United States most grain acreage is for feed grains. Soybeans are the largest U.S. oil crop, compared with sunflower seeds in the FSU.

During the period from 1950 to 1995, Soviet livestock numbers increased substantially. The low livestock numbers of 1950 are a reflection

---

[8] Fletcher Pope, Jr., Valentine Zabijaka, and William Ragsdale, "Agriculture in the United States and the Soviet Union," *Foreign Agricultural Economic Report*, No. 92, Economic Research Service, USDA, Revised January 1977, p. 2.

**TABLE 15-5**   U.S.–FSU Distribution of Sown Acreages for Selected Crops

|  | 1950 | | 1995 | |
|---|---|---|---|---|
|  | U.S. | FSU | U.S. | FSU |
|  | *Million acres* | | | |
| Wheat | 61.6 | 95.1 | 69.2 | 120.3 |
| Rye | 1.8 | 58.3 | 1.6 | 16.0 |
| Corn | 72.4 | 11.9 | 71.2 | 7.0 |
| Oats | 39.3 | 40.0 | 6.3 | 22.4 |
| Barley | 11.2 | 21.3 | 6.7 | 52.4 |
| Potatoes | 1.7 | 21.2 | 1.4 | 14.1 |
| Vegetables | 1.6 | 3.2 | 3.4 | 3.5 |
| Tobacco | 1.6 | 2.5 | 0.7 | NA |
| Cotton | 17.8 | 5.7 | 16.9 | 6.5 |
| Forage crops | 85.1 | 51.1 | 59.8 | NA |
| Fruit, berries, nuts | 3.3 | 3.4 | 3.7 | NA |
| Sugar crops | 1.3 | 3.2 | 2.4 | 6.8 |
| Soybeans | 13.8 | 0.0 | 62.6 | 1.8 |
| Sunflower seeds | 0.0 | 8.9 | 3.5 | 15.9 |
| Total | 336.4 | 364.9 | 309.4 | NA[a] |

SOURCE: A Compendium of Papers, Joint Economic Committee, Congress of the United States, "Soviet Economic Prospects For The Seventies," 93rd Congress, 1st Session, June 27, 1973; U.S. Department of Agriculture, *Agricultural Statistics*, U.S. Government Printing Office, Washington, D.C., 1992; and U.S. Department of Agriculture, *Former USSR Situation and Outlook*, Washington, D.C., 1996.

[a] NA means not available.

of the losses due to the forced collectivization in the 1930s and the effects of World War II. As can be seen from Table 15-6, the FSU produces more sheep, but fewer cattle and hogs than does the United States. Most of the cattle in the FSU are dual purpose, being raised for both milk and meat production. The FSU has very few high-grade beef cattle such as are raised in the United States for meat purposes. Most "beefsteak" purchased in Soviet restaurants is of poor quality. Also, while the Soviets produce more milk, the output per animal is less because of the breeds of animals used. Livestock inventories in the FSU have declined about 30 percent from 1992 to 1996 because of the increased feed costs and reduced demand for livestock products.

## Levels of Technology

Soviet technology has lagged far behind that of the United States. Soviet agriculture is typical of a developing nation's agriculture, that is, it is char-

**TABLE 15-6**  Livestock Numbers

|  | 1950 | | 1995 | |
| --- | --- | --- | --- | --- |
|  | U.S. | FSU | U.S. | FSU |
|  | | Million head | | |
| Cattle | 78 | 58 | 103 | 91 |
| Hogs | 59 | 22 | 60 | 49 |
| Sheep | 30 | 86 | 9 | 74 |

SOURCE: Economic Research Service, USDA.

acterized by a high percentage of the labor force employed in farming and by low levels of mechanization and mineral fertilizers used. This is a result of the low priority placed on agriculture over the past years and the inadequate agricultural background of most past Soviet leaders. Agriculture has received increasing attention as food problems have developed, but much more investment is needed to increase the productivity of agriculture.

## Foreign Trade

Soviet foreign trade had been increasing very rapidly over the 2 decades prior to 1992. In the past, the Soviet's need for Western technology and food has increased trade with the West. The FSU exported crude oil and petroleum products, coal, coke, wood and wood products, metals, mineral fertilizers, and machinery and equipment to the West. Their imports from the West included chemical equipment, automotive manufacturing equipment, steel and pipe, industrial consumer goods, as well as wheat and feed grains.

The major events of 1972 were the proposed United States–Soviet Union trade agreement that was later rejected and the normalizing of relations with the People's Republic of China. The U.S.–FSU trade agreement called for a tripling of trade between 1973 and 1975. In 1972, the United States exported $547 million worth of all goods to the FSU and imported $95 million.

U.S. agricultural trade with the FSU before 1972 was relatively unimportant—exports to the FSU consisted primarily of hides, skins, and almonds. However, with the easing of relationships, the removal of U.S. flag shipping requirements, and poor weather, the FSU increased imports of feed grains (mostly corn), wheat, and soybeans. The value of U.S. agricultural exports to the FSU totaled less than $10 million per year up to 1970, increased to $459 million in 1972, and amounted to more than $1.3 billion in 1995 (Table 15-7). In 1989 U.S. agricultural exports to the FSU exceeded $3.5 billion, with corn shipments alone amounting to over $2.1 billion. The FSU also imported large amounts of wheat, corn, and raw

**TABLE 15-7**   U.S. Agricultural Trade with the Former USSR

| Commodity | 1971 | 1975 | 1978 | 1985 | 1995 |
|---|---|---|---|---|---|
| | | | Million dollars | | |
| U.S. Exports to FSU[a] | | | | | |
| Wheat | 0.7 | 672.7 | 355.8 | 162.3 | 104.3 |
| Coarse grains[b] | 26.3 | 457.8 | 1109.4 | 1540.7 | 16.2 |
| Corn | 24.5 | 452.6 | 1109.4 | 1540.7 | 16.2 |
| Other coarse grains | 1.8 | 5.2 | 0.0 | 0.0 | 0.0 |
| Soybeans[c] | — | 2.9 | 222.1 | 0.0 | 68.3 |
| Cattle hides | 10.9 | 5.2 | 8.1 | 0.0 | 1.7 |
| Fruits, nuts, and berries | 1.5 | 6.1 | 16.8 | 67.7 | 3.5 |
| All other | 5.2 | 25.6 | 52.9 | 137.1 | 1138.0 |
| Total | 44.6 | 1170.3 | 1765.1 | 1907.8 | 1332.0 |
| U.S. Imports from FSU | | | | | |
| Casein and glue | — | 1.7 | 2.4 | 0.1 | 23.9 |
| Fur skins | 2.7 | 3.5 | 8.8 | 7.8 | 4.0 |
| All other | 0.2 | 1.0 | 1.1 | 0.7 | 23.4 |
| Total | 3.0 | 7.2 | 12.3 | 8.6 | 51.3 |

SOURCE: Economic Research Service, *U.S. Foreign Agricultural Trade Statistics Reports*, USDA, annually.

NOTE: — means, negligible or none.

[a] Includes transshipments through Canada, Belgium, the Netherlands, and West Germany.

[b] Includes corn, rye, barley, oats, and sorghum.

[c] Includes seeds, oilcake, and meal.

sugar from countries other than the United States. Current FSU imports are mostly consumer food products.

American agricultural imports from the FSU are relatively insignificant. They amounted to only $51 million in 1995. The primary imports are furs and skins, casein, and glue. Sugar, cheese, fruit juices and essential oils are also imported from the FSU.

## USA–Former USSR Grain Agreements

The governments of the FSU and the United States entered into a grain agreement in 1975. The purpose of the contract was to assure the FSU a supply of grain and to indicate to U.S. producers the amount of grain the FSU would be purchasing. The agreement allowed the FSU to purchase 6 million metric tons of wheat and corn in about equal proportions in each 12-month period between October 1, 1976, and September 30, 1981. The sales of U.S. grain would be by private U.S. commercial traders at prevailing market prices. The FSU could purchase an additional 2 million tons in any 12-month period unless the total grain supply in the United States fell below 225 million metric tons. Purchases of grain in excess of 8 million tons per year could only be made after consultation with the U.S. government. This agreement covered only wheat and corn, permitting the FSU to purchase additional quantities of other U.S. grains such as barley, oats, and rice if needed.

In response to the FSU's 1979 invasion of Afghanistan the president of the United States on January 4, 1980, placed an embargo on the shipment of agricultural commodities to the FSU, except for the 8 million tons of wheat and corn that the United States was committed to deliver under the fourth year of the grain agreement. The partial embargo was left in effect for 15 months. During the embargo, the Soviet Union was forced to purchase grain at higher prices elsewhere, primarily from Argentina.

On July 28, 1983, the United States and the FSU signed another 5-year grain agreement. The agreement required the FSU to purchase a minimum of 8 million tons of grain (about one-half wheat and one-half corn) per year. In addition, they were to purchase another 1 million tons of wheat and/or corn, or 500,000 tons of soybeans or soybean meal. The FSU could also buy an additional 3 million tons of wheat or corn without consultation with the U.S. government. The FSU failed to meet their purchase requirements for wheat under the Long-Term Grain Agreement for 1984–85 and 1985–86. The FSU contended that U.S. prices were not competitive with world prices, while the United States maintained that the agreement stipulated prevailing prices on the U.S. market.

A third 5-year grain agreement was approved in 1990, effective beginning in 1991. Under the new pact, the Soviets were required to purchase 10 million metric tons of U.S. feed grains, wheat, and soybeans per year. They could purchase up to 14 million tons of wheat and feed grains without prior consultation. Of the 10 million tons required annually, 4 million tons was to be feed grains, 4 million wheat, and the remaining 2 million metric tons may be in grains, soybeans, or soybean meal. This long term grain agreement, however, became ineffective with the breakup of the Soviet Union. More recent grain exports to the former Soviet Union have been made under various food assistance and credit programs.

# TRANSITION TO A "FREE-MARKET" ECONOMY[9] _____

## Relationships Between Member Countries

The Communist party in the former USSR was abolished following a failed coup attempt against President Gorbachev in August 1991. The

---

[9] Economic Research Service, USDA, "Former USSR, Agriculture and Trade Report," *Situation and Outlook Series,* Washington, D.C., May 1992 and May 1993.

Union itself also broke up and was officially dissolved in December 1991 along republic lines. Eleven of the fifteen republics in the FSU formed a very loose association known as the *Commonwealth of Independent States* (CIS) as can be seen in Figure 15-1. These eleven CIS republics are Azerbaijan, Armenia, Byelarus, Kazakhstan, Krgyzstan, Moldova, Russia, Tajikistan, Turkmenistan, Uzbekistan, and the Ukraine. The three Baltic States and Georgia elected to be nonparticipating republics. The Baltic countries are Lithuania, Latvia, and Estonia. Russia accounts for about half the population and about three-fourths of the area of the former USSR.

There was also a collapse of the former Council for Mutual Economic Assistance (CMEA), formed in January 1949, which included Bulgaria, Czechoslovakia, Hungary, Poland, Romania, Cuba, Vietnam, and Mongolia. East Germany was a member of the CMEA prior to unification with West Germany. The CMEA group involved trade and other concessions provided by the former USSR to these countries. CMEA countries received oil and other natural resources at subsidized prices from the USSR; the former USSR was obligated to receive generally inferior capital and consumer goods in return. There is still some trade activity between former CMEA members and the former USSR; however, trade has declined significantly in recent years. The subsidization of former USSR exports to these countries has been curtailed as well as to the former USSR republic members. The former USSR, especially Russia, is expected to benefit in the future because of the improvement in terms of trade with former CMEA countries and republic members. However, in the short run, the rapid collapse of central planning in the former Soviet bloc countries, as well as the breakup of CMEA and the former USSR has caused serious disruption in trade and other economic relations between these countries. The ruble was formerly accepted as a common currency among the republics, but many are now planning their own currencies.

Many of the former republics have become more dependent on food imports from outside the former USSR with concessionary terms including agricultural aid and credit as a result of the recent shift from an administrative to market system. Western nations provided food related assistance of $20 billion from 1990–1993, with more than half of this amount in the form of export credit guarantees. Food imports have been particularly important to Russia, the Transcaucasus republics, and the republics of Central Asia. Food aid and export credits, however, have been reduced since the early 1990s.

## Internal Adjustments Within Countries

Russia has been the leader in progressive market reforms in the former USSR and was among the first of the republics to liberalize prices, to initiate *privatization*, to make the ruble more convertible, and to integrate

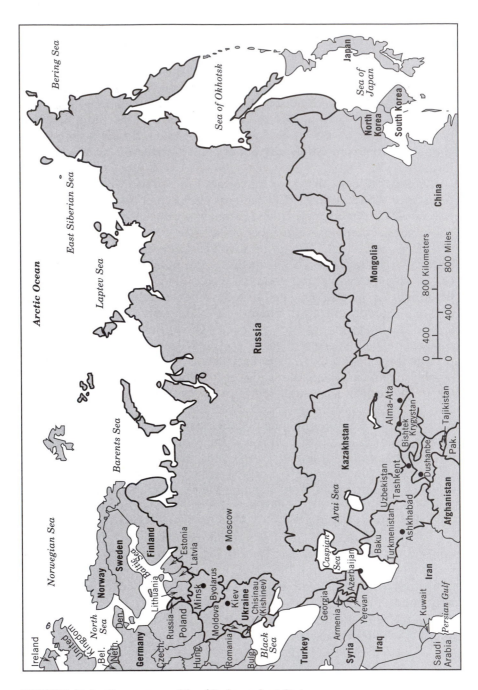

**FIGURE 15-1**   Commonwealth of Independent States.

with the world economy. However, the transition from a planned economy to a free-market economy has been a painful experience for all republics, requiring tremendous shifts in the use of resources to restructure the economy. The shift has temporarily idled some resources including labor and land, for example, in the switch from collective and state ownership to private ownership of farms. The cost of restructuring was exacerbated because of the large producer and consumer subsidies in use prior to the break-up of the former USSR. The use of these subsidies caused major distortion in resource allocation among farmers, processors, and distributors.

Along with the problem of continued subsidy distortions, most of the newly independent States have experienced inflationary problems that have contributed to increased food shortages. Per capita money income had increased rapidly in recent years as well as state budget deficits and expansion in the money supply. Prices of food and most other consumer goods were restricted from rising to repress inflation. This long-term macroeconomic imbalance contributed to further serious problems, including hoarding and bartering consumer goods and a major decline in industrial and agricultural output in all former USSR republics. The macroeconomic imbalance and the associated weakness of the ruble resulting from continued poor management practices of the command system were important factors in strengthening the post "coup" resolve of the republics to attain independence and to dissolve the Union. The republics had become unwilling to continue accepting the ruble and had lost faith in the central government's ability to maintain fiscal and monetary discipline.

Currently economic reforms are being introduced to abandon the command system and adopt a free-market system in each of the republics. The Russian Republic led the way in a radical reform program to create the institutional base for an effective market capitalist system including increased privatization, and integrating the Russian economy into the world economy.

Prices have been liberalized in Russia in stages starting in January 1992; however, economic conditions are not yet stable (Table 15-8). Other republics have generally followed Russia's lead in liberalizing, or at least raising consumer prices. To do otherwise would have meant that Russian consumers would have purchased goods from other republics at lower prices. Many of the republics had been dependent on Russia for fuel, other natural resources, and heavy industrial goods. This dependence provided an opportunity for them to coordinate their economic reforms and policies with Russia; however, many have chosen to separate themselves economically and to resist restructuring in order to protect existing enterprises and jobs and to avoid domination by Russia. The reform movement has been delayed in both Russia and other republics as the governments continue to expand credit for existing enterprises so as to protect

**TABLE 15-8**   Economic Indicators for selected FSU Countries

| Country/Item | 1992 | 1993 | 1994 | 1995 |
|---|---|---|---|---|
| Russian Federation: | | | | |
| GDP | (14.5) | (8.7) | (12.6) | (4.0) |
| Industrial Production | (18.0) | (14.1) | (20.9) | (3.3) |
| Agricultural Production | (9.4) | (4.4) | (12.0) | (8.0) |
| Consumer Prices | 2,564 | 840 | 215 | 131 |
| Budget Deficit (% GDP) | 3.2 | 4.6 | 10.3 | 2.4 |
| Ukraine: | | | | |
| GDP | (13.7) | (14.2) | (23.0) | (12.3) |
| Industrial Production | (6.4) | (8.0) | (27.3) | (11.5) |
| Agricultural Production | (8.0) | 2.0 | (16.7) | (10.0) |
| Consumer Prices | 2,000 | 10,160 | 400 | 180 |
| Budget Deficit (% GDP) | NA | 5.1 | 9.4 | 5.2 |
| Belarus: | | | | |
| GDP | (9.6) | (10.6) | (15.8) | (10.0) |
| Industrial Production | (9.4) | (7.4) | (17.1) | (11.5) |
| Agricultural Production | (9.0) | 4.0 | (14.0) | (6.0) |
| Consumer Prices | 970 | 2,000 | 1,956 | 245 |
| Budget Deficit (% GDP) | 2.0 | 5.6 | 3.3 | NA |
| Kazakhstan: | | | | |
| GDP | (13.0) | (12.9) | (24.6) | (8.9) |
| Industrial Production | (13.8) | (14.8) | (28.5) | (7.9) |
| Agricultural Production | 1.0 | (5.0) | (20.1) | (21.0) |
| Consumer Prices | 1,510 | 2,170 | 1,160 | 62 |
| Budget Deficit (% GDP) | (1.4) | (1.9) | (0.5) | (0.3) |

SOURCE: Economic Research Service, USDA, *Former USSR Situation and Outlook Report*, WRS-96-1, Washington, DC. 1996.

NA means not available

( ) means a negative value

jobs and the economy's current structure. The reform movement will continue to be a difficult process because of the economic hardship imposed on the population. In fact, Russian leaders have made concessions that increase the inflationary pressure that threatens further progress on price liberalization, ruble convertibility, privatization, and other key reform measures. Inflation has been reduced in 1994 and 1995 as a result of stricter monetary policy and declining budget deficits.

## Privatization of Farms in the Former USSR

Prior to the collapse of Communism in the former USSR, the private agricultural sector was limited to the household plots of farm employees and small gardens of city dwellers. Independent family farms had been abolished by the 1940s by forced collectivization. However, new legislation was passed in late 1990 to reestablish the individual family farm as a legal entity and to allow farmers to receive land from a special redistribution reserve. The reserve was developed mainly from land that was considered "poorly farmed" by collective and state farms. Overseeing the land distribution process was a State Committee for Land Reform and Support of the Peasant Farm that established local land reform committees throughout Russia. By January 1, 1995, there were nearly 694,300 private family farms that had been registered occupying an area of nearly 21 million hectares. Farm sizes ranged from as small as 1 hectare for truck crops near cities to 346 hectares in the republic of Kazakhstan. The average of all family farms was 30 hectares (about 74 acres) (Table 15-9).

Although the number of private farms has been growing rapidly, they are still relatively minor in terms of the area farmed. Only about 5 percent of FSU's plowed land and only 5 percent of the total agricultural land was operated by individual farmers and farm or livestock cooperatives at the beginning of 1995. However, the number of private farms registered in the Russian Federation had increased to 285,600 by the end of 1994. Currently only a small percentage of the state and collective farms have divided themselves into private farms.

The current Yeltsin government in Russia appears to be committed to extending private agriculture and has taken measures to continue the implementation of restructuring and privatization. State and collective farms may continue, but individuals have a right to leave, with a share of the property. The new draft of the Land Code prohibits private ownership, sale of land, and mortgaged lending. It does allow for long-term leasing and inheritance. Other new measures include the assessment of land taxes on state and collective farms, the liquidation of nonprofitable farms, and the proposed creation of private farm supply and marketing sectors in Russia. The new enterprise sectors will allow citizens to engage in middlemen, retail, or wholesale trading activities, where it was previously forbidden, and the agriservice industry functions will no longer be legally controlled by the state and collective farms.

Despite the encouraging measures being taken to increase the number of private farms and to promote privatization, the new farm operators are facing tremendous difficulties. About 6000 farms in Russia are reported to have failed in the last two years. Principal reasons cited for the failures include: the lack of clear land laws; state farms unwillingness to turn over lands; the absence of markets for input suppliers and continued depen-

**TABLE 15-9**  Number, Average Size, and Growth of Private Farms, Selected Former Soviet Union Republics

| Republic | Jan. 1, 1992 No. of Farms | Avg. Size | Jan. 1, 1995 No. of Farms | Avg. Size | Growth in No. of Farms | Area as Portion of Agricultural Land in Enterprises |
|---|---|---|---|---|---|---|
| | 1,000 units | Hectares | 1,000 units | Hectares | Percent | Percent |
| Azerbaijan | 0.1 | 44 | 0.8 | 25 | 700 | 0.2 |
| Armenia | 164.5 | 1 | 305.0 | 1 | 85 | 4.4 |
| Belarus | 0.7 | 22 | 2.9 | 21 | 314 | 0.5 |
| Kazakhstan | 3.3 | 242 | 21.0 | 346 | 536 | 3.2 |
| Kyrkyzstan | 4.1 | 25 | 21.7 | 29 | 429 | 7.7 |
| Moldova | 0.005 | 0 | 12.7 | 3 | 2,539 | 0.2 |
| Russian Fed. | 49.0 | 43 | 285.6 | 41 | 483 | 1.7 |
| Tajikistan | 0.004 | 25 | 0.2 | 131 | 49 | 0.0 |
| Turkmenistan | 0.1 | 11 | 0.3 | 8 | 200 | 0.0 |
| Uzbekistan | 1.9 | 7 | 12.8 | 14 | 574 | 0.2 |
| Ukraine | 2.1 | 19 | 31.3 | 22 | 1,390 | 1.2 |
| Total | 225.9 | 15 | 694.3 | 30 | 207 | NA |

SOURCE: Statkom SNG.
NA means not available.

dence on state farms for most farm inputs; the unavailability of small scale farm equipment; the lack of alternate channels (other than state farms) to market produce; and inadequate finances and bank credit for private farmers.

Overall, while private farms are increasing in numbers, the former USSR has not changed the structure of agriculture to make any significant difference in productive capacity or efficiency. Also the infrastructure has not been adequately developed and management has not provided adequate incentives for change.

# THE PEOPLE'S REPUBLIC OF CHINA _____

Prior to 1978, China had a command agricultural system. The production and distribution of almost all of China's agricultural commodities were governed by rigid central planning. Compulsory quotas for most products were set at prices well below the world market. China also had a subsidized food policy under which it distributed centrally processed agricultural commodities to urban residents at very low prices. The net effect of the planning system was to support industrial development by taxing agriculture.

Under the pre-1979 commune system, production incentives were low, and as a result, agricultural production remained low. Since 1979, the government introduced a series of international and market reforms to stimulate agricultural productivity. The introduction of individual decision making on farms, called the *household responsibility system* (HRS), allowed shifting production decisions from collectives. Important reforms were also made in the highly controlled areas of procurement, distribution, and marketing of agricultural production.

A state monopoly, controlled by the ministry of foreign trade, handles all international transactions. China trades with about 100 countries, but its largest accounts are with Hong Kong and Japan. American trade with China has been increasing, and some experts believe that the United States has a potential to export large quantities of chemicals, grains, and fertilizers.[10] Basically, China is attempting to become self-sufficient in food production so the Chinese people will not have to rely on outside sources for much of their food supply.

Food production is increasing at a more rapid rate than population. Multiple cropping is expanding, new varieties are being planted, and larger quantities of chemical fertilizers are applied to grain crops. However, prices paid to farmers for grain are generally less than prices paid

---

[10] U.S. Congress, Joint Economic Committee, *Economic Developments in Mainland China*, 92nd Congress, 3rd Session, Washington, D.C., 1972.

for other commercial crops. Subsidies are provided to farmers for fertilizer, tractors, and fuels to encourage agricultural development.

China has a cropland area approximating that of the United States, and much of the farmland is the productive equal of the land cropped in this country. The two areas, in terms of climate, rainfall, and growing conditions, are not greatly dissimilar. In general, China's yields of food grains and oil-bearing crops were more than U.S. yields of those crops. With these rapid increases in yields since the late 1970s, one must wonder how much of China's agricultural inefficiency can be attributed to the command structure of the economy.

In 1958, under the leadership of Chairman Mao Tse-tung, China launched its "Great Leap Forward" program that was to produce a dramatic growth in its industrial and agricultural output. Following their "Cultural Revolution" that began in 1965, heavy reliance was given to ideology as the primary incentive for greater human effort toward the economy's development. The result was, instead, an unwieldy bureaucracy, political unrest, and stifling disincentives for its people.

In a surprising display of pragmatism, China completely restructured its rural economy, encompassing rural industry as well as agriculture.[11] Begun experimentally in 1978, certain more impoverished areas in Anhui Province were permitted to eliminate their production teams in favor of the household responsibility system. The results were so striking that major reforms in the rural economy were launched for the rest of the nation, permitting free markets and free-market pricing and individual specialization according to their comparative advantage. Although not officially sanctioned until late 1981, 45 percent of China's production teams had already been displaced by HRS by that time; by the end of 1983, more than 94 percent of the nation's production teams had been so switched.[12]

Land ownership is unchanged, but each household may now contract with *economic cooperatives* (the former production teams) to farm a specific tract of land. Under the new farming system, households have much more freedom to decide what to produce, how to produce it, and how to market their products. With quotas and product prices set by the government for contracted household deliveries to the state, households are free to market the remainder of their production through free markets, if they wish to do so. These markets now exceed 72,000 and are the major outlets for vegetables, fruits, meats, and certain specialty products.

---

[11] Much of the information and specific data on the restructuring of China's rural areas is to be found in the continuing series by the Economic Research Service titled *China Situation and Outlook Report*, available from the U.S. Department of Agriculture, Washington, D.C.
[12] Justin Yifu Lin, "The Household Responsibility System Reform in China: A Peasant's Institutional Choice," *American Journal of Agricultural Economics*, Vol. 69, No. 2, May 1987, pp. 410–415.

Table 15-10, with 1980 as a bench-mark year, shows how significant has been the stimulation for China's agriculture and the resulting changes in land use and output rates, as both producers and consumers are more free to exercise their preferences through the free marketing system.

With their efforts rewarded on the basis of their market-determined value productivity, farmers have greatly increased their output of food grains and other crops, and some of these total output increases have come in spite of considerable reductions in the amount of land devoted to producing those crops. Urban Chinese have increased their consumption of wheat and rice relative to the less desired coarse grains and have doubled their per capita consumption of vegetable oils. Producers have responded to the changing market demand by increasing their use of fertilizers and other yield-increasing inputs, as well as their own labor and management, so that yields have risen sharply. For all crops (except cotton and oilseeds) shown in the table, total acres used in production declined from 1980 to 1995 by 6.2 percent, yet total output increased by 46 percent because of a 56 percent increase in the yields of those crops. The output of cotton rose by 67 percent, even though the land area devoted to cotton production increased by only 12 percent, the end product of a 49 percent increase in cotton yields.

Such marked changes have not come about without raising other serious problems, however. Given certain restrictions on individual choices, households are now able to shift their productive efforts to such nonagricultural pursuits as food processing, light manufacturing (mainly handicrafts), construction, transportation, and services. As a consequence, large numbers of rural industries have been established, boosting industrial output. Retail sales outside the government-operated commercial system, for example, grew from 8.4 billion yuan in 1980 to 273.2 billion yuan in 1994, 41 percent of the nation's total retail sales. With a great many families shifting from agriculture (where they typically provided for their own food needs from their agricultural output) to more urban lives, and a rapid growth of consumer incomes resulting from those choices, has come a sharply increased demand for market-provided meats, milk, eggs, and other preferred foods.

The nation's administration is subsidizing many of the system's economic activities (some heavily), with a special focus on developing a more efficient, flexible, and market-oriented system for moving agricultural commodities from producers to consumers and into the international markets.

U.S. agricultural trade with the People's Republic of China resumed in late 1971 after 20 years of a trade embargo. In 1972, U.S. agricultural exports of grains and vegetable oil totaled $64 million. Our exports were needed to supplement domestic production and imports from other countries.

**TABLE 15-10** China's Changing Agriculture

| Crop | Sown Area | | | Production | | |
|---|---|---|---|---|---|---|
| | 1980 | 1995 | Percent Change | 1980 | 1995 | Percent Change |
| | Million hectares | | | Million metric tons | | |
| Wheat | 29.2 | 28.8 | -1.4 | 55.2 | 102.0 | 84.8 |
| Rice | 33.9 | 30.7 | -9.4 | 140.0 | 185.2 | 32.3 |
| Coarse grains[a] | 31.0 | 27.2 | -12.3 | 84.2 | 126.3 | 50.0 |
| (Corn) | (20.4) | (22.8) | (11.8) | (62.6) | (112.0) | (78.9) |
| Potatoes[b] | 10.2 | 9.5 | -6.9 | 28.7 | 32.1 | 11.8 |
| Other[c] | 13.0 | 13.6 | 4.6 | 12.5 | 20.9 | 67.2 |
| Total | 117.2 | 109.9 | -6.2 | 320.6 | 466.6 | 45.5 |
| Oilseeds[d] | 15.2 | 26.3 | 73.0 | 15.6 | 43.3 | 177.6 |
| Cotton | 4.9 | 5.5 | 12.2 | 2.7 | 4.5 | 66.7 |

SOURCE: Economic Research Service, *China Situation and Outlook Report*, July 1996.

[a] Includes corn, sorghum, millet, barley, and oats.

[b] Converted to grain-equivalent weight.

[c] Includes soybeans for beans, pulses, and other miscellaneous grains.

[d] Includes peanuts, cottonseed, rapeseed, sesame seed, sunflower, oil-bearing flax seed, and castor beans.

**TABLE 15-11**  Trade in Major Agriculture Commodities, People's Republic of China

| Item | 1971 | 1973 | 1982 | 1985 | 1991 | 1995 |
|---|---|---|---|---|---|---|
| | | | Thousand metric tons | | | |
| Total grain imports | 3128 | 7645 | 15,546 | 6059 | 13,412 | 19,691 |
| Wheat imports | 3021 | 5987 | 13,258 | 5626 | 12,367 | 11,585 |
| From: | | | | | | |
|    Argentina | — | — | 94 | 815 | 391 | 256 |
|    Australia | 33 | 768 | 2102 | 1214 | 1364 | 436 |
|    Canada | 2988 | 2398 | 3526 | 2370 | 4504 | 4861 |
|    United States | — | 2815 | 6870 | 816 | 4586 | 3840 |
|    Other | — | 6 | 666 | 351 | 1522 | 2192 |
| Coarse grain imports | 107 | 1626 | 1967 | 120 | 751 | 6455 |
| From: | | | | | | |
|    Argentina | 107 | 126 | 155 | 5 | 0 | 195 |
|    United States | — | 1500 | 1591 | 0 | 31 | 5181 |
| Rice exports | 924 | 2142 | 500 | 1019 | 690 | 50 |
| Soybean exports | 460 | 310 | 137 | 1140 | 1110 | 380 |
| Soybean imports | — | 255 | 299 | 0 | 0 | 294 |
| Sugar imports | 464 | 736 | 2160 | 1909 | 1014 | 2572 |
| Cotton imports | 151 | 377 | 109 | 0 | 371 | 742 |
| Corn exports | — | — | — | 6340 | 7780 | 110 |

SOURCE: Economic Research Service, USDA.

NOTE: — means none or negligible.

[a] Includes transshipments through Canada.

[b] Raw sugar equivalent.

As shown in Table 15-11, China has imported large amounts of wheat and coarse grains (mostly corn), with nearly all the wheat coming from the United States, Canada, Australia, and the EU. In the years 1960–65 China imported about 4.5 million tons of wheat compared to the 13.3 million tons imported in 1982 and 11.6 million tons in 1995.

The United States and China signed a 4-year agreement in October 1980, providing for the export of 6 to 8 million metric tons of grain to China each year, beginning January 1, 1981. China could purchase an additional million tons of grain without prior notification. The agreement specified that 15 to 20 percent of the purchases must be corn, and the remaining 80 to 85 percent wheat.[13] Because of its greatly increased grain output,

---

[13] Economic Research Service, *Agriculture Outlook,* Washington, D.C.: U.S. Government Printing Office.

China does not currently need to import large quantities of grain, and thus has not renegotiated long-term grain agreements.

Normalizing relationships between the United States and China has brought increased trade. In 1982, China imported 8.5 million tons of wheat and corn from the United States, but that declined to less than 1 million tons in 1986, and rose again to 19.7 million tons in 1995. All U.S. exports to China amounted to $11.7 billion in 1995, while U.S. imports from China in that year were $45.6 billion. The volume of trade between these nations in future years will depend on their ability to continue to develop export industries and their ability to encourage economic reforms to provide equal access for U.S. products.

China shifted from a net exporter of 12 million tons of grain in 1994 to a net importer of 13.8 million tons in 1995. This 26 million ton swing in world trade sent commodity prices soaring on world markets.

# SUMMARY

Most economies of the world are mixed. They have some characteristics of both capitalism and socialism. Capitalism is an unplanned, self-regulating system with no government involvement in private economic decisions. On the other hand, state socialism is an economic system with collective ownership of all productive resources with state-directed decision making.

Soviet agriculture includes collective, state, and private farms. Collective and state farms are large compared to U.S. farms. The private plots are very small, but produce one-fourth of the total Soviet agricultural output. Total agricultural production in the former USSR is not planned at all. What has been planned is the amount of agricultural products that the state will buy from the farms at specific prices. When farms meet their delivery quotas, they can receive bonuses for additional output.

The Soviet Union was officially dissolved in December 1991 along republic lines. Eleven of the republics formed a very loose association called the *Commonwealth of Independent States*.

New legislation passed in late 1990 reestablished individual family farms. At present, however, only about 5 percent of the total agricultural land is operated by individual farmers. Only 4 percent of the state and collective farms have divided themselves into private farms.

Russia appears to be committed to extending private agriculture and has taken the steps to implement restructuring and privatizing the economy. Many of the republics have become dependent on food imports on concessionary terms as they attempt to restructure their economies to a free-market system.

United States agricultural trade opened with the People's Republic of China in 1971. China is importing grain, cotton, wool, sugar, and vegetable oils, and is an exporter of broilers, oilseeds, cotton yarn and raw silk.

# CHAPTER HIGHLIGHTS

1. The political philosophy of a nation determines the economic system within which agriculture operates.
2. The two general types of economic systems are capitalism and socialism.
3. The most prevalent system in the world is a mixture of capitalism and socialism.
4. The FSU and China have been typical of socialistic or command economies. Now they are slowly transforming their economies to mixed systems that are more market oriented.
5. Most of the agricultural output of the FSU comes from large-scale farms.
6. The FSU has three types of farms. These are collective, state, and private farms.
7. Private-plot farms in the FSU accounted for only a small portion of total agricultural land, yet they produce about one-fourth of the total agricultural output. Since 1990 the former USSR has been slowly moving to a private farm-free market type of agriculture.
8. Increased attention is being given to agriculture in the FSU because of their inability to meet domestic food desires.
9. The FSU and China used 5-year plans to provide production goals for agriculture.
10. The FSU and China used quotas in planning the amounts of agricultural products their farms must deliver to the state.
11. FSU collective and state farms are huge by American standards.
12. FSU farms use much more labor per acre and are much less mechanized than American farms.
13. The performance of the FSU agricultural system is poor as compared with American agriculture.
14. The U.S. imports of FSU agricultural commodities are small, but our exports to them have grown to more than $1.3 billion.
15. U.S. exports of grain to the former USSR were governed by long-term grain agreements. Now U.S. grain exports are in the form of food aid, credit programs, or market transactions.

16. The People's Republic of China had patterned its agricultural industry after that of the FSU but, since 1978, has adopted a household agriculture that is market oriented.
17. Current Chinese agricultural trade is primarily with the United States, Canada, Australia, Japan, Hong Kong, Cuba, and the EU.

## KEY TERMS AND CONCEPTS TO REMEMBER

Capitalism

Collective farm

Commonwealth of Independent States

Household responsibility system

Privatization

Socialism

State farm

## REVIEW QUESTIONS

1. What is the difference between capitalism and socialism? Do most countries have a pure economic system?
2. What is a collective farm? What is a state farm? How do collective and state farms in the FSU differ from family farms in the United States?
3. Is Soviet agriculture planned? Describe the structure of the Soviet planning system. Why is the former USSR moving to a private farm free-market type of agriculture?
4. Why is the FSU importing American grain? Have its priorities changed?
5. The FSU has been interested in increasing foreign trade with Western countries. What does it want? What do we get in exchange?
6. What were the provisions of the U.S.–FSU grain agreement? Was this grain agreement of any benefit to the United States?
7. Describe the structure of Chinese agriculture. Does it resemble the structure of agriculture in the United States?
8. Why did the People's Republic of China stop purchasing grain from the United States in 1975 and 1985? Why have they stopped signing long-term wheat agreements with the United States, Canada, Australia, and Argentina?

## SUGGESTED READINGS

Bornstein, Morris, and Daniel R. Fusfeld, eds. *The Soviet Economy: A Book of Readings,* 4th ed. Homewood, Ill.: Richard D. Irwin, 1974, Parts 1, 2, and 5.

Brandow, G. E. "Impressions of Chinese Agriculture: Viewing Flowers from Horseback in Winter," Department of Agricultural Economics and Rural Sociology, The Pennsylvania State University, A. E. 110, State University, Pa., March 1974.

Diamond, Douglas B., Lee W. Bettis, and Robert E. Ramsson, "Agricultural Production," Office of Economic Research, Central Intelligence Agency, October 1980.

Economic Research Service. *China: World Agriculture Regional Supplement* (annual), USDA, Washington, D.C., June 1983.

Economic Research Service. *USSR: World Agriculture Regional Supplement* (annual), USDA, Washington, D.C., May 1983.

International Conference of Agricultural Economists. *Agriculture of the Soviet Union.* Moscow: Mir Publishers, 1970.

Joint Economic Committee, Congress of the United States, "Soviet Economic Prospects for the Seventies," 93rd Congress, 1st Session, June 27, 1973.

Johnson, D. Gale, "Prospects For Soviet Agriculture in the 1980s," Paper No. 81:10, Chicago: The University of Chicago, April 1981.

McConnell, Campbell R. and Stanley L. Bruce, *Economics*, 12th ed. New York: McGraw-Hill, 1993, Chapter 40.

Nove, Alex. *The Soviet Economy.* New York: Frederick A. Praeger, 1965.

# AN OUTSTANDING CONTRIBUTOR

**D. Gale Johnson**   D. Gale Johnson is Eliakim Hastings Moore Distinguished Service Professor in the Department of Economics at the University of Chicago.

Johnson was born on a farm near Vinton, Iowa. He earned his B.S. degree at Iowa State College in 1938, and the M.S. degree at the University of Wisconsin in 1939. Following 2 years of further graduate work at the University of Chicago, he completed the work for a Ph.D. degree at Iowa State College and was awarded that degree in 1945.

Dr. Johnson was appointed assistant professor at Iowa State College in 1941. He left there in 1944 to accept an appointment in the Department of Economics at the University of Chicago, where he has served continuously since then. He has held a number of responsible administrative offices at the university, engaged as associate dean and dean of the Division of Social Sciences, vice president and Dean of Faculties, and Provost.

Johnson's primary research interest has been in the areas of agricultural policy and world agricultural production and trade. He has written extensively on the agricultural development of the Soviet Union. Johnson has written or edited 12 books and contributed chapters or sections to an even larger number of other books, as well as numerous professional articles in a variety of U.S.

and foreign journals dealing with world food and resource allocations between agriculture and other sectors of the economy. About a dozen of his writings have also been published in Japanese, Italian, Spanish, and Portuguese.

The expertise of Johnson has been utilized by many government and other groups. He has served as consultant or advisor to the Office of Price Administration, Department of State, Department of the Army, Tennessee Valley Authority, Office of Price Stabilization, RAND Corporation, President's Committee to Appraise Employment and Unemployment Statistics, agricultural advisor to the Special Representative for Trade Negotiations, National Advisory Commission of Food and Fiber, Department of State Policy Planning Council, National Commission on Population Growth and the American Future, U.S. Council on International Economic Policy, and the National Academy of Sciences.

Johnson has served as vice president (1953) and president (1965) of the American Agricultural Economics Association. He was elected a Fellow of that association in 1967.

He is married to the former Helen Wallace, and they have two children.

CHAPTER **16**

# International Economics

International trade is the exchange of goods and services between countries. It occurs because a country is able to purchase goods abroad more cheaply than it can produce them at home. The result of trade is to increase a country's level of living.

Many people have difficulty in understanding international trade, but if we start from the premise that, in principle, international trade is no different from interregional trade (trade within a nation), its workings and values are more easily understood. There are some differences, of course, such as currencies, monetary systems, language, space, and national policies, but basically international and interregional trade are the same.

Within the United States, people take for granted trade between states or regions. For instance, trade with California or Florida for citrus products or with Michigan for automobiles is accepted. Most people recognize that a higher level of living is possible when each area specializes in producing those items in which it is relatively proficient and then exchanges some of its output for some of the output of other areas. For example, Montana exchanges wheat, barley, and beef calves for processed food from California and farm inputs from the Midwest.

On the other hand, when it comes to the international exchange of goods and services, many people fail to see any advantage to specialization and trade. However, because of differences in a country's resources, it is possible to produce products at different costs, which is crucial to trade and allows for increased levels of living for all participants.

Endowments of factors of production differ substantially from country to country and are significant in determining the types of economic activities in various areas of the world. The classical factors of production are land, labor, capital, and management. Since every country has varying amounts of these factors, each country has different advantages in producing products. Therefore, products requiring large amounts of labor in the production process should be produced in countries where labor is abundant, relative to other factors of production, and where wage rates are low, relative to the cost of other factors of production. One would expect that country to export labor-intensive goods to other countries where labor is relatively scarce and wage rates are relatively high. This thesis holds true for other products that use substantial amounts of other factors of production as well.

# BASIS FOR FOREIGN TRADE _____

The underlying proposition of foreign trade is that consumers and producers benefit from the exchange of goods and services. Specialization increases the supply of goods and services available. To illustrate this point and to show that trade is beneficial, let's examine a simple example.

Assume a two-country world—the United States and the European Union (EU)—and two commodities, wheat and wine. Also, assume that all factors of production are categorized under one input called labor, of which the total quantity is fixed, with the following marginal outputs obtained from a short segment of a curvilinear production possibilities curve.

|  | Marginal Output (Amount Produced) | | |
|---|---|---|---|
|  | Wheat (bu) |  | Wine (gal) |
| A U.S. day's labor will produce | 20 | or | 5 |
| A EU day's labor will produce | 4 | or | 12 |

In this example it is obvious that America can produce wheat more cheaply than the EU because it gets a larger output with a day of labor service (20 bushels compared to 4). The EU can produce wine more cheaply because with the same amount of labor input they can produce 12 gallons of wine compared with 5 gallons for the United States. Therefore, one can say that the United States has an absolute advantage in the production of wheat and the EU has an absolute advantage in wine production.

If the United States and EU practice self-sufficiency, dividing their labor service in the production of both commodities equally, the United States could produce 10 bushels of wheat and 2.5 gallons of wine, while the EU could produce 2 bushels of wheat and 6 gallons of wine. The total world output would be 12 bushels of wheat and 8.5 gallons of wine. But if both countries specialize, they could produce 20 bushels of wheat and 12 gallons of wine, a gain of 8 bushels of grain and 3.5 gallons of wine. For these countries to benefit, they must trade.

It is easy to show that trade is beneficial if each country has an absolute advantage. What happens if one country has an absolute advantage in the production of all goods and services? Trade is still beneficial to both countries. Given the same assumption as above, but with outputs changed as follows:

|  | Marginal Output (Amount Produced) | | |
|---|---|---|---|
|  | Wheat (bu) | | Wine (gal) |
| A U.S. day's labor will produce | 18 | or | 3 |
| A EU day's labor will produce | 2 | or | 1 |

From this example, one can see that the United States has an *absolute advantage* in the production of wheat (18:2 or 9: 1) and in wine (3:1). The example also shows that even though America has an *absolute advantage* in the production of both commodities, it has a *relative* (or *comparative*) *advantage* in wheat (9:1) as compared to wine (3: 1).

Assume that the United States specializes in wheat and the EU in wine. Before trade opens between the two areas, six bushels of wheat are worth a gallon of wine in the United States (18:3 or 6:1); in the EU, two bushels of wheat exchange for a gallon of wine. The domestic exchange ratio in the United States is 6:1; in the EU it is 2:1. We know the United States will produce wheat; one bushel is worth ⅙ gallon of wine in the U.S. If the bushel of wheat was sold to the EU, it would procure ½ gallon of wine. Therefore, rather than getting ⅙ gallon of wine without trade, it is possible to get ½ gallon of wine with trade, and thus trade makes the U.S. consumers better off.

Because trade must be beneficial to both participants in order for it to occur, let's look at the situation from the viewpoint of the EU. With a gallon of wine, the EU can sell it domestically for two bushels of wheat, or, if they engage in trade with the United States, for six bushels of wheat. Therefore, they will produce wine. You can see that it pays the EU to specialize in wine production and America to specialize in wheat, and then trade, even though the United States has an absolute advantage in the production of both commodities.

It is *comparative advantage* that provides the basis for foreign trade. The *law of comparative advantage* states that "a country will gain economically if it concentrates its efforts in those economic activities where it has the greatest relative advantage or the least relative disadvantage, and then trades with other countries."[1] In the foregoing example, the United States had its greatest relative advantage in the production of wheat, and the EU had its least relative disadvantage in wine production.

---

[1] Walter Krause, *International Economics*, Boston: Houghton Mifflin, 1965, p. 11.

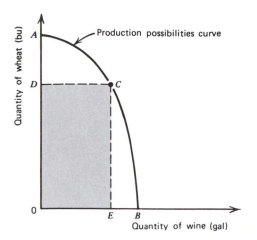

**FIGURE 16-1** Production possibilities curve.

## Putting Trade Theory Together

Land, labor, capital, and management can be used to produce wheat or wine along what is called the *production possibilities curve*, which is shown in Figure 16-1.

If all resources are used to produce wheat, the United States could produce $0A$ bushels, whereas, if they produced only wine they could produce $0B$ gallons. It is also possible to produce some wheat and some wine at point $C$ (or at any point along the curve $AB$). At this point, the United States can produce $0D$ bushels of wheat and $0E$ gallons of wine. It can be seen that the resources of the United States are better suited to producing wheat. The production possibilities curve, however, only shows what is *possible* to produce, given the resource base and existing technology. How much is actually produced of each commodity also depends on the demand for these products. It is the demand and supply for commodities that determine their prices, and hence, comparative advantage. Before we analyze demand, however, let us look at the producers' position.

## Maximizing Profits to Producers

Producers are in business to make a profit. If we assume a competitive industry, prices are fixed to the individual producers. The producer accepts the market price as given. If the market price of wheat is $4 per bushel and the price of wine is $8 per gallon, the relative prices may be represented by isorevenue lines (Figure 16-2). These lines show the various combinations of wheat and wine that give the producer the same total revenue.

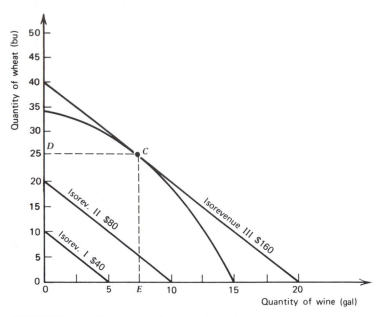

**FIGURE 16-2**   Isorevenue lines and production possibilities.

Isorevenue line I shows the various combinations of wheat and wine that will produce a revenue of $40. A revenue of $40 is possible by selling 10 bushels of wheat at $4 per bushel; or 5 gallons of wine at $8 per gallon; or 5 bushels of wheat at $4 per bushel *and* 2½ gallons of wine at $8 per gallon. Since market prices are fixed, isorevenue lines parallel and to the right of I, such as II, represent larger total revenues. Isorevenue line II represents a revenue of $80. Producers want to maximize their revenue for a given production possibilities curve, and that point is shown at C where Isorevenue line III is tangent to the production possibilities curve, producing 0D bushels of wheat and 0E gallons of wine.

Changes in the market price of commodities will cause changes in the amount of each product that will be produced. For example, in the previous illustration, the price of wine was $8 per gallon and the price of grain was $4 per bushel. If the market price of wine drops to $4 per gallon, the slope of the isorevenue line becomes flatter (Figure 16-3). A decrease in the price of wine would cause producers to increase their production of wheat, the higher priced product, and to reduce their production of wine, as shown in Figure 16-4. In the initial equilibrium, firms were producing 0A bushels of grain and 0B gallons of wine. With the decrease in the price of wine, the isorevenue line becomes flatter (grain

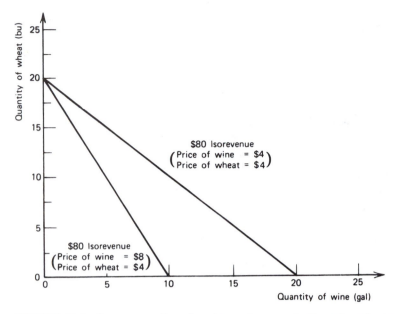

**FIGURE 16-3**  Isorevenue line showing a decrease in the price of wine.

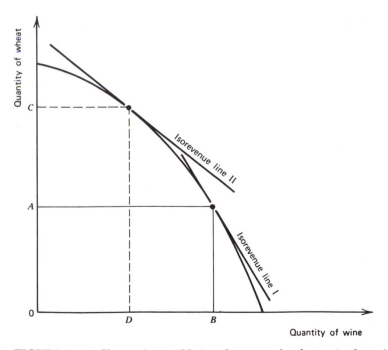

**FIGURE 16-4**  Change in equilibrium because of a change in the price of wine.

becomes relatively more profitable), so producers will increase their grain output from 0A to 0C and decrease their production of wine from 0B to 0D.

## Prices Determine Production and Consumption

As producers attempt to maximize profits from their expenditures in the production process, consumers also attempt to maximize utility or satisfaction given the amount of money income that they earn from selling their labor services, or from rent from their owned resources. Economics represents consumer tastes and preferences by indifference curves, which show the various combinations of goods or services that will yield equal satisfaction. An individual consumer has an entire indifference map, with indifference curves farther out from the origin showing higher levels of utility. Referring to Figure 16-5, $I_0$ may represent 50 units of utility, with $I_1$ representing 80 units of utility. The steepness or the flatness along the indifference curve indicates the preferences of consumers. Where the curve is steep, the consumer is willing to give up a large amount of grain for a unit of wine; and where the curve is flat, the consumer is willing to give up only a small amount of grain for a unit of wine because the consumer has a lot of wine and relatively little grain.

A consumer maximizes utility, given money income, in much the same way a producer maximizes profits. We represent the consumer's spend

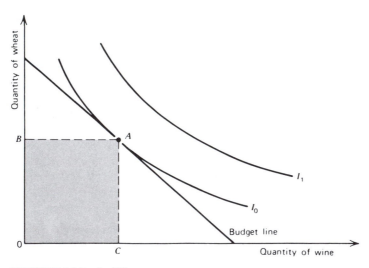

**FIGURE 16-5**   Indifference curves.

ing possibilities with a *budget line,* in which the slope is determined by the market prices of the goods being bought (Figure 16-5). The budget line shows all the combinations of two goods the consumer may buy with the given money income.

Let's use the same pair of goods as before (wheat and wine) to demonstrate the principle of optimizing satisfactions. Our consumer may buy only wheat, only wine, or any combination of wheat or wine along the budget line. Given money income and market prices, the consumer tries to get to the highest indifference curve possible. Consumers are in equilibrium when the budget line is tangent to the highest indifference curve at point *A.* In equilibrium, the consumer is buying 0*B* of wheat and 0*C* of wine, and has reached a maximum of satisfaction from that person's spendable income.

To arrive at a general equilibrium solution as to what will be produced and consumed, we must go from the individual to an indifference curve for a nation. This change causes problems because the utility that people derive from consuming different amounts, or even the same amounts, of the same commodity varies from individual to individual. If we trade grain for wine, the wine drinkers are made better off, but the cereal lovers are made worse off. We simplify the exposition by assuming the tastes of a community can be described in the same way as the tastes of an individual.

Under these assumptions, a nation without trade is in equilibrium where the community indifference curve $I_0$ is tangent to the production possibilities curve and the isorevenue line is tangent to both, at point *A* in Figure 16-6. The nation is in equilibrium producing and consuming 0*B* bushels of wheat and 0*C* gallons of wine.

## How Trade Makes Nations "Better Off"

Now if we assume the United States has a comparative advantage in wheat and the EU in wine, the production possibilities will have the shape as in Figure 16-7. Before trade, the isorevenue line (*BT*) will be steep in the United States, reflecting the high price of wine relative to grain because the United States can produce a large amount of grain. The isorevenue line (*BT'*) in the EU will be relatively flat, reflecting the low price of wine as wine is plentiful.

After trade, ignoring transportation and handling charges, the prices of grain and wine should equalize in the two areas. Prices tend to equalize because the United States will export grain and its price will increase because total demand for U.S. grain has increased. Supplies of grain have increased in the importing country, so therefore the price of grain should fall. Also, the EU will export wine to the United States, which will increase its price in the EU and decrease its price in the United States. The

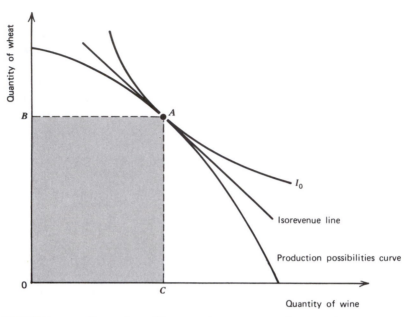

**FIGURE 16-6**   General equilibrium of a country without trade.

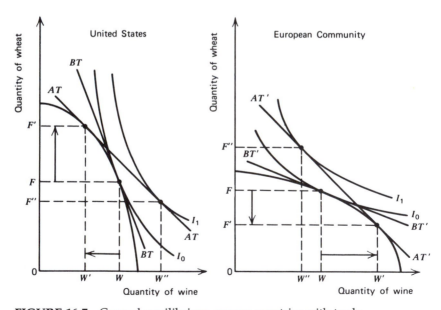

**FIGURE 16-7**   General equilibrium among countries with trade.

after-trade isorevenue lines are shown by $AT$ and $AT'$, which have the same slope. Because market prices change after trade opens, the production of grain increases in the United States from $OF$ to $OF'$ and the production of wine falls from $OW$ to $OW'$. In the EU, the production of grain falls from $OF$ to $OF'$ and the production of wine increases from $OW$ to $OW'$.

Consumers were in equilibrium on indifference curve $I_0$ before trade, consuming all that was originally produced. Now that trade has taken place, consumers reach a higher indifference curve, $I_1$ rather than $I_0$ (Figure 16-7). Consumers in the United States are consuming $OF''$ of grain after trade and $OW''$ gallons of wine, while the EU is consuming $OF''$ bushels of grain and $OW''$ gallons of wine. After trade, the United States produces $OF'$ and consumes only $OF''$ bushels of grain, so it exports $F'$ to $F''$ bushels of grain to the EU. The EU produces $OF'$ and consumes $OF''$ bushels of grain. To do this it imports $F'$ to $F$ bushels of grain, which equals the amount the United States exported. Thus, exports equal imports. The EU produces $OW'$ gallons of wine, but consumes only $OW''$ gallons of wine and exports the difference ($W'$ to $W''$) to the United States. The United States consumes $OW''$ gallons of wine but produces only $OW'$ and thus imports $W'$ to $W''$ gallons from the EU.

The important concept is that international trade occurs because of differences in the prices of products in the two countries before trade. Consumers benefit by being able to consume more goods and services than the resources of a country could produce, allowing them to attain a higher level of satisfaction. It is important to remember that the goal of any production is to increase satisfaction (utility).

Many countries set up barriers to trade which they intend to "protect" a domestic industry, but which really negate the benefits of trade. Most barriers are detrimental to consumers but beneficial to special interest groups that use political pressure to maintain their market position. Because tariffs are determined by Congress, the tariff system in the United States reflects a variety of political interests.

## What's Wrong with Trade Barriers?

The major purpose of a trade barrier is to reduce the volume of imports into a country. Two general types are used widely in the world: *tariffs* and *import quotas*. A tariff is a tax levied on a commodity when it crosses a national boundary. An import quota restricts the volume of imports by setting a maximum amount that can be imported.

In the early days, tariffs were imposed primarily as a source of revenue to finance federal government operations, but as mentioned, they are presently used to protect domestic industries from foreign competition. Two types of tariffs are used. One is an *ad valorem duty* and the second is a *specific duty*. The ad valorem tax is a fixed percentage of the value of the commodity. For example, poultry liver in the EU is taxed at 14 percent of its value. A specific duty is a tax that is a fixed sum of money per unit of commodity, such as $50 per imported motorcycle.

Import quotas have been employed in U.S. agricultural programs. For example, under these programs, the U.S. can restrict the volume of wheat imported. Such limits are, of course, below what would be imported under free market conditions, or there would be no need to impose a quota. The United States has developed the legal authority to use quotas only to protect those agricultural industries that have been involved in domestic agricultural programs; other countries use similar devices. Most underdeveloped countries use quotas on all products, but the only developed country where quotas are prevalent is Japan.

## WHAT IS THE ECONOMIC EFFECT OF TARIFFS AND QUOTAS? _____

Most students believe that if a tariff is imposed by an importing country, this tax is passed on to consumers in the form of a price increase for the commodity. This is only partially true. Depending on the price elasticities of demand and supply for the commodity, part of the tax is paid by the consumers in the importing country in terms of higher prices and part is absorbed by the exporting country in terms of a lower world price.

As an example, let us assume that the United States puts a $0.10 per pound specific tariff on bananas from Central and South America, as shown in Figure 16-8. This figure shows that at a price of $0.20 per pound, U.S. consumers will purchase a billion pounds. Once the tariff is imposed, the price of bananas increases to $0.25 per pound because the supply curve shifts from $SS$ to $S'S'$.[2] Given the demand curve for bananas, the increase in price forces consumers to consume a smaller quantity. They will switch to substitute products such as oranges or apples, while cutting consumption of bananas. This decision causes a reduction in quantity demanded and a reduction in the world price to $0.15 per pound. In this case where the supply and demand elasticities are approximately equal, about $0.05 of the $0.10 tariff is passed on to consumers and $0.05 passed back to banana producers in terms of a lower world price. Thus, the exporters are forced by the market system to pay part of the tax the United States imposed.

The distribution of the burden between importing and exporting countries is not the same in all cases because it depends on the elasticities of supply and demand and the market power of the country imposing the tariff. In the example just mentioned, the world price fell because the

---

[2] The demand curve could have been shifted instead. The economic results are the same.

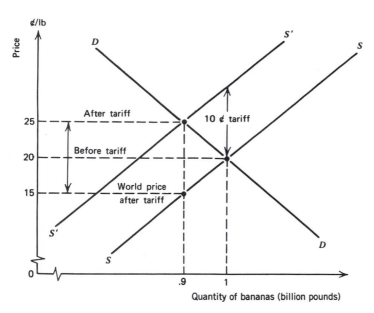

**FIGURE 16-8**   The economic impact of a tariff.

United States consumes a large part of the bananas from Central and South America. If a small nation accounting for only a small share of world imports imposes a tariff, then only a slight reduction in consumption would result, and there would be only a small impact on the world price of the commodity.

The economic effects of an import quota are shown in Figure 16-9. In this illustration assume that all of this product is being supplied by foreign producers. The demand and supply[3] curves intersect at a price of $P_0$ and the equilibrium quantity taken of $Q_0$. If the quota is effective and sets an absolute limit on imports, the new supply curve ($S'$) is vertical at that physical limit ($0Q$) and the price increases to $P_1$. Quotas restrict the amount of a commodity consumers can purchase and, thus, force consumers to pay a higher price. This action causes consumers to switch to less desirable substitute commodities, while domestic producers of substitute commodities expand their output under the quota protection, using resources drawn from more efficient industries.

## Domestic Effects of a Tariff or Quota

To analyze the domestic effects of a tariff imposed by a country, let us make two assumptions. First, assume the country is small so that it does

---

[3] This supply curve ($S_f$) includes only foreign supplies.

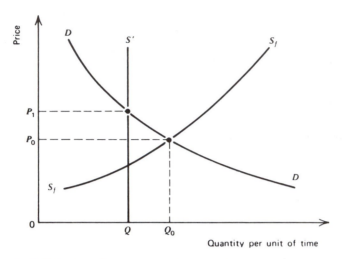

**FIGURE 16-9**   The economic impact of a quota.

not influence world prices by its actions; and second, assume this country can import all the products it wants at the prevailing world market price (Figure 16-10).

Before trade occurs, this country is producing quantity $Q_0$ at a market price of $P_0$ where its demand and domestic supply curves intersect. Af-

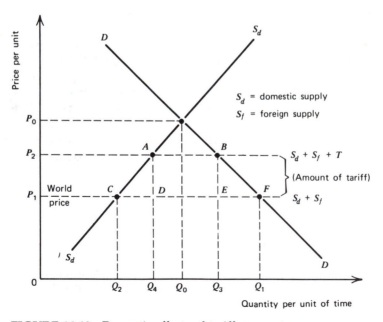

**FIGURE 16-10**   Domestic effects of tariffs or quotas.

ter free trade, the new supply curve is $S_d + S_f$ because at the world price this country can purchase all it wants. The demand curve and the "new" supply curve intersect at quantity $Q_1$ so consumers in this importing country increase their consumption under free trade from $Q_0$ to $Q_1$ and purchase this larger quantity at price $P_1$. At price $P_1$, domestic producers supply only $0Q_2$ of the market and foreign suppliers provide $Q_2$ to $Q_1$. Thus the benefits of free trade are obvious. The consumers get more of the product at lower prices through trade than they would get through domestic production.

Assume a tariff is imposed in the amount of $T$ (or $P_1$ to $P_2$). This tax raises the free trade price by the amount of the tariff to $S_d + S_f + T$. Now the supply curve intersects the demand curve at $P_2$, so the price increases by the amount of the tariff. Consumers reduce consumption to $Q_3$ and domestic producers increase output to $Q_4$. This tariff reduces imports from ($Q_2$ to $Q_1$) to ($Q_4$ to $Q_3$).

Because of the tariff, the loss to consumers is area $P_2BFP_1$. Area $P_1P_2AC$ is redistributed to producers. This increase in producer earnings is why you read about industries wanting tariff protection; American cattle producers want to keep out New Zealand and Australian beef, and most major manufacturing firms also wish to restrict foreign competition. This redistribution of income from consumers to producers is called the *redistribution effect* of the tariff.

Area *ADEB* is called the *revenue effect* of a tariff. This loss to the consumer goes to the government in the form of tax collections. Area *ADEB* is equal to the amount of tariff ($P_1$ to $P_2$) times the amount imported, which is quantity ($Q_4$ to $Q_3$).

Triangle *ACD* is called the *protective effect* of a tariff. It is the loss to the consumer due to the fact that producers must withdraw resources from other sectors of the economy to increase output from $Q_2$ to $Q_4$. Using these resources in this protected sector causes a reduction in productive efficiency, and it is a real loss to the economy. The same is true for triangle *BEF*. This *consumption effect* is also a loss to the consumer that is a residual not accounted for by the other effects.

The *domestic effect* of a quota is the same as that given for a tariff if the quota is $Q_4$ to $Q_3$. Rather than having a revenue effect, area *ADEB* does not go to the government, but to the importing firm. This revenue is referred to as monopoly profit.

With respect to economic efficiency, quotas are much more harmful than tariffs in general. With tariffs, the domestic price can never exceed the world price by more than the tax. With a quota, on the other hand, any increase in domestic demand must cause the domestic price to rise because the amount of imports cannot be increased. Thus, there is no limit to the difference that can exist between the domestic and world price of that commodity.

## Arguments for Protection

Many arguments are given as reasons why Congress should protect some special interest groups. These arguments run from national security, balance of payments, infant industry, to cheap foreign labor, and numerous other reasons. The most reasonable argument is the infant industry case that states that when a young industry starts business, it takes a few years to grow and develop its markets internationally and to exploit its economies of a scale so that it can be competitive. This argument has appeal, but the U.S. steel industry unsuccessfully tried this argument on Congress in the 1960s.

## A Move Toward Freer Trade

The issue of tariffs in America has always been mixed with national politics, and its economic aspects have been confused by the activities of pressure groups. From 1789 until the Hawley–Smoot Tariff of 1930, the U.S. tariffs were very protective. Since 1934, the United States has been attempting a program of orderly tariff reductions, except for the mixed mood of the country because of the balance of payments problems beginning in 1971.

Two main approaches have been used in the last 40 years to liberalize trade. One is the international approach under the General Agreement on Tariffs and Trade (GATT), and the World Trade Organization (WTO), and the second is a regional approach.[4] This latter approach is illustrated by organizations such as the EU, where a number of countries agree to liberalize trade among themselves, but maintain common and uniform tariffs against others.

## International Approach

GATT, an international organization, aimed at lowering trade barriers and eliminating discriminatory treatment, came into being in 1947. For the United States to participate in reciprocal tariff negotiations under GATT, enabling legislation must pass through Congress. In the past, the U.S. has participated in GATT negotiations under the authority of the Trade Agreements Act of 1934 and its extensions, the Trade Expansion Act of 1962, the Trade Act of 1974, and the Omnibus Trade and Competitiveness Act of 1988.

---

[4] Mordechai E. Kreinin, *International Economics: A Policy Approach*, 2nd ed. New York: Harcourt Brace Jovanovich, 1975, p. 307.

GATT's membership presently includes more than 120 nations. These nations are responsible for most of the world's trade. The basic purpose of GATT is to serve as a vehicle to handle multilateral tariff negotiations among its members by establishing rules, regulations, and principles to govern the conduct of trade. Its three main features are (1) provision for equal treatment among all parties (all countries receive unconditional most-favored-nation treatment, which means that any concession one grants to another it must also grant to all other GATT countries); (2) provision for eliminating quantity restrictions to trade except when a country is having balance of payments difficulties, and (3) provision for members to hold meetings to discuss and settle trade problems.

The United States has concluded its eighth round of GATT negotiations. The most recent trade rounds were named. In 1961, the United States participated in what was called the Dillon Round; the Kennedy Round lasted from 1962 to 1967; from 1975 to 1979, it was the Tokyo Round. Members of GATT opened the last round of multilateral trade negotiations in Uruguay in September 1986, and the Uruguay Round Agreement was finally signed in April 1994.

Major accomplishments in successive GATT rounds have included signing an antidumping agreement in the Kennedy Round (1964–67), addressing nontariff as well as tariff barriers to trade in the Tokyo Round (1974–79), and further addressing nontariff as well as tariff barriers including new areas of agriculture, services, intellectual property, and strengthened procedures for dispute settlement in the Uruguay Round (1986–94).

Major features of the completed Uruguay round agreement included the following:

1. A 34 percent average reduction in industrial products tariffs.
2. Agreement to convert quotas and other trade restraints to tariffs, to cut export and domestic subsidies, and to require minimum market access commitments for agricultural commodities.
3. Elimination of quotas on textile and clothing imports over a 10-year period.
4. Extension of Most Favored Nation (MFN) treatment, national treatment, and other principles to service sectors in which countries make specific market-opening commitments.
5. Recognition and strengthening of trade obligations for intellectual property such as patents, trademarks, and copyrights.
6. Establishment of the World Trade Organization (WTO) as a single umbrella for trade agreements in goods, services, intellectual property, and other areas. The WTO will implement the Uruguay Round provisions,

provide an international dispute settlement body, and provide a trade policy review mechanism for the future.

The U.S. Department of Agriculture reports on its analysis of the results that can be expected from agricultural trade liberalization among all the world's industrialized economies. World price increases would be greatest in dairy products, sugar, wheat, rice, coarse grains, and ruminant meats.[5] Total annual gains in real world income were estimated at $30 billion. The EU was projected to gain $14 billion, the U.S. $9 billion, and Japan $6 billion. Developing countries could expect to lose $4 billion, and the centrally planned economies about $1 billion.[6] Most of the benefits in the EU and Japan accrue through lower domestic prices for consumers. U.S. benefits would show up primarily in the form of federal budget savings. The tariff reductions in the Uruguay Round Agreement are estimated at $744 billion over the next 10 years. Agricultural producers could be the major losers if government support programs are eliminated, but direct income payments do not violate GATT if they are trade neutral.

The Uruguay Round for the first time brings agriculture, a sector that accounts for 13 percent of world trade, under international rules. Nontariff barriers are being converted to their tariff equivalents, and the resulting tariffs must be reduced by a minimum of 15 percent in each tariff line and by an average of 36 percent overall. Countries are also required to grant minimum market access in products where there has been little or no trade. This includes the end of bans by Japan and Korea on rice imports and commitments by all countries to increase wheat, corn, rice, and barley imports by a total of 3.5 million metric tons. Export subsidies are to be reduced by 36 percent in value from 1986–90 levels over 6 years, and the volume of subsidized exports is to be reduced by 21 percent.

## Regional Approach

Customs unions, common markets, and free trade areas are types of regional integration. A *customs union* is an economic and political organization between two or more countries abolishing trade restrictions among themselves and establishing a common and uniform tariff to outsiders. A *free trade area* is the same as a customs union except there is no common and uniform tariff imposed on those excluded from the union. Likewise, a *common market* is the same as a customs union, except that it includes the free mobility of factors of production.

---

[5] Vernon O. Roningen and Praveen M. Dixit, "Economic Implications of Agricultural Policy Reforms in Industrial Market Economies," *ERS Staff Report*, No. AGES 89–36, Washington, D.C.: USDA, August 1989.

[6] Ibid., p. 28.

Some of the customs unions are located in Europe and Central and South America. The EU was established as a customs union in 1958 and originally included West Germany, Italy, France, Belgium, the Netherlands, and Luxembourg. It was expanded to include Great Britain, Ireland, Denmark, Greece, Portugal, and Spain by 1986.

In 1995, the EU was expanded to cover 15 countries by adding Austria, Finland, and Sweden. Other countries have been given "associate member" status, including Bulgaria, the Czech Republic, Estonia, Hungary, Latvia, Lithuania, Poland, Romania, Slovakia, and Slovenia. The EU also has various bilateral free trade and customs union arrangements with other nearby countries such as Switzerland, Norway, Iceland, Malta, Cyprus, Morocco, Turkey, and Israel. About 70 African, Caribbean, and Pacific (ACP) countries receive economic assistance and preferential tariff treatment from the EU including virtually all of Africa.

Other preferential trade arrangements have been established throughout the world. The European Free Trade Association (EFTA) is a free trade agreement between Austria, Finland, Iceland, Liechtenstein, Norway, Sweden, and Switzerland formed in 1960. Some members have left EFTA since joining the EU. Three customs unions were formed in Latin America, including the Central American Common Market (CACM), the Andean Pact, and the Southern Cone Common Market (Mercosur). The CACM members are Costa Rica, El Salvador, Guatemala, Honduras, Nicaragua, and Panama. The Andean Pact includes Bolivia, Columbia, Ecuador, and Venezuela.

Customs unions established in Africa include the West African Economic Community (CEAO), the Economic and Customs Union of Central Africa (UDEAC), the Economic Community of West African States (ECOWAS), the Economic Community of Central African States (CEEAC), and the Arab Meghreb Union (AMU). Most of the African customs unions allow free movement of labor and capital between member countries.

Two free trade agreement regions have been established in Asia. The Association of Southeast Asian Nations (ASEAN) formed in 1967 included Brunei, Indonesia, Malaysia, the Philippines, Singapore, and Thailand. The Australia–New Zealand Closer Economic Relations Trade Agreement (ANZCERT) was founded in 1983.

The North American Free Trade Agreement (NAFTA) was completed in 1994. This Agreement was created to establish a free trade area between the United States, Canada, and Mexico. Tariff reductions in NAFTA are to be phased in over 10 to 15 years. NAFTA does not include any agreement to form common foreign policies, stabilize exchange rates, or coordinate welfare or immigration policies. The trade preferences only apply to goods made in NAFTA countries, defined to be at least 62.5 percent domestic parts used to manufacture the product.

Although regional integration reduces tariff barriers between those countries inside the organization, it does not necessarily provide economic benefits to these countries. The economic effects of integration on resource efficiency of the nations involved may be positive or negative.

The benefits of a regional organization are called *trade creation*. Trade creation is generated when the output of any given product shifts from a country, in which the cost to produce that item is high, to a participating country with lower production costs. The benefit to the consumers in the regional association is that they are now able to buy more imports at a lower price. *Trade diversion* may also occur within an association if a country changes its trade to purchase a particular good from a high-cost intraunion source rather than from a low-cost external source as before. To determine whether an economic union makes a country better off, one must determine whether trade creation offsets trade diversion. Each case must be investigated separately.

# AGRICULTURAL TRADE OF THE UNITED STATES _____

Because of excess capacity in agriculture in the United States, the export market has been important for the agricultural sector. In the past, the export market has been a place to dump surplus production, but with the recent changes in the demand for agricultural commodities it appears that agribusiness firms will have an opportunity to sell more of their output in foreign markets. This market is especially relevant if worldwide food shortages continue to exist and production controls on U.S. agriculture are not needed to maintain farm income.

In 1995, agricultural exports amounted to 10 percent of all merchandise exports of the United States. In the last 20 years, the United States has made great strides in developing foreign markets for feed grains, soybeans, and wheat. The large fluctuations in the export demand for agricultural commodities make it difficult for American farmers to establish consistent production patterns. Since 1960, the value of U.S. agricultural exports increased from $5.9 billion to $43.3 billion in 1981, then declined to $26.1 billion in 1986, and increased back to $56 billion in 1995. With the very inelastic demand and supply curves in the short run for farm products at the farm level, large fluctuations in demand cause large changes in agricultural prices and income. When wheat exports fell between 1976 and 1977, prices received by farmers dropped 70 cents per bushel. This decrease in price was a direct result of the increase in wheat production in the United States that occurred in an attempt to meet expected export demand.

There has been much discussion about a strategic food reserve system to stabilize grain flows to the market and to reduce price instability. This is being considered by several food importing and food exporting countries. In addition, a farmer-owned reserve plus a small strategic food reserve has been used by the United States. The major problem with this type of arrangement concerns the guidelines established to determine when reserves are to be released. Reserves are adjusted to dampen price fluctuations. Producers and consumers disagree on the price level at which stocks should be placed on the market. Much debate over the issue of reserves will continue for years to come.

## Exports

Table 16-1 shows U.S. agricultural exports from 1970 to 1995, by individual products. Feed grains, wheat, and soybeans are the largest components of exports. Wheat and flour ranks second, soybeans third, and meats and meat products fourth of the major agricultural exports. Cotton and tobacco, once very important, are still significant, but grains and preparations, oilseeds, fruits and vegetables, animals, and animal products are all important exports.

At present, 25 percent of U.S. farm cash receipts are derived from the export market. In terms of 1995 crop production, the United States exported about 54 percent of the wheat crop, 43 percent of the soybean crop, 21 percent of the corn crop, 38 percent of the grain sorghum crop, and about 49 percent of the rice crop (Figure 16-11).

Asia and Western Europe are the largest markets for U.S. agricultural exports (Table 16-2). Asia has been increasing its imports of rice, wheat, soybeans, and feed grains. In 1995, Japan became our first $10 billion agricultural customer, the largest single importer of U.S. farm products. Also, 1973 was the first year in which Asia took significantly more products than Western Europe.

As seen by Table 16-2, trade with the former Soviet Union, China, and Eastern Europe has fluctuated considerably in recent years. Even with severe monetary and social problems the FSU and Eastern Europe spent $1.6 billion, primarily on grains. In the past, the centrally planned economies have wanted to maintain their self-sufficiency for reasons of national security and to insulate themselves from capitalistic systems, but it appears that as they become more market oriented, longer-term trade relationships are emerging.

Foreign trade in command economies has been nationalized and conducted by state trading organizations. These countries have conducted trade to import what they consider to be essential. Thus, exports are not considered to be an end in and of themselves, but a financial means to obtain foreign exchange with which to purchase imports. When the FSU

**TABLE 16-1**   U.S. Agricultural Exports: Value by Commodity, Selected
Calendar Years

| Commodity | 1970[a] | 1980 | 1982 | 1986 | 1992 | 1995[b] |
|---|---|---|---|---|---|---|
| | | | Million dollars | | | |
| Animals and animal products | | | | | | |
| Dairy products[c] | 127 | 175 | 347 | 438 | 726 | 711 |
| Fats, oils, and greases | 247 | 769 | 663 | 411 | 525 | 827 |
| Hides and skins, excluding | | | | | | |
| fur skins | 187 | 694 | 1022 | 1521 | 1346 | 1621 |
| Meats and meat products | 132 | 890 | 978 | 1113 | 3339 | 4522 |
| Poultry products[d] | 56 | 603 | 515 | 496 | 1211 | 2345 |
| Other | 101 | 660 | 410 | 530 | 778 | 907 |
| Total animals and animal | | | | | | |
| products | 850 | 3791 | 3935 | 4509 | 7925 | 10,933 |
| Cotton, excluding linters | 372 | 2864 | 1955 | 786 | 1999 | 3681 |
| Fruits and preparations | 334 | 1335 | 1376 | 2040 | 2732 | 3240 |
| Grains and preparations | | | | | | |
| Feed grains, excluding products | 1064 | 9759 | 6444 | 4330 | 5737 | 8153 |
| Rice, milled | 314 | 1289 | 997 | 621 | 725 | 996 |
| Wheat and flour | 1111 | 6586 | 6927 | 3279 | 4675 | 5734 |
| Other | 107 | 357 | 273 | 398 | 3035 | 3654 |
| Total grains and preparations | 2596 | 17,991 | 14,641 | 8629 | 14,172 | 18,537 |
| Oilseeds and products | | | | | | |
| Cottonseed and soybean oil | 244 | 915 | 692 | 343 | 432 | 786 |
| Soybeans | 1228 | 5880 | 6218 | 4321 | 4380 | 5400 |
| Protein meal | 358 | 1654 | 1447 | 1302 | 1398 | 1701 |
| Other | 91 | 944 | 784 | 493 | 980 | 1036 |
| Total oilseeds and products | 1921 | 9393 | 9141 | 6459 | 7190 | 8923 |
| Tobacco, unmanufactured | 517 | 1334 | 1547 | 1209 | 1651 | 1400 |
| Vegetables and preparations | 206 | 1188 | 1174 | 1024 | 2871 | 3889 |
| Other | 463 | 3360 | 2853 | 1349 | 4389 | 5211 |
| Total exports | 7259 | 41,256 | 36,622 | 26,064 | 42,929 | 55,814 |

SOURCE: Economic Research Service, *Foreign Agricultural Trade of the United States*, USDA,
January/February/March 1996.

[a] Beginning in 1970, export values include small amounts of commodities formerly classified as nonagricultural.

[b] Preliminary.

[c] Includes some additional commodities beginning in 1971.

[d] Includes live poultry beginning in 1971.

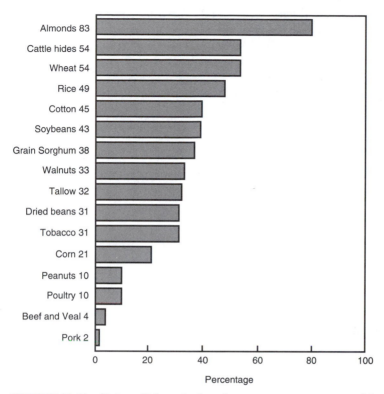

**FIGURE 16-11**  Sixteen U.S. agricultural exports, as percentages of their total production.

has a crop failure, it occasionally tightens its belt. Recently the FSU, instead, has come into the market, as it did in 1972 when it purchased 400 million bushels of wheat from the United States, popularly called the "Russian Grain Deal." Also, the People's Republic of China imported about 6.2 million tons of grain in 1972–73, including 5.4 million tons of wheat and 800,000 tons of corn. In 1992–93, China was expected to import 9 million tons of grain, mostly wheat. China has had long-term wheat import agreements with Canada, Australia, and the United States, but because of its improved production has chosen not to renew those contracts. China's method of payment has been in cash from its exports. China has been a major grain importer in 1995, contributing to an upsurge in international grain prices.

## Imports

The United States increased its agricultural imports in 1995 (Table 16-3). Imports rose 22 percent from 1992 to $30 billion. The major imports were

**TABLE 16-2**   U.S. Agricultural Exports by Regions, Selected Years

| Region[a] | 1973 | 1980 | 1982 | 1986 | 1992 | 1995[b] |
|---|---|---|---|---|---|---|
| | | | Million dollars | | | |
| Western Europe | 5605 | 12,351 | 11,414 | 7026 | 7804 | 9003 |
|   Enlarged EU-15 | 4526 | 9256 | 8398 | 6604 | 7290 | 8537 |
|   Other Western Europe | 1079 | 3095 | 3016 | 423 | 514 | 466 |
| Eastern Europe and FSU[c] | 1498 | 3285 | 2718 | 1091 | 2663 | 1639 |
|   FSU | 920 | 1130 | 1871 | 658 | 2346 | 1346 |
|   Eastern Europe | 577 | 2155 | 847 | 433 | 317 | 293 |
| Asia | 6509 | 15,037 | 13,675 | 10,537 | 17,923 | 27,939 |
|   West Asia | 521 | 1418 | 1428 | 1334 | 1782 | 2478 |
|   South Asia | 564 | 734 | 800 | 450 | 629 | 1019 |
|   SE Asia, excluding Japan and PRC | 1851 | 4281 | 4387 | 3586 | 1545 | 3012 |
|   Japan | 2998 | 6331 | 5555 | 5106 | 8437 | 10,957 |
|   People's Republic of China (PRC) | 575 | 2273 | 1505 | 61 | 545 | 2633 |
| Latin America | 1692 | 6176 | 4438 | 3641 | 6669 | 7926 |
| Canada, excluding trans-shipments | 1034 | 1905 | 1805 | 1533 | 4902 | 5738 |
| Canadian trans-shipments | 677 | 12 | 15 | 14 | 0 | 0 |
| Africa | 583 | 2303 | 2287 | 1997 | 2570 | 3070 |
|   North Africa | 307 | 1244 | 1223 | 1340 | 1461 | 2144 |
|   Other Africa | 276 | 1059 | 1064 | 657 | 1109 | 926 |
| Oceania | 83 | 188 | 270 | 224 | 398 | 499 |
|   Total | 17,680 | 41,256 | 36,622 | 26,064 | 42,929 | 55,814 |

SOURCE: Economic Research Service, *Foreign Agricultural Trade of the United States*, USDA, January/February/March 1996.

[a] Not adjusted for transshipments.

[b] Preliminary.

[c] Includes Yugoslavia.

coffee, meat and meat products, vegetables, grains, and beverages. Most of the U.S. import trade has been with Canada, Mexico, Colombia, and Brazil.

# AGRICULTURAL TRADE BALANCE _____

Although the trade balance is not always an important figure, you will find it in many publications, and its meaning should be understood. The **balance of trade** is the value of exports (goods) minus value of imports (goods). If this figure is positive, it is said a country has a *favorable bal-*

**TABLE 16-3**  U.S. Agricultural Imports: Value by Commodity, Selected Years

| Commodity | 1970 | 1980 | 1982 | 1986 | 1992 | 1995[a] |
|---|---|---|---|---|---|---|
| *Complementary*[b] | | | Million dollars | | | |
| Bananas, fresh | 188 | 416 | 581 | 749 | 1097 | 1140 |
| Cocoa beans | 201 | 395 | 704 | 1111 | 1080 | 1134 |
| Coffee, green | 1160 | 3873 | 2903 | 4544 | 1706 | 3263 |
| Drugs, crude | 25 | 159 | 151 | 187 | 313 | 346 |
| Essential oils | 32 | 100 | 75 | 109 | 195 | 273 |
| Fibers, unmanufactured | 19 | 35 | 27 | 23 | 42 | 49 |
| Rubber crude, excluding allied gums | 231 | 720 | 530 | 615 | 770 | 1629 |
| Spices | 55 | 141 | 152 | 353 | 262 | 334 |
| Tea, crude | 53 | 130 | 129 | 133 | 179 | 190 |
| Wool carpet | 31 | 45 | 30 | 39 | 29 | 43 |
| Other complementary | 166 | 978 | 41 | 23 | 5 | 5 |
| Total complementary products | 2161 | 6992 | 5323 | 7886 | 5678 | 8406 |
| *Supplementary*[c] | | | | | | |
| Animals and animal products | | | | | | |
| Cattle, dutiable | 111 | 226 | 469 | 426 | 1245 | 1413 |
| Dairy products | 125 | 488 | 612 | 809 | 857 | 1089 |
| Hides and skins | 110 | 87 | 198 | 210 | 185 | 199 |
| Meats and meat products | | | | | | |
| (nonpoultry) | 1010 | 2259 | 2037 | 2344 | 2638 | 2305 |
| Wool, apparel | 59 | 70 | 104 | 126 | 140 | 164 |
| Other animals and animal | | | | | | |
| products | 146 | 630 | 200 | 539 | 615 | 820 |
| Total animals and animal | | | | | | |
| products | 1561 | 3760 | 3620 | 4454 | 5680 | 5990 |
| Cotton, raw, excluding linters | 6 | 4 | 13 | 2 | 0 | 10 |
| Fruits and preparations | 146 | 564 | 975 | 861 | 1408 | 1631 |
| Grains and preparations | 70 | 283 | 474 | 678 | 1586 | 2362 |
| Nuts, edible, and preparations | 100 | 231 | 225 | 370 | 467 | 485 |
| Coconut oil | 77 | 270 | 184 | 174 | 273 | 307 |
| Other oilseeds and products | 87 | 324 | 291 | 412 | 946 | 1508 |
| Sugar and molasses | | | | | | |
| Sugar, cane, and beet | 725 | 1995 | 798 | 625 | 642 | 683 |
| Other | 43 | 87 | 204 | 319 | 497 | 572 |
| Tobacco, unmanufactured | 139 | 392 | 342 | 602 | 1353 | 550 |
| Vegetables and preparations | 298 | 864 | 1134 | 1579 | 2184 | 3103 |
| Wines | 145 | 692 | 783 | 1006 | 1087 | 1153 |
| Malt beverages | 32 | 367 | 466 | 784 | 864 | 1166 |
| Other supplementary vegetable | | | | | | |
| products | 141 | 541 | 115 | 1299 | 1959 | 2067 |
| Total supplementary products | 3608 | 10,374 | 9909 | 13,165 | 18,946 | 21,587 |
| Total agricultural imports | 5770 | 17,366 | 15,232 | 21,051 | 24,624 | 29,993 |

SOURCE: Economic Research Service, *Foreign Agricultural Trade of the United States*, USDA, January/February/March 1996.

[a] Preliminary.

[b] Complementary agricultural import products consist of all products that are not commercially produced in the United States.

[c] Supplementary agricultural import products consist of all products similar to agricultural commodities produced commercially in the United States, together with all other agricultural products interchangeable to any significant extent with such U.S. commodities.

*ance of trade*; if this figure is negative (imports greater than exports), a country is said to have an *unfavorable balance of trade*. This unfortunate terminology came from the Mercantilist period (approximately 1500 to 1750). During this era, countries thought it was desirable to have an export trade surplus in order to accumulate gold and other precious metals that they thought were measures of wealth. It was Adam Smith who pointed out in his *Wealth of Nations* in 1776 that goods rather than gold are the true wealth of a nation. Smith demonstrated that through division of labor, specialization, and free trade, benefits in terms of increased output are possible and the gains are divided among nations through international trade. However, Smith did not argue for an import balance either. He realized that exports must equal imports in the long run.

In most years throughout its history, our nation has maintained a surplus of exports over imports. Following World War II, the balance of trade averaged almost $5 billion per year. A negative trade balance began to occur in the early 1970s, averaged about $–30 billion from 1977 through 1980, then peaked near $–160 billion in 1987. The trade balance improved somewhat since then, with a balance of $–140 in 1995.

The balance of trade surplus in agricultural commodities ran about $1 to $2 billion a year until 1972. In that year, the balance of trade was positive by almost $3 billion, jumped to $7 billion in 1981, then declined to $5 billion in 1986. Agricultural exports improved each year through the remainder of the 1980s, reaching a surplus of $25.8 billion for 1995 and helping to offset the nonagricultural trade balance of $–166 billion.

## SUMMARY

International trade is the exchange of goods and services between countries. People recognize that through trade a higher level of utility is reached when one specializes in those products where one is superior and then exchanges some of this output for products of another.

The basis for foreign trade is the law of comparative advantage. This principle states that a country (or individual) will gain economically if it concentrates its efforts where it has the greatest relative advantage or the least relative disadvantage and then trades with other countries (or individuals).

The major international trade barriers are tariffs and import quotas. The sole purpose of these barriers is to reduce the volume of imports, thus reducing foreign competition. As a consequence, domestic consumers pay a higher price for the product and are forced to purchase less, *ceteris paribus*.

Two approaches have been used in recent years to move the world toward freer trade. One is the international approach under GATT and the World Trade Organization (WTO), and the second is the formation of customs unions called the regional approach.

Customs unions and preferential trade groups have been formed throughout the world. Customs unions are typified by the West African Economic Community European Union, the Andean Pact, and the Central American Common Market.

International trade is very important to U.S. agriculture. We export large amounts of feed grains, wheat, soybeans, fruits and vegetables, and animal products. Our primary agricultural imports are coffee, meat, vegetables, grains, and beverages.

## CHAPTER HIGHLIGHTS

1. International trade is the exchange of good and services between countries.
2. A country engages in international trade because it increases that nation's level of living.
3. Specialization and trade are made possible by a nation's comparative (relative) advantage.
4. International trade permits a nation to consume beyond its production possibilities.
5. Tariffs and quotas are trade barriers that reduce the gains from trade.
6. Special interest groups favor trade barriers because of their redistribution effects.
7. International and regional associations can be effective in eliminating trade barriers.
8. U.S. agricultural commodity exports totaled about $56 billion in 1995, 10 percent of all U.S. exports.
9. Exports of agricultural commodities constitute 25 percent of U.S. farm cash receipts.
10. Major export commodities are feed grains, wheat, soybeans, fruits and vegetables, and meats and meat products.
11. Our best (regional) agricultural customers are Asia and Western Europe.
12. U.S. agricultural imports were $30 billion in 1995, less than 5 percent of all U.S. imports.
13. Major import commodities are coffee, meat and meat products, vegetables, grains, and beverages.
14. Agriculture provided a trade surplus of $26 billion in 1995.

# KEY TERMS AND CONCEPTS TO REMEMBER

Absolute advantage

Balance of trade

Common market

Comparative advantage

Customs union

Free trade area

Import quota

International trade

Law of comparative advantage

Tariff

# REVIEW QUESTIONS

1. Explain comparative advantage. In what commodities do we have a comparative advantage? Can America have a comparative advantage in the production of every commodity?
2. Show graphically the gains to a country from being involved in international trade. (Assume a two-country and two-commodity world.)
3. Using a demand–supply diagram, show and explain the economic effects of tariffs and quotas.
4. What are the two major approaches that are being used to liberalize international trade?
5. What is meant by the terms "trade creation" and "trade diversion"? How are these concepts used to determine whether or not a regional organization would make a country "better off"?
6. Why are export markets important to U.S. agriculture? What impact do these markets have on agricultural prices?
7. What countries are the primary importers of U.S. agricultural products? Has the make-up of these countries changed much over time?
8. What agricultural products are we exporting and importing? Why do we export and import some of the same commodities?
9. How is the balance of trade computed? What does a "favorable" balance of trade mean?

# SUGGESTED READINGS

*Food and Agricultural Policy*, Washington, D.C.: American Enterprise Institute for Public Policy Research, 1977, Part 3.

Caves, Richard F., Ronald W. Jones, and Jeffrey A. Frankel. *World Trade and Payments: An Introduction*, 5th ed. Glenview, IL: Scott, Foresman and Co., 1989.

Husted, Steven and Michael Melvin. *International Economics*, 3rd ed. New York: Harper Collins, 1995.

Ingram, James C., and Robert M. Dunn. *International Economics*, 3rd ed. New York: John Wiley & Sons, 1992.

Kindleberger, Charles P., and Peter H. Lindert. *International Economics*, 6th ed. Homewood, Ill.: Richard D. Irwin, 1978, Chapters 2 and 3.

Snider, Delbert A. *Introduction to International Economics*, 6th ed. Homewood, Ill.: Richard D. Irwin, 1975, Chapters 13–15.

Wexler, Imanuel. *Fundamentals of International Economics*, New York: Random House, 1968, Chapters 13, 16, and 17.

Storey, Gary G., Andrew Schmitz, and Alexander H. Sarris, eds. *International Agricultural Trade*, Boulder, Colo.: Westview Press, 1984.

# AN OUTSTANDING CONTRIBUTOR

**Andrew Schmitz**   Andrew Schmitz is currently Ben Hill Griff professor in the Department of Food and Resource Economics at the University of Florida.

   Schmitz was born and raised on a farm near Central Bufte, Saskatchewan, Canada. He continues to maintain an active interest in the operation and management of the family farm. Schmitz was granted the B.S.A. degree in agricultural economics at the University of Saskatchewan in 1963. His M.Sc. degree in agricultural economics followed in 1964—this thesis winning for him the 1964 Master's Thesis Award for the best thesis in agricultural economics in Canada. Graduate work in economics at the University of Wisconsin, specializing in international trade, earned him an M.A. degree in 1966 and the Ph.D. degree in 1968. His Ph.D. thesis won the University of Wisconsin's Harold Groves Doctoral Dissertation Award in 1968.

   Professor Schmitz, held the Robinson Chair at Berkeley and the Van Vliet Chair at the University of Saskatchewan, served as Chairman of his Berkeley department during the years 1989 to 1993. Schmitz has published widely in more than a dozen U.S. and foreign professional journals including the *American Economic Review, Journal of Political Economy*, the *Economic Journal* and the *Quarterly Journal of Economics*. He has co-authored over 10 books and has contributed more than 50 chapters or sections in several books edited by others.

   During his tenure at the University of California, Dr. Schmitz has been called as consultant to agencies and organizations, including the U.S. Tariff Commission; U.S. Department of Agriculture; U.S. Department of Transportation, Central Inteffigence Agency; International Food Policy Research Institute, Teknekron, Inc.; Ford Foundation; International Trade Commission; National Grain and Feed Association; U.S. Agency for International Development; Agriculture Canada; and numerous other private groups or firms including several law firms.

In addition to his dissertation awards, Schmitz was honored early by the American Agricultural Economics Association with two awards for Published Research (1970 and 1978), their Award for Outstanding Journal Article (1970), Quality of Communication Award (1979), Publication of Enduring Quality Award (1981), and their Quality of Research Discovery Award (1982). He was elected Fellow of the American Agricultural Economics Association in 1985. Later he received several new awards including another Enduring Quality award for "Concept of Economic Surplus and its Use in Economic Analysis," This publication has also been nominated as a classic by the English Classic Society. He has been serving as project leader of the Fulbright Bulgarian Project.

Dr. Schmitz is married to the former Carol Andersen, and they have five children.

# CHAPTER 17

# World Population and Food Supply

One of the basic needs of human life is, of course, food. Since the beginning of time, our activities and very existence have been governed by a quest for food. Only relatively recently have developed countries been able to provide adequate food supplies for their populations. Many less-developed countries still suffer from constant shortages and from frequent famines caused by drought and other natural disasters.

Current estimates of the situation in the world today indicate that 786 million people either do not consume enough calories per day, and are considered undernourished, or are not able to consume the nutrients needed for an adequate diet and are therefore malnourished (Table 17-1). Of these undernourished and malnourished people, about one-half are children who are chronically hungry or actually starving. These figures include only those people who suffer from food shortages; they do not include those persons who have the required foods available, but are malnourished because of a lack of education in regard to basic nutritional needs.

## World Diets

Each of us consumes different kinds of foods. Individual tastes and preferences are such that some people like steak, others like chicken, and still others like fruits and vegetables. In the United States we take the availability of food for granted. It is plentiful and cheap. In many other countries, however, food supplies are not plentiful, nor do they have the range of fresh and processed products that are offered in our supermarkets; the choice of products is much more limited. Their marketing systems are un-

**TABLE 17-1**  Estimated Number of People with Insufficient Protein/Energy Supply by Regions

| Region | Number of Undernourished (Million) | Percentage of Population |
|---|---|---|
| Africa | 168 | 33 |
| Far East | 528 | 19 |
| Latin America | 59 | 13 |
| Near East | 31 | 12 |
| Total | 786 | |

SOURCE: United Nations, Food and Agriculture Organization, March 1996.

derdeveloped, and many cannot respond to consumer desires. They lack the legal, transportation network, and pricing mechanisms to form an effective market system. The world's food supplies are not canned ham, frozen vegetables, nor cake mixes, as we know them. Available products for most of the world's people are commodities that are purchased in raw form with the family unit transforming these commodities into usable foods.

The world's daily diet is composed primarily of cereal products. These food grains are wheat, rice, corn (maize), and sorghum, accounting for about 50 percent of the caloric intake per day. Wheat is an important food in North and South America, Europe, parts of Asia and the Middle East; rice is the major food grain in Asia; whereas maize is a significant food in Africa and Latin America.

Meat and animal products make up 14 percent of the world's daily diet per capita in terms of calories. Consumption of meat is highest in the developed countries. Milk is widely used throughout the world.

Roots and tubers account for 5 percent of the world diet; fats and oils 10 percent; sugar for 9 percent; pulses (dry beans and peas), nuts, and oilseeds for 6 percent; fruits and vegetables for 4 percent; and 1 percent are other foods.

When we speak of the world's food, we are talking basically about cereals, primarily wheat and rice and to a lesser extent maize, barley, and sorghum.

## World Population Increases

Estimates show that there are over 5 billion people in the world today. Censuses of population were not kept before the 1800s, but scientists have attempted to estimate populations based on interpretations of archeological evidence.[1] It is estimated that around 8000 B.C. there were only 5 million people. During the period from 8000 B.C. to 1600 A.D., the population doubled approximately six times. In other words, it took on the average, 1500 years for the population to double during this period. Between 1600 and 1850, it doubled again. Population then doubled again by 1930 and again by 1975. Thus, as one can see (Table 17-2), population has been increasing at an ever increasing rate.

The world population increased 19 percent between 1975 and 1985, reaching a total of about 4.8 billion people by 1985 (Table 17-3). It has been estimated that by the year 2000 there will be more than 6 billion people on this planet. In 1975, the less-developed countries accounted for 73 percent of world population, and by the year 2050 they will account for 86

[1] Paul R. Ehrlich and Anne H. Ehrlich, *Population, Resources, Environment,* 2nd ed. San Francisco, Calif.: Freeman, 1972, p. 6.

**TABLE 17-2** Estimated World Population and the Years It Took to Double the Population

| Date | Estimated World Population | Doubling Time |
|---|---|---|
| 8000 B.C. | 5 million | |
| 1650 A.D. | 500 million | 1500 years |
| 1850 A.D. | 1 billion | 200 years |
| 1930 A.D. | 2 billion | 80 years |
| 1975 A.D. | 4 billion | 45 years |

SOURCE: Paul R. Ehrlich and Anne H. Ehrlich, *Population, Resources, Environment,* 2nd ed., San Francisco, Calif.: Freeman, 1972, p. 6.

**TABLE 17-3** World Population Estimates and Projections to Year 2050 for Regions and Countries

| Area | 1970 | 1975 | 1985 | 2000 | 2050 |
|---|---|---|---|---|---|
| | | | Millions | | |
| World | 3704 | 4085 | 4850 | 6241 | 10713 |
| Europe | 460 | 474 | 492 | 490 | 486 |
| Western Europe | 352 | 361 | 371 | 390 | 358 |
| Central Europe | 108 | 113 | 121 | 127 | 128 |
| USSR (Former) | 243 | 255 | 279 | 312 | 386 |
| North America | 226 | 239 | 264 | 305 | 419 |
| United States | 205 | 216 | 238 | 275 | 383 |
| Canada | 21 | 23 | 25 | 30 | 36 |
| Oceania | 19 | 21 | 25 | 31 | 44 |
| Australia/New Zealand | 16 | 17 | 19 | 23 | 28 |
| Middle East | 115 | 133 | 177 | 280 | 821 |
| Asia | 1996 | 2232 | 2670 | 3424 | 5165 |
| Japan | 104 | 112 | 121 | 128 | 109 |
| China | 820 | 918 | 1050 | 1303 | 1655 |
| India | 555 | 623 | 770 | 1018 | 1623 |
| Indonesia | 123 | 137 | 172 | 221 | 341 |
| Latin America | 286 | 323 | 403 | 531 | 848 |
| Mexico | 53 | 61 | 77 | 103 | 170 |
| Brazil | 96 | 109 | 138 | 181 | 281 |
| Africa | 358 | 408 | 541 | 842 | 2543 |

SOURCE: Economic Research Service, U.S. Department of Agriculture, April 1993.

TABLE 17-4   Percentage of Total World Population by Selected Countries and Regions, 1975–2000

| Region | Percentage of World Population | | | |
|---|---|---|---|---|
| | 1975 | 1985 | 2000 | 2050 |
| Asia | 55 | 55 | 55 | 48 |
| China | 22 | 22 | 21 | 15 |
| India | 15 | 16 | 16 | 15 |
| Africa | 10 | 11 | 13 | 24 |
| Latin America | 8 | 8 | 9 | 8 |
| Middle East | 3 | 4 | 4 | 8 |
| All underdeveloped world | 73 | 75 | 78 | 86 |
| Europe | 12 | 10 | 8 | 5 |
| FSU | 7 | 6 | 5 | 4 |
| North America | 6 | 5 | 5 | 4 |
| Oceania | 1 | 1 | 1 | 1 |
| All developed world | 27 | 25 | 22 | 14 |

SOURCE: Economic Research Service, U.S. Department of Agriculture, April 1993.

percent (Table 17-4). Those countries with the greatest food problems are also expected to have the greatest increases in populations. Africa's population is projected to increase over 200 percent between 2000 and 2050, Asia's population by 51 percent, Middle East 193 percent, and Latin America's 60 percent. Smaller increases are anticipated in North America, and the FSU. An absolute decrease in population is expected in Europe.

# WORLD FOOD NEEDS ———————————————

Because population and income are both major factors determining food consumption, one must take account of these variables in order to make projections of world food demand. Between 1970 and 2000, world food demand is expected to grow at a rate of 2.0 percent per year. About 1.8 percent of this increase will be caused by the increase in population and about 0.2 percent from the rising purchasing power of consumers. In the poorest countries, food demand may grow at 2.5 percent for population and 0.55 percent per year for income. Former centrally planned food production over this same period is expected to increase at 1.9 percent per year (Table 17-5). In the African developing economies, food supplies are estimated to increase at 2.0 percent per year, while population is increasing at 2.8 percent per year.

In the Western Hemisphere, food problems regularly arise in Central America (particularly in El Salvador, Honduras, and Nicaragua), two An-

**TABLE 17-5**  Extrapolated Growth Rates of Food Production and Population, 1969–71 to 2000

| Area | Food Production[a] | Population |
|------|------|------|
| | Percent per year | |
| Developed countries | 1.50 | 0.69 |
|   United States | 2.08 | 0.64 |
|   Western Europe | 0.97 | 0.47 |
|   Japan | 0.92 | 0.81 |
| Centrally planned (former) | 1.86 | 1.30 |
|   Eastern Europe | 2.04 | 0.63 |
|   USSR (former) | 1.84 | 0.81 |
|   China | 1.76 | 1.54 |
| Developing countries | 3.03 | 2.48 |
|   Latin America | 3.55 | 2.77 |
|   North Africa/Middle East | 3.14 | 2.88 |
|   Other Developing Africa | 2.01 | 2.80 |
|   South Asia | 2.40 | 2.26 |
|   Southeast Asia | 3.32 | 2.27 |
|   East Asia | 2.98 | 2.03 |
| World | 2.01 | 1.84 |

SOURCE: Economic Research Service, USDA.

[a] Projections assume relatively high petroleum prices.

dean countries (Bolivia and Peru), and some of the poorer parts of the Caribbean (Haiti being an extreme case). Several of India's neighbors are at risk as well, including Afghanistan, Bangladesh, Nepal, Pakistan, and Sri Lanka. The circumstances of several Arab countries, including Algeria, Morocco, Tunisia, and Yemen, are not much better.

Food problems are especially severe in Sub-Saharan Africa. As has been mentioned already, the region is experiencing rapid population growth, since poverty, illiteracy, and other factors related to high human fertility are all present. Furthermore, increases in production have stayed even with mounting demand in only a handful of countries.

Adverse natural conditions, including poor soil quality and sparse and erratic precipitation, explain part of agriculture's poor performance in Sub-Saharan Africa. As in other parts of the world where crop and live-stock production falls far short of its potential, misguided public policies also make a negative contribution.

Many African governments have prevented agricultural commodity markets from functioning effectively. To satisfy urban populations, which want cheap food, and also to exert downward pressure on urban dwellers' wages and salaries, prices have been controlled. Where those prices have been set below equilibrium levels, as has often been the case, markets have failed to clear. In addition, commodity exports have been taxed explicitly or implicitly (e.g., when exporters have been obliged to convert foreign exchange earnings into the domestic currency at unfavorable rates).

In addition to governmental interference with market forces, agricultural development has been hampered by inadequate investment in roads and other infrastructure and also in the sector's scientific and technological base. Where infrastructure is deficient, farmers find it difficult to market their output; in particular, they often find themselves selling crops and livestock at low prices. A strong agricultural research and extension system is needed to develop and to acquaint farmers with new varieties and farming techniques. Where such a system is not in place, opportunities for agricultural intensification are lost.

Short-run food needs are especially apparent in Algeria, Morocco, Tunisia, Central African Republic, Lesotho, Mozambique, Swaziland, Zambia, Zimbabwe, Burkina Faso, Cape Verde, Chad, Gambia, Ghana, Guinea, Guinea Bissau, Mali, Mauritania, Senegal, Togo, Somalia, Sudan, Tanzania, Costa Rica, Haiti, Jamaica, and Nicaragua. Food import needs are based on nutritional requirements that are estimated at 27 million metric tons of cereal grains.

This projection shows the food gap between separate estimates of demand and supply indicators. In reality, effective demand must equal effective supply at some equilibrium price and output of food. At this equilibrium, it still is possible to have either plenty or shortages of food.

This economic meaning is not to be confused with the discussions of *nutritional needs.* The calculations of biological necessities to meet a specific reference individual has nothing to do with whether or not food will be produced or if consumers have enough purchasing power to make their wants effectively felt in the market place. Estimates such as these are useful in calculating quantities of needed food aid. But we must realize that there is a big difference between needing something and being able to purchase it; the food hunger problem is directly related to purchasing power. It is the poor people of the world who suffer from food shortages, not the rich (even in the poorest of countries).

The world food problem is one of an energy-intake deficiency. Food energy and protein recommendations made in 1973 by the World Health Organization and the Food and Agricultural Organization indicate it is likely that if energy requirements are sufficient, the diet would also be sufficient in protein. More food intake alone, even from cereals and pulses, will normally be adequate to meet both energy and protein needs. An exception to this, however, is for young children who have a limited ability to consume more food.

The amount of additional food that is required to close the food gap, so that all people have a nutritionally adequate diet, is impossible to cal-

**TABLE 17-6**   World and Regional Caloric Food Supplies Per Capita

| Country/Region | Recommended Minimum Caloric Intake | Food Supplies[a] 1961–63 Average | Food Supplies[a] 1988–90 Average | 1988–90 Intake as Percentage of Recommended Minimum |
|---|---|---|---|---|
| | | Calories/cap/day | | |
| Developed countries | 2550 | 3031 | 3404 | 133 |
| United States | 2387 | 3067 | 3642 | 153 |
| France | 2328 | 3288 | 3593 | 154 |
| Japan | 2257 | 2513 | 2921 | 129 |
| USSR (Former) | 2362 | 3146 | 3380 | 143 |
| Developing countries | 2255 | 1940 | 2473 | 110 |
| Bangladesh | 1861 | 1976 | 2037 | 109 |
| China (PRC) | 2096 | 1658 | 2641 | 126 |
| Nepal | 1981 | 1914 | 2205 | 111 |
| Samoa | 2188 | 2040 | 2695 | 123 |
| Ghana | 2047 | 2030 | 2144 | 105 |
| Mali | 2012 | 2167 | 2259 | 112 |
| Uganda | 1945 | 2293 | 2178 | 112 |
| Antigua | 2187 | 2120 | 2307 | 105 |
| Haiti | 2128 | 2028 | 2005 | 94 |
| Peru | 2077 | 2223 | 2037 | 98 |
| World | 2355 | 2287 | 2697 | 115 |

SOURCE: Food and Agriculture Organization, United Nations.

[a] Caloric food supplies are calculated as the quantity of food available per capita for human use, measured in terms of calories per day, at the retail level after provision is made for changes in food stocks, quantities traded or fed to livestock, and quantities used as seed, for industrial purposes, or lost in collecting, processing, or marketing.

culate with the data available. The annual average food energy requirements determined for 1961–63 and 1988–90 give a clue to the amount of deficiency. As shown in Table 17-6, the total world food situation improved significantly in all regions of the world. All developed countries in 1988–90 had more than adequate food energy supplies. They were actually 33 percent above requirements. The developing countries as a whole met their energy needs, but the lowest income countries were below minimum requirements. Parts of Africa, Asia, and Latin America are most in need of additional food supplies.[2] If we assume that 786 million people need from 200 to 600 more calories per person per day, that would amount

---

[2] The extent of malnutrition in parts of Asian centrally planned economies is unknown, but the volume of their net grain imports would suggest that their agriculture is unable, under current conditions, to provide adequately for their people.

to from 18 to 55 million tons of cereals per year. At $250 per ton, the cost would range from about $4.5 billion to $13.7 billion.

The USDA currently estimates world food needs at approximately 27 million metric tons of cereals, at a cost of about $6.7 billion. To maintain current per capita status quo consumption, total annual food aid requirements are estimated at 15.1–16.7 million mt in 1996 and 27.1–32.4 million mt in 2005. To meet minimum nutrition needs, total annual food aid requirements are estimated at 34.5–38.7 million metric tons in 1996 and 39.8–48.3 million metric tons in 2005. The variance in estimated food aid requirements is due to varying assumptions regarding the future growth of the recipient country's exports, foreign exchange earnings, and capacity for commercial food imports. The estimated food aid requirements for status quo food consumption levels almost double from 1996 to 2005 under both growth scenarios.

## World Food Crises of 1972–75

Food crises occur every year in one part of the world or another. Agricultural production is affected by variable weather patterns, insects, diseases, and natural disasters such as floods, earthquakes, and tornadoes. These relatively small disasters are handled by food aid programs.

In 1972 world grain production fell by 33 million tons, the largest drop in 20 years. This decline was caused by grain crop failures in the FSU and in the rice-producing areas of Asia. World stocks were drawn on, which stripped America and Canada of the large stocks that were on hand. Then in 1974, North American stocks were not restored because of a poor corn crop. These short supplies drove grain prices far above their pre-1972 levels. A similar situation occurred in 1979–80.

In addition, the energy crisis and the unstable world monetary system aggravated conditions. The energy crisis made it difficult for many farmers in underdeveloped countries to purchase oil and gasoline in order to farm. Also, the devaluation of the dollar made American grain cheaper than it would have been in the world market, leaving less grain available for food aid. These latter two factors were not large, but did make a bad situation worse.

The 1972–75 experience has shown that the world is ill equipped to handle food shortages and their resultant price impacts. It is evident that some sort of arrangement is needed to coordinate world product flows and reserves in a way to cope with both surpluses and shortages.

Dependence on North American and Australian cereals has been increasing for some time. Asia, Africa, and Latin America import mainly food grains, whereas Western and Eastern Europe, Japan, and the former USSR need feed stuffs (Table 17-7). The United States alone has about 32 percent of the international trade in wheat, 68 percent of the coarse grain

TABLE 17-7  Net Exports (+) and Net Imports (−) For All Cereals, by Region

| Region | 1934–38 | 1960–63 | 1973/74 | 1980/81 | 1992/93 | 1995/96[a] |
|---|---|---|---|---|---|---|
| | | | Million metric tons | | | |
| North America[b] | + 5 | +43 | +86 | +137 | +111 | +103 |
| Latin America | + 9 | + 1 | − 1 | − 8 | − 14 | − 12 |
| Western Europe | −23 | −26 | −22 | − 7 | + 28 | + 13 |
| Eastern Europe and USSR (Former) | + 4 | 0 | −10 | − 55 | − 30 | − 6 |
| Africa and Middle East | + 1 | − 4 | −11 | − 27 | − 46 | − 48 |
| Asia | + 2 | −16 | −40 | − 41 | − 57 | − 76 |
| Oceania (Australia and New Zealand) | + 3 | + 7 | + 9 | + 14 | + 14 | + 17 |

SOURCE: Economic Research Service, *World Agricultural Situation*, USDA, monthly series.

NOTE: Minor imbalances in a given period are due to rounding or to variations in reporting methods used by certain countries.

[a] Estimated from ERS data.

[b] The United States accounted for the following portions of these totals (in million metric tons): 1973/74 = 72.9; 1980/81 = 118.6; 1989/90 = 91.0; 1992/93 = 85.0; and 1995/96 = 96.9.

market, and 17 percent of the rice market. Because of the large percentage of the world market the United States has for food, some people have suggested during food shortages that the United States use its "agripower" to force and keep farm prices up. Cartel-type arrangements probably would not work in the longer run because of the increased production worldwide that would result from the higher prices. Both currently and in the long run, U.S. agriculture has been able to produce much more than we can generally sell for domestic and export purposes.

## Improvements Made Since the 1974 World Food Conference

Because of widespread food shortages, a global World Food Conference was held in Rome, Italy, in November 1974. From this conference has come new international institutional improvements that should be helpful in organizing a global perspective on farm and food policies. The major organization formed was the International Fund for Agricultural Development that was established in 1977 (Figure 17-1). The purpose of this fund is to provide additional funds to increase food production in developing countries.

Other organizations under the Food and Agriculture Organization are (1) the global information and early warning systems, and (2) committee on world food security. These institutions are a movement in the proper direction, but it will take much more effort and willpower to solve the hunger problem.

The United States has been the most active of all countries in providing food aid and attempting to increase food production in the developing countries. In fact, before the 1972 food crisis it provided almost all the world's food aid.

Even though food production is not a problem in the world, the distribution of food is concentrated in the hands of a few developed countries. Some food-deficit countries can afford to pay for imports. These include the industrialized countries such as Japan, most European countries, and the OPEC countries that can pay for food with their oil earnings. However, there are many less-developed countries that are unable to pay for their food imports. Therefore, it is the policy of the United Nations to attempt to increase food production in these countries and decrease their reliance on the few major grain exporters.

Since the goal of the 1974 World Food Conference to eradicate hunger and malnutrition "within a decade" was not accomplished, a new World Food Summit arranged by FAO was scheduled to take place in Rome in the fall of 1996. The 1996 World Food Summit will provide an international forum not only to assess the evaluation of the food situation since the World Food Conference 20 years earlier, but will also lay the basis for

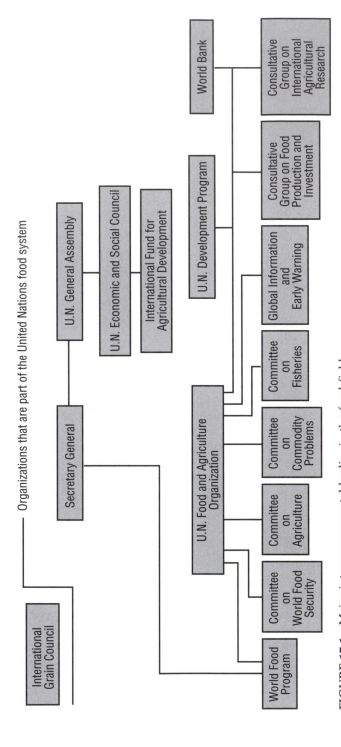

**FIGURE 17-1**  Major intergovernmental bodies in the food field.

SOURCE: Martin M. McLaughlin, "World Food Insecurity: Has Anything Happened Since Rome?" Overseas Development Council, No. 27, 1975, updated 1996.

a new global campaign. There is renewed concern about the food problem by FAO as the present world population of 5.5 billion is projected to increase by 3 billion by the year 2030.

The number of chronically undernourished people in Sub-Saharan Africa alone is expected by FAO to reach 0.3 billion by 2010.

## Food Aid Convention

The Food Aid Committee was established by the Food Aid Convention of the International Grains Arrangement in 1967 to administer the Convention on a periodic basis. The objective of the Convention was to secure, through a joint effort by the international community, the annual distribution of food aid to developing countries in the form of grain suitable for human consumption. The annual distribution target of the 1995 Food Aid Convention was 10 million tons of total food aid. Minimum annual contributions of donors in wheat equivalent agreed to for 1995/96 (July–June) in the 1995 Convention were: Argentina (35,000 mt), Australia (300,000 mt), Canada (400,000 mt), EU (1,755,000 mt), Japan (300,000 mt), Norway (20,000 mt), Switzerland (40,000 mt), and the United States (2.5 million mt) for a total contribution of 5,350,000 mt. The 1995 Convention will remain in force until 1998.

The World Food Program (WFP) became fully operational in 1963, the same year that the Kennedy Round trade negotiations started. The first Food Aid Convention (FAC) was organized by the UN and FAO during these negotiations to establish minimum donor commitments. Twelve countries, including Japan, the United States, Canada, Australia, Britain, and the EU, made commitments in the first FAC to supply at least 4.5 million metric tons of food aid or cash equivalents of food. The FAC was renewed in subsequent years with gradual increases in the minimum annual contributions to a total of 7.8 million metric tons by 1980. The total minimum contribution was finally raised to 10 million metric tons in the 1974 World Food Conference.

The principal purposes of the WFP have been to use food aid to stimulate and advance economic and social development in developing countries, and to provide emergency food relief in disaster areas. It was conceived as a program to utilize food surpluses of donor nations stemming from their national domestic agricultural policies—principally the United States, the EU, Canada, and Australia.

The 1996 Federal Agricultural Improvement and Reform (FAIR) Act continues food aid activities of previous U.S. farm bill legislation such as: (1) the Food for Progress Program to provide commodities to governments of developing countries and emerging democracies or to private voluntary organizations (PVO's) to introduce elements of free enterprise; (2) the P.L. 480 Program with increased maximum funding to

$28 million under Title II, with the addition of the World Food Program as an eligible organization to receive P.L. 480 support, and extension of the authority for the Food Aid Consultative Group through 2002; (3) amending the Food Security Wheat Reserve to include other commodities for a 4-million-ton reserve to provide humanitarian relief in disasters; (4) continuing the use of short-term and medium-term supplier credit guarantees with a mandated annual program level of $5.5 billion through 2002 for both types of credit guarantees; and (5) continuing the emerging markets program requiring the CCC to make at least 1 billion dollars of direct credit or credit guarantees available to "emerging markets" with growth potential for U.S. agricultural exports during fiscal 1996–2002 plus up to $10 million annually for technical assistance.

## Historical Food Production

Total world food production has been increasing much more rapidly in less-developed countries than in developed countries (Table 17-8). Since 1989–91, the food production index in both areas of the world increased 9 percent; up 20 percent in the developing countries, and down 5 percent in the developed countries. However, on a per capita basis, food production in 1995 decreased 7 percent in the developed countries, but increased 9 percent in the less-developed countries. This means that even with larger populations their per capita food supply has increased substantially since the 1989–91 period. Per capita food production in the developing nations has been increasing slowly over the last 30 years, but not enough to satisfy food requirements.

# INCREASING THE WORLD'S FOOD SUPPLIES _____

There are two primary ways of increasing agricultural production: yields per unit of land can be increased, or the land area under production can be expanded. Other important factors are the weather and an agricultural policy that provides incentives to producers and food merchants.

## Crop Yields

There is a wide disparity in yields between the developed and the less-developed countries. For example, the 1995 yield of wheat per hectare in Pakistan was 2.1 metric tons, less than one-third that of West Germany's

**TABLE 17-8**  Indexes of World Food Production

| Calendar Year | World Total | World Per Capita | Developed Countries Total | Developed Countries Per Capita | Developing Countries Total | Developing Countries Per Capita |
|---|---|---|---|---|---|---|
| | | | 1989–91= 100 | | | |
| 1965 | 55 | 86 | 66 | 80 | 45 | 78 |
| 1970 | 63 | 90 | 74 | 87 | 54 | 82 |
| 1975 | 71 | 92 | 83 | 92 | 61 | 83 |
| 1980 | 80 | 95 | 90 | 96 | 71 | 87 |
| 1981 | 82 | 96 | 91 | 97 | 74 | 89 |
| 1982 | 85 | 97 | 94 | 100 | 76 | 90 |
| 1983 | 85 | 96 | 91 | 96 | 79 | 92 |
| 1984 | 89 | 99 | 97 | 101 | 82 | 93 |
| 1985 | 91 | 99 | 98 | 101 | 85 | 95 |
| 1986 | 93 | 100 | 98 | 101 | 88 | 96 |
| 1987 | 93 | 98 | 98 | 100 | 89 | 95 |
| 1988 | 95 | 98 | 96 | 97 | 94 | 98 |
| 1989 | 98 | 100 | 100 | 101 | 97 | 98 |
| 1990 | 101 | 101 | 101 | 101 | 100 | 100 |
| 1991 | 101 | 99 | 98 | 98 | 103 | 101 |
| 1992 | 103 | 100 | 98 | 97 | 107 | 103 |
| 1993 | 103 | 98 | 95 | 93 | 111 | 104 |
| 1994 | 107 | 100 | 96 | 94 | 116 | 107 |
| 1995 | 109 | 100 | 95 | 93 | 120 | 109 |

SOURCE: Food and Agriculture Organization, United Nations.

6.9 tons per hectare.[3] For corn, India's production was 1.6 tons, about one-fifth the U.S. yield of 7.1 metric tons per hectare. These high yields in the developed countries come essentially from improved seeds, fertilizers, plant protection, irrigation water, and a sound financial system. Applying these agricultural input packages continues to offer opportunities to increase yields in developing countries.

The introduction of high-yielding varieties (HYV) of wheat and rice to Asia has been called the *green revolution*. Planting these short-stalked, nitrogen responsive, and photoperiod insensitive plants has doubled and even tripled yields per hectare over traditional plant varieties. About 51 million hectares of HYV wheat were planted in Asia, Africa, Near East,

---

[3] One hectare equals 2.471 acres.

**TABLE 17-9**   Land Area of the World

| Continent | Total | Potentially Arable | Percent of Total | Cultivated | Percent of Total |
|---|---|---|---|---|---|
| | | Billion acres | | | |
| Africa | 7.46 | 1.81 | 24 | .45 | 6 |
| Asia | 6.76 | 1.55 | 23 | 1.12 | 17 |
| Australia and New Zealand | 2.03 | .38 | 19 | .11 | 5 |
| Europe | 1.18 | .43 | 36 | .35 | 30 |
| North America | 5.21 | 1.15 | 22 | .60 | 12 |
| South America | 4.33 | 1.68 | 39 | .40 | 9 |
| FSU | 5.52 | .88 | 16 | .57 | 10 |
| Total | 32.49 | 7.88 | 24 | 3.59 | 11 |

SOURCE: President's Science Advisory Committee, *The World Food Problem*, Vol. 2, U.S. Government Printing Office, Washington, D.C., 1967, p. 434, and Francis Urban and Thomas Vollrath, "Patterns and Trends in World Agricultural Land Use," *Foreign Agricultural Economic Report*, No. 198, Economic Research Service, USDA, April 1984.

and Latin America, and about 73 million hectares of HYV rice.[4] To date the green revolution covers 79 percent of the wheat area of Asia and 45 percent of the rice area, while in Africa HYV are 51 percent of the wheat area and 5 percent of the rice area. Additional emphasis on improved plants and the agricultural inputs complementary to them should show large dividends in future years.

## Land Area

As one looks at a map of the earth's surface, large areas of water, polar ice, and mountains are evident. The report of the president's Scientific Advisory Committee estimated that the land surface is only about 25 percent of the total global area. This amounts to about three times as much land as is actually harvested in a year. Africa and South America have the largest areas of potentially arable land (Table 17-9). Probably three billion acres could be developed for cultivation. Presently Africa is farming roughly 20 percent of its potentially arable land and South America 10 percent. The United States, Canada, Australia, and New Zealand have more than a billion acres of land that could be farmed. There appears to be substantially less land left to farm in Asia, Europe, and the FSU. In

[4] Dana G. Dalrymple, "Development and Spread of High-Yielding Varieties of Wheat and Rice in Developing Countries," Washington, D.C.: Agency for International Development, 1986, pp. 108–110.

Asia there is little land remaining to be cultivated without water development.

Two broad classifications of land are used in differentiating between land actually in use as opposed to land that could be used in the future. *Arable land* is the land that currently is being cropped or is capable of being cropped without incurring additional costs of development. *Potentially arable land*, on the other hand, is land that could be brought into production after undergoing sufficient physical development to make it capable of producing food and fiber for human use.

The world is using 44 percent of its 7.88 billion acres of arable land. Of this arable land, only 550 million acres is irrigated. If HYV are to be grown in Asia, it will be necessary to increase water projects. The Food and Agriculture Organization estimates land development costs in Asia at $440 per hectare, and to develop irrigated land at $1500 per hectare. The cost of bringing in 222 million acres of additional cultivated land into production is estimated at $90 billion. These estimates should give some idea that it is not yet economically feasible to develop much additional land.

For the short term, it might make more economic sense to increase yields per unit of land. Land will only be developed when it becomes economical to do so, at least in the capitalist countries. Command economies may appear able to disregard their costs and returns, but it is only because they can disguise the consequences of uneconomic development.

Through 1990, the world's arable land area expanded rather slowly (Table 17-10). The increase in tilled land area alone is not very indicative

**TABLE 17-10**   Arable and Potentially Arable Land by Region

| Area | Arable Land | | | | Potentially Arable |
|---|---|---|---|---|---|
| | 1975 | 1980 | 1985 | 1990 | |
| | Million hectares | | | | |
| World | 1394 | 1417 | 1431 | 1444 | 3190 |
| Europe | 142 | 141 | 140 | 139 | 174 |
| FSU | 232 | 232 | 232 | 230 | 356 |
| NSC America[a] | 358 | 375 | 379 | 388 | 1106 |
| North America[b] | 268 | 274 | 274 | 274 | — |
| South America | 90 | 101 | 105 | 114 | — |
| Oceania | 44 | 46 | 49 | 51 | 154 |
| Asia | 448 | 450 | 453 | 456 | 628 |
| Africa | 169 | 173 | 177 | 182 | 733 |

SOURCE: Food and Agriculture Organization, United Nations, Rome, Italy, 1993.

[a] North, South, and Central America.

[b] North and Central America.

of total food production capability, however. While world population increased by 30 percent from 1975 to 1990, the land area used to support that population increased by only 3.6 percent. The fact that tilled land increased at a much lower rate than population increased suggests the increased productivity of land and other resources devoted to human food production.

Potentially arable land is an even more tenuous basis for projecting future food production. The classification as "potential" is based on the physical capability of presently undeveloped land being able to produce for human satisfaction in some future period. But that land can become a part of the economic supply of land only if its value of output can cover the costs of bringing that land into production. Thus, income levels, effective demand, and the productivity of land and its allied resources, all will influence the portion of potentially arable land that is brought into production in any future period. About 35 percent of the world's 3190 million hectares of potentially arable land lies in the Western Hemisphere, an area in which land development for economic use is strongly influenced by market forces.

# WORLD FOOD, 2000 _____

The U.S. Department of Agriculture has projected world demand, production, and trade in grains to 2000, given two different sets of conditions (Table 17-11). Alternative I is a baseline projection that assumes median population and income growth rates through the year 2000. Technological advances and weather assumptions are compatible with historic trends of the last two decades. Agricultural and trade policies will continue as they have been in the past. Food-deficient countries will attempt to limit trade, whereas food-surplus nations will attempt to maximize their exports. Alternative I also has petroleum prices at their 1974–76 real price highs for the year 2000, while Alternative II assumes that real prices of petroleum will double between 1974–76 and 2000, with only a slight increase in the real price of grains.

In the first situation, world grain production increases at an annual rate of 2.3 percent between 1985 and 2000. Developed countries would expand grain at a rate of 1.78 percent and developing countries at a rate of 2.94 percent. In the second situation, both regions of the world would reduce grain production. The developed countries would expand grain production at 1.72 percent per year, and the developing countries by 2.74 percent per year. The productive capacity of the major grain-producing countries would be sufficient to meet any foreseeable increase in import demand levels to 2000.

**TABLE 17-11**   World Annual Output of Wheat, Coarse Grains,[a] and Rice

| Country/Region | 1961–71 | 1973–75 | 1985 | Year 2000 Alternatives I | II |
|---|---|---|---|---|---|
| | | Million metric tons | | | |
| **United States** | | | | | |
| Production | 208.8 | 228.7 | 304.0 | 418.0 | 402.0 |
| Consumption | 169.0 | 158.2 | 210.9 | 290.0 | 272.4 |
| Net exports | 39.8 | 72.9 | 93.1 | 128.0 | 129.6 |
| **Other developed exporters** | | | | | |
| Production | 58.6 | 61.2 | 93.0 | 121.9 | 106.1 |
| Consumption | 33.2 | 34.3 | 47.1 | 68.1 | 65.2 |
| Net exports | 28.4 | 27.7 | 45.9 | 53.8 | 40.9 |
| **Centrally planned (Former)** | | | | | |
| Production | 401.0 | 439.4 | 567.0 | 722.0 | — |
| Consumption | 406.6 | 472.4 | 596.0 | 758.5 | — |
| Net imports | 5.2 | 24.0 | 29.0 | 36.5 | — |
| **USSR (Former)** | | | | | |
| Production | 165.0 | 179.3 | 230.0 | 290.0 | — |
| Consumption | 161.0 | 200.7 | 242.5 | 305.0 | — |
| Net trade | +3.9 | −10.6 | −12.5 | −15.0 | — |
| **China (PRC)** | | | | | |
| Production | 163.9 | 176.9 | 227.0 | 292.0 | — |
| Consumption | 166.9 | 180.8 | 235.0 | 302.0 | — |
| Net imports | 3.0 | 3.9 | 8.0 | 10.0 | — |
| **Less-developed countries** | | | | | |
| Production | 306.5 | 328.7 | 471.7 | 728.0 | 735.6 |
| Consumption | 326.6 | 355.0 | 526.0 | 784.8 | 767.4 |
| Net imports | 18.5 | 29.5 | 54.3 | 56.8 | 31.8 |
| **World** | | | | | |
| Production | 1109.2 | 1202.8 | 1608.2 | 2175.5 | — |
| Consumption | 1107.5 | 1202.0 | 1608.2 | 2175.5 | — |
| Net trade | +1.7 | +0.8 | 0.0 | 0.0 | — |

SOURCE: Economic Research Service, USDA.

[a] Coarse grains include corn, barley, rye, oats, sorghum, and millet.

Under the first alternative, U.S. grain import demand is projected at 128 million tons; however, U.S. export supply would also be larger because they would include private grain stocks and those in the farmer-owned reserve. An even larger margin of production over consumption is projected under Alternative II. The reason for this margin is that U.S. production is responsive to world prices and adjustments in human and feed use in livestock are possible. Animal feeding is a very flexible grain user. Between 1972 and 1974, because of the unfavorable beef-to-grain price ratio in the United States, feed grains fed to livestock and poultry declined 41 million tons.[5] Thus, during periods of high grain prices, economic adjustments are possible in the livestock industry that greatly increase cereal availability for human consumption. Also, livestock is a food stock that can be slaughtered for meat. It is true they consume more energy than they produce, but many of the calories for livestock are derived from forage and feed grains that are not used for human consumption.

# WORLD CROPS AND LIVESTOCK _____

The major grains for food are wheat, the coarse grains, and rice. Recent years' world output of these grains and the leading producing areas are summarized in Table 17-12.

## Wheat

Worldwide wheat production for the 1995/96 crop year was estimated at 534 million metric tons.[6] This output was below the previous record crop of 588 million tons in 1990/91.

The world's major wheat producers are the People's Republic of China, the former USSR, the United States, India, France, Canada, and Australia. China produces 19 percent of the world's supply, the FSU 11 percent, and the United States produces about 11 percent. Major exporters of wheat are the United States, Canada, the EC, and Australia. China, Eastern and Western Europe, the FSU, and Italy are large wheat importers. Wheat is a staple food grain in almost all developed countries.

## Coarse Grains

Coarse grains include corn, barley, oats, rye, millet, and sorghum. Production in 1995/96 dropped about 76 million tons to 777.2 million metric tons. Important producers are the United States, the former USSR, and the People's Republic of China. The United States is the largest producer,

---

[5] Henry B. Arthur and Gail L. Cramer, "Brighter Forecast for the World's Food Supply," *Harvard Business Review*, May–June 1976.

[6] One metric ton equals 1000 kilograms, which amounts to 2204.62 pounds (or 36.74 bushels of 60-pound wheat).

**TABLE 17-12**  World Production of Wheat, Coarse Grains, and Rice

| Crop and Producing Areas | 1973/74 | 1975/76 | 1980/81 | 1983/84 | 1992/93 | 1995/96 |
|---|---|---|---|---|---|---|
| | | | Million metric tons | | | |
| **Wheat** | | | | | | |
| China (PRC) | 30.1 | 40.0 | 55.2 | 81.4 | 101.6 | 100.0 |
| Canada | 16.2 | 17.1 | 19.1 | 26.6 | 29.9 | 25.4 |
| Australia | 12.0 | 12.0 | 11.0 | 19.0 | 15.4 | 16.6 |
| Argentina | 6.6 | 8.6 | 7.8 | 11.5 | 9.2 | 8.6 |
| Western Europe | 50.8 | 48.5 | 63.9 | 69.0 | 88.3 | 87.3 |
| USSR (Former) | 109.8 | 66.2 | 98.1 | 85.0 | 87.9 | 58.9 |
| Eastern Europe | 31.5 | 28.4 | 34.5 | 33.3 | 26.6 | 35.3 |
| India | 24.7 | 24.1 | 31.6 | 42.5 | 55.1 | 65.5 |
| Others | 43.7 | 41.7 | 55.7 | 50.7 | 77.1 | 77.4 |
| Total non-U.S. | 325.2 | 292.1 | 376.9 | 419.0 | 491.0 | 475.0 |
| United States | 46.4 | 58.1 | 64.5 | 65.5 | 66.9 | 59.5 |
| World total | 371.6 | 350.1 | 441.4 | 484.5 | 558.0 | 534.5 |
| **Coarse Grains** | | | | | | |
| Canada | 20.4 | 20.0 | 21.6 | 20.8 | 19.5 | 24.1 |
| Australia | 4.7 | 5.6 | 5.5 | 8.5 | 8.3 | 9.1 |
| Argentina | 17.9 | 12.4 | 18.8 | 17.6 | 15.4 | 13.1 |
| South Africa | 11.9 | 7.7 | 11.4 | 10.2 | 9.1 | 10.7 |
| Thailand | 2.5 | 3.3 | 3.5 | 4.3 | 3.8 | 3.7 |
| Brazil | 16.9 | 18.5 | 20.5 | 24.3 | 28.8 | 31.8 |
| Western Europe | 84.1 | 81.5 | 95.2 | 84.9 | 91.5 | 92.3 |
| USSR (Former) | 101.0 | 65.8 | 80.7 | 103.0 | 92.3 | 59.8 |
| Eastern Europe | 55.7 | 59.6 | 61.5 | 64.7 | 42.9 | 50.9 |
| Others | 158.8 | 175.1 | 197.0 | 203.3 | 263.5 | 272.3 |
| Total non-U.S. | 473.9 | 449.5 | 515.7 | 541.6 | 575.1 | 567.8 |
| United States | 186.6 | 184.9 | 198.7 | 139.5 | 277.7 | 209.4 |
| World Total | 660.5 | 634.5 | 714.4 | 681.1 | 852.8 | 777.2 |
| **Rough Rice** | | | | | | |
| Bangladesh | 17.6 | 18.9 | 21.3 | 22.5 | 27.4 | 27.0 |
| Burma | 8.6 | 9.2 | 12.2 | 14.0 | 13.5 | 17.2 |
| India | 66.1 | 74.3 | 81.1 | 82.6 | 108.0 | 118.5 |
| Indonesia | 21.5 | 22.6 | 28.7 | 33.8 | 47.3 | 49.5 |
| Japan | 15.2 | 16.5 | 12.2 | 13.7 | 13.2 | 13.4 |
| Republic of Korea | 5.8 | 6.5 | 5.0 | 7.6 | 7.3 | 6.4 |
| Pakistan | 3.7 | 3.9 | 4.7 | 5.3 | 4.6 | 5.7 |
| PRC | 112.9 | 119.0 | 141.5 | 154.0 | 185.0 | 190.0 |
| Thailand | 14.3 | 15.2 | 18.0 | 17.7 | 19.8 | 21.8 |
| Subtotal | 256.8 | 286.0 | 324.7 | 351.2 | 426.1 | 449.5 |
| EC | 1.1 | 1.0 | 1.0 | 1.1 | 2.2 | 2.0 |
| Australia | 0.4 | 0.4 | 0.6 | 0.7 | 1.1 | 1.1 |
| Argentina | 0.3 | 0.3 | 0.3 | 0.3 | 0.5 | 0.9 |
| Brazil | 6.5 | 8.5 | 9.7 | 9.5 | 10.5 | 9.9 |
| All others | 46.0 | 51.4 | 53.9 | 57.8 | 70.2 | 76.5 |
| Total non-U.S. | 320.1 | 347.7 | 390.2 | 420.6 | 510.6 | 539.9 |
| United States | 4.2 | 5.8 | 6.6 | 4.6 | 8.1 | 7.9 |
| World total | 324.3 | 353.5 | 396.8 | 425.3 | 518.7 | 547.8 |
| All grains total | 1356.4 | 1338.1 | 1552.6 | 1590.8 | 1929.5 | 1859.5 |

SOURCE: Foreign Agricultural Service, *Foreign Agricultural Circular*, Washington, D.C.: USDA, May 1996.

growing about 27 percent of the world's harvest. In the United States, we normally think of coarse grains as feed grains for livestock, but these grains are food grains as well for many countries in Africa, Asia, Latin America, and the Middle East. The leading exporters of coarse grains are the United States, France, Argentina, Germany, and China. Western Europe, the former USSR, and Japan import most of the coarse grains for livestock feed. Coarse grains, like wheat, are grown primarily in temperate climatic zones.

## Rice

Rice is the major food crop in the world and is consumed mainly in countries of the Far East. The world rice crop totaled 548 million metric tons in 1995/96. Of this total, Asian countries produced about 91 percent, and the People's Republic of China alone accounts for 35 percent. Asian rice exports amount to about 13 million metric tons annually. And, a surprising statistic for many, the United States is the world's third largest rice exporter, shipping more than 2 million tons a year. Thailand is the largest supplier, exporting more than 5 million tons per year.

## Livestock

The beef cattle industry is concentrated in Europe and North and South America. In 1993, the United States had an inventory of 104 million head, and Brazil 152 million head (Table 17-13). The former USSR, the EU, and Argentina accounted for another 201 million head.

Beef is a significant food in the American diet. The United States produced over 11 million metric tons of beef and imported about 701,000 tons or roughly 6 percent of production. Each American now consumes about 45.1 kilograms (99 pounds) of beef and veal per year. Major beef-importing countries are the United States, the EU, and Japan. Australia, New Zealand, the United States, and the EU are the leading beef exporters.

Pork is a popular meat in Western and Eastern Europe and China. Pork consumption per capita is higher in those regions than is beef. Major hog producers are China, the EU, the FSU, the United States.

Mutton, lamb, and goat meat are produced primarily in the FSU, Australia, New Zealand, and the EC. The leading exporters of sheep and lambs are New Zealand and Australia. Per capita consumption of sheep and goat meat is 17 kilograms per year in Australia, 25 in New Zealand, and 15 in Greece. Consumption in the Russian Federation averages only 1.7 kilograms per capita.

Total world beef production is about 45 million metric tons, whereas total world poultry meat production is about 44 million metric tons. Major poultry producers are the United States, the EU, and China.

TABLE 17-13   Numbers of Livestock in Specified Countries and Areas

| Type of Livestock and Producing Areas | 1975 | 1980 | 1987 | 1990 | 1993 | 1996 |
|---|---|---|---|---|---|---|
| | | | Million head | | | |
| **Cattle and buffalo** | | | | | | |
| United States | 131.8 | 111.0 | 102.0 | 98.2 | 100.9 | 103.8 |
| FSU | 109.1 | 115.0 | 122.1 | 118.3 | 101.5 | 65.1 |
| Brazil | 94.0 | 93.0 | 97.0 | 130.9 | 129.4 | 152.1 |
| EU | 79.2 | 77.9 | 82.1 | 85.9 | 79.8 | 82.9 |
| Argentina | 59.6 | 59.5 | 51.7 | 57.3 | 56.9 | 53.5 |
| Mexico | 28.7 | 29.5 | 33.6 | 31.7 | 30.7 | 28.1 |
| Australia | 32.8 | 26.2 | 23.5 | 24.7 | 25.6 | 26.0 |
| World[a] | 724.5 | 949.1 | 1037.8 | 1070.5 | 1050.1 | 1049.7 |
| **Hogs** | | | | | | |
| China (PRC) | 232.8 | 325.1 | 337.2 | 352.8 | 385.0 | 441.4 |
| United States | 55.1 | 67.0 | 50.9 | 53.8 | 59.8 | 60.2 |
| FSU | 72.3 | 73.7 | 79.5 | 78.8 | 60.0 | 36.8 |
| EU | 69.7 | 75.7 | 103.6 | 113.9 | 108.1 | 114.8 |
| Brazil | 43.5 | 36.8 | 31.7 | 33.2 | 32.3 | 32.5 |
| World[a] | 618.5 | 756.5 | 748.3 | 760.0 | 771.6 | 787.5 |
| **Sheep** | | | | | | |
| FSU | 145.3 | 143.7 | 142.2 | 146.7 | 117.0 | 53.0 |
| Australia | 151.7 | 135.2 | 158.8 | 177.8 | 147.2 | 120.7 |
| New Zealand | 55.3 | 64.0 | 69.2 | 60.6 | 54.3 | 47.3 |
| EU | 44.4 | 48.0 | 84.2 | 100.5 | 99.4 | 97.4 |
| World[a] | 697.3 | 732.5 | 696.3 | 907.3 | 819.3 | 912.4 |

SOURCE: Foreign Agricultural Service, *Foreign Agricultural Circular*, Washington, D.C.: USDA.
[a] Total for selected countries and areas only.

Poultry consumption per capita is 47 kilograms in the United States and Hong Kong, and 45 kilograms in Israel. Poultry consumption has been increasing rapidly in the United States. In fact, per capita poultry meat consumption surpassed beef consumption in 1993.

# FOOD FROM THE SEA _____

Aquatic products are an especially important source of food in Japan, Norway, Spain, Iceland, Portugal, and Southeastern Asia. The world's oceans and other bodies of water have provided more than 110 million metric tons of fish. About 75 million tons of this amount was used for human food, the balance being used mainly in the manufacture of animal feeds. Fish and fish products consumed by humans were just 2.5 percent of total food by weight, only 1 percent of all food calories, and 6 percent of our protein from all food sources.

To suggest the seas as a food source that can solve the world's hunger problem is little more than wishful thinking. With present management and technology, the long-run potential fish harvest has been estimated to be somewhere in the range of 100 to 150 million metric tons. But even tripling or quadrupling fish harvested for food would still leave this as a relatively minor source of calories and protein for the world's people.

# THE MALTHUSIAN DILEMMA[7] _____

The Reverend Thomas R. Malthus, writing in 1798, gained fame for what he called the *principle of population.* The basic significance of his proposition was that the food supply can only be expected to expand arithmetically, while population will tend to expand geometrically, until starvation or some equally fatal deterrent intervenes.

> It may be fairly pronounced, therefore, that considering the present average state of the earth, the means of subsistence, under circumstances the most favourable to human industry, could not possibly be made to increase faster than in an arithmetical ratio.[8]

The arithmetic nature of food expansion rested primarily on the economic premise of diminishing returns on a fixed supply of land as more and more labor is applied to it. The geometric expansion of population, meanwhile, could not be maintained in the longer run because it would rapidly encounter the arithmetic food barrier. Thus, hunger and other population checks (most of them related to hunger and poverty) would keep the unhappy balance. The logically pessimistic conclusion derives from the fundamental premises on food production and population, earning for Malthus the label of "gloomy dean."

The fact of the matter is that during the almost two centuries since Malthus wrote his treatise, both food supplies and population have persistently followed not the arithmetic expectation of his theory, but a pronounced geometric pattern. Hunger, never absent in this imperfect world, has been relatively less widespread and less acute in the past few de-

---

[7] Henry B. Arthur and Gail L. Cramer, "Brighter Forecast for the World's Food Supply," *Harvard Business Review*, May–June 1976.

[8] Thomas R. Malthus, *The Principle of Population*, 1878 edition. London: Reeves and Turner, London, pp. 5–6.

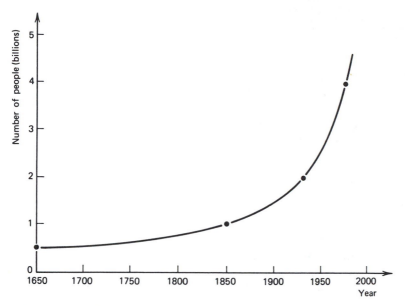

**FIGURE 17-2** World population estimates, selected years.

cades than in any half-century of the world's history. (We say this for perspective, not as cause to celebrate a "victory" over the hunger problem.)

With the world's population estimated at about one-quarter billion at the time of Christ's birth, growth has been very slow, with occasional sharp declines from severe famines and epidemics. Rapid growth as we know it today is only a recent phenomenon. A sketch of the population data from Table 17-2 for the years beginning with 1650, as in Figure 17-2, presents a striking picture of what Malthus meant by a *geometric* population growth potential.

But what about production? With some (relatively) minor variations, can it have been anything other than the same trend line as for population? Obviously, it can't have been less than that for population or there would have been an excess of people over the food needed for their support. Thus food production must also have exhibited a very similar geometric growth.

A variety of circumstances have (so far) prevented the predicted catastrophe: (1) For the first hundred years or so following Malthus' prediction, the opening of new lands provided the world with greater quantities of food; (2) beginning in the 1920s, mechanization in important food production areas increased yields, releasing many millions of acres formerly used to provide feed and forage for draft animals; and (3) the application of science to agriculture over the past 40 years or more (commercial fertilizers, pesticides, genetics, etc.) sharply increased the output of many important foods.

Does this all mean that Malthus was right at least on the population side of his principle? In one sense, yes. The laws of nature provide the *potential* for abundant human propagation, as it does for almost all natural species. Moreover, there is no doubt that food requirements of the world are so massive, and the machinery for producing and delivering these requirements is so complex, that we cannot complacently assume that adequate supplies will be forthcoming as needed. The dilemma is there, even if improvements in world food production have prevented Malthus' predicted outcome. The best information we have seems to call for a reappraisal, however. The two horns of the dilemma may not be as terrifying as was feared, either independently or together.

Births are undoubtedly becoming more discretionary, both in terms of technology and in cultural acceptance, despite the drags of ignorance and tradition. In many of the developing countries, as well as in the FSU, Japan, some European countries, the United States, and Canada, birth rates are declining. Estimated world population figures show that the global total birth rate has fallen, especially in the past 20 years.

The continuing growth of population results mostly from longer life spans due to medical technology and better diets. This factor of population increase still has a distance to go in many countries, but it is a one-time shift, not a continuing growth factor; its effect tapers off as age-spans approach a (present) biological limit.

We have seen that the food side of the dilemma has in fact been able to attain a geometric growth pattern in the past 150 years—not of course matching the biological *potential* population expansion, but nonetheless geometric in rate, and very impressive in actual volume. The assumed ceiling with which Malthus dealt—the result of limited and unexpandable land resources—has been breached by revolutionary changes in technology and in the use of capital to produce dramatic increases in per-acre yields. The world's available food has actually outrun population growth so that *in the aggregate* the world has been eating better (per capita) than at any time in history.

This is not the place to predict how far into the future we can expect the situation to continue. Unlike many others, however, we see no indication of the exhaustion of the kind of unforeseeable ingenuity that has found answers in the past. New developments, however, are no guarantee that the rate of expansion of the food supply will match what may be needed by the world's people.

In other words, the dilemma persists, if only in the sense of the ominous potential on either side. To date, we seem *not* to have pressed against the food barrier as Malthus contemplated and we are even a little ahead of the game on a global basis.

Such a statement as this is no comfort at all to the millions who are ill-fed and poverty stricken in many parts of the world. But it does seem most clear that theirs is as much a problem of food distribution as a problem of production.

# LOOKING TO THE FUTURE ————————————

During the two centuries that have passed since Malthus acquired fame for his views on human procreation, hunger, and related topics, exponential growth of our species has proven not to be a permanent phenomenon. The world has come to understand that as economic growth occurs, populations undergo a transition from a traditional equilibrium of high fertility and mortality to the sort of equilibrium that the world's richest countries now enjoy. In the latter equilibrium, natural increase does not occur, but with the important difference that births as well as deaths are much less common events.

Global agriculture has responded more than adequately to the challenge presented by demographic transition. So the crucial question to be faced in the decades to come really is not whether food needs can be met—they can be. Instead, the important matter to deal with is how food issues will be resolved.

Some of the options are truly dismal. The Director General of the International Food Policy Research Institute (IFPRI) points out that demand can continue to be held back because of widespread poverty. If this is our future course, though, a high price will end up being paid in terms of tens of millions of children who fail to reach their full mental and physical potential resulting from inadequate nutrition.

Another gloomy possibility is that increased demand for agricultural commodities will be satisfied through widespread use of current uncultivated land. The species extinction and other environmental impacts resulting from the conversion of forests and other natural habitats into cropland and pasture would be enormous.

There is a better option, expressed in IFPRI's *Vision 2020*. Raising crop and livestock yields can continue to result from productivity-enhancing investment. In addition to arresting encroachment on the domains of other forms of life that inhabit the planet, agricultural intensification can also be pursued in ways that do not damage soil resources, or result in agricultural chemical pollution, or expose human beings to harmful levels of pesticides. With adequate support for education and public health and nutrition programs, people worldwide will be able to earn enough to satisfy their own nutritional needs and, in the case of rural populations, to produce the crops and livestock that will be demanded.

# SUMMARY

There are at least 786 million people suffering from hunger who do not have the income nor means to purchase enough food. These people live in Asia, Latin America, and Africa. The size of the food gap is around 12 to 37 million tons of cereals per year.

The world food crises of 1972–75 and 1979–80 were caused by crop failures, a decrease in world stocks, the energy crisis, and an unstable monetary system. These factors came together at about the same time. It was apparent that the world was not prepared to handle the food shortages and the resultant price increases.

Since the World Food Conference, several new international institutions have been formed to handle food problems. These include the International Fund for Agricultural Development, and a committee system under the Food and Agricultural Organization. To date, these organizations have not been very effective in reducing the number of hungry people in the world.

For much of the time since the Second World War, global demand for food has been growing at historically unprecedented rates. All parts of the developing world have embarked on a demographic transition that eventually will take them to an equilibrium featuring low mortality and fertility. But until that equilibrium is reached, natural increase will continue, and the number of mouths needing to be fed will grow. Demand for agricultural commodities is also expected to rise as living standards improve.

Through the early 1900s, demand growth was met almost always by clearing additional land for crop and livestock production. The development of agriculture in the American midwest, the Argentine Pampas, and other places where Europeans were settling during the nineteenth century are examples, on a grand scale, of this response. Opportunities for developing additional land still exist in various places. However, land improvement expenses are high in many places, and the environmental consequences of habitat loss have provoked widespread concern. Intensification, or increasing yields, which has become the main response to demand growth, is a much more appealing alternative.

As a whole, farmers and ranchers have done a commendable job of keeping up with demand. Production of grain, livestock, and other commodities has reached new heights in just about every corner of the world. Where agriculture's performance has been disappointing, the fault usually lies with natural disasters or misguided governmental policies, including the suppression of market forces and inadequate investment in roads and agricultural research and extension. Usually, famine occurs not because agriculture has failed, but because food distribution networks have been disrupted by armed conflict.

# CHAPTER HIGHLIGHTS

1. There are at least 786 million people in the world suffering from hunger and malnutrition.

2. World population will increase to more than 6 billion people by the year 2000. The developing countries have about 78 percent of the world's population. By the year 2050 they will have 86 percent of the world's population.

3. The cereals comprise about 50 percent of the caloric intake per day of the average person on earth. Food grains are primarily wheat and rice, although corn, barley, and sorghum are important foods in many developing nations.

4. Between 1985 and 2000, population will grow about 1.8 percent per year and food production will increase about 2 percent per year.

5. There is enough food available worldwide to feed all people. However, many people in Asia, Latin America, and Africa do not have the income to purchase it.

6. The primary food problem is that most of the hungry people are not getting enough calories. It is likely that most of these people would attain adequate supplies of protein if they ate enough staples to meet energy requirements.

7. It would take from 18 to 55 million tons of cereals to close the food gap. This amount of food would probably cost $4.5 to $13.7 billion per year.

8. The 1972–75 and 1979–80 world food crises occurred because of crop failures in various parts of the world, a draw-down in North American food stocks, and the depreciation of the dollar.

9. Institutional improvements made since the world food crises include the formation of (1) the International Fund for Agricultural Development, and (2) the new committees under the Food and Agriculture Organization.

10. Per capita food production has decreased 7 percent from the 1989–91 average production in the developed nations, but increased 9 percent in the developing countries.

11. World dependence on North America and Oceania for cereals has increased greatly over the past 50 years.

12. Both the U.S. Department of Agriculture and the United Nations project adequate world food supplies to the year 2000.

13. Animals are flexible grain users. Livestock derives much of its feed from forage and feed grains that have little human food value.

14. Many of the developing nations associated with OPEC can afford to pay for their food imports.

15. It has been U.S. policy to help food-deficit nations increase their food supplies rather than rely on food aid. Increased food supplies are likely to come from increased yields per unit of land and expanded acreage.

16. Food from the sea will not add much to the world's energy or protein supply in the foreseeable future, but it is an important source of food.

17. The United Nations is attempting to increase food output in developing countries. It also is working on a world food security system.

18. The United States provides much of the world's food aid, but in addition the United States is trying to use the land grant college concept to increase food production in developing countries.

19. Malthus stated that food supply could only be expected to expand arithmetically while population would expand geometrically.

20. Both food supplies and population have followed a pronounced geometric growth pattern over the past 150 years.

21. Economic development will lower mortality and fertility so that the world population may stabilize at around 12 to 16 billion people during the next century.

## KEY TERMS AND CONCEPTS TO REMEMBER

Arable land                              Principle of population
Potentially arable land

## REVIEW QUESTIONS

1. What is the world food problem? Is it one of protein, calories, food production, or income distribution?

2. World population has been increasing at a very rapid rate. What factors influence the rate of population growth? Do you think the rate of population growth will decrease in the years ahead? Why?

3. What is the difference between effective demand and nutritional needs for food?

4. The food crises of 1972–75 were very severe. Why did they occur? How can we prevent future food crises?

5. During the food crises, some people were arguing that Americans should reduce their meat intake so that more grain would be available for the hungry world. If Americans cut their meat consumption, would that grain saved as a result be shipped to the hungry in the developing countries? Discuss.

6. A number of new institutions have been formed in order to improve the information flow about world food problems. Why are they needed? Will these institutions be able to handle the next large food crisis?

7. Aquatic products are an important source of food. Can the world's food problems be solved by developing only this source of food?

8. Malthus thought that the food supply could be expected to expand arithmetically while population would tend to expand geometrically. Was he right or wrong? Why was he right or wrong?

9. What efforts are being made to increase food production in the developing countries? What impact will these efforts have on the export market for U.S. products?

# SUGGESTED READINGS

Avery, Dennis T. *Saving the World with Pesticides and Plastic*, Indianapolis: Hudson Institute, 1995.

Brown, Lester. *State of the World 1995*, Washington: Worldwatch Institute, 1995.

Bulatao, Rodolfo A., Eduard Bos, Patience W. Stephens, and My T. Vu. *World Population Projections, 1989–90 Edition: Short- and Long-Term Estimates*, Baltimore: Johns Hopkins University Press, 1990.

Carter, Harold O., et al. *A Hungry World: The Challenge to Agriculture*, Davis, Calif.: University of California, July 1974.

Cramer, Gail L., and Eric J. Wailes, 2nd ed. *Grain Marketing*, Boulder, CO: Westview Press, 1993.

Economics, Statistics, and Cooperatives Service. *Global Food Assessment, 1980*, Washington, D.C.: USDA, Foreign Agricultural Economic Report, No. 159, July 1980.

Economic Research Service. *The World Food Situation and Prospects to 1985*, Foreign Agricultural Economic Report, No. 98. Washington, D.C.: U.S. Government Printing Office, 1974.

Economic Research Service. *World Food Needs and Availabilities 1986–87*, Washington, D.C.: USDA, August 1986.

Ehrlich, Paul R., and Anne H. Ehrlich. *Population, Resources, Environment*, 2nd ed. San Francisco, Calif.: Freeman, 1972, Chapters 2 and 3.

Foster, Phillips, *The World Food Problem*, Boulder, CO: Lynne Rienner Publishers, Inc., 1992.

International Food Policy Research Institute. "Feeding the World, Preventing Poverty, and Protecting the Earth: A 2020 Vision," Washington, 1996.

Johnson, D. Gale. "World Food Problems and U.S. Agriculture," *Lectures in Agricultural Economics,* Economic Research Service, USDA, Washington, D.C., June 1977.

*Proceedings: The World Food Conference of 1976, June 27–July 1,* Ames, Iowa: The Iowa State University Press, 1977.

Woods, Richard G., ed. *Future Dimensions of World Food and Population,* Morrilton, Ark.: Winrock International, 1981.

Wheeler, R. O., Gail L. Cramer, K. B. Young, and Enrique Ospina. *The World Livestock Product, Feedstuff, and Foodgrain System,* Morrilton, Ark: Winrock International, 1981.

# AN OUTSTANDING CONTRIBUTOR

**G. Edward Schuh**   G. Edward Schuh is Dean of the Hubert H. Humphrey Institute of Public Affairs at the University of Minnesota, a position he has held since December 1987.

Schuh was born and raised on a vegetable farm just outside Indianapolis, Indiana, graduating from Ben Davis High School in 1948. He went to Purdue University, from which he graduated with distinction in 1952 with a B.S. degree in agricultural education. He did graduate work at Michigan State University and received his M.S. degree in agricultural economics in 1954. Following 2 years of military service, most of it spent in South Korea and Japan, he went to the University of Chicago, where he earned the M.A. degree (1958) and the Ph.D. degree (1961) in economics.

Dr. Schuh joined the faculty at Purdue University as Instructor in the Department of Agricultural Economics in 1959. He advanced to assistant professor in 1961, to associate professor in 1962, and to professor in 1965. He went to Brazil on a Purdue University project, where he was visiting professor at the Federal University of Vicosa. In Brazil he helped begin the first graduate program in agricultural economics in Latin America. Later (1966 to 1972), he served as Program Advisor in Agriculture to the Ford Foundation (while remaining on the faculty at Purdue University), and helped develop other graduate programs in Brazil. He was head of the Department of Agricultural and Applied Economics at the University of Minnesota from 1979 to 1984, and Director of Agriculture and Rural Development at the World Bank from 1984 to 1987.

Schuh has taught graduate courses in agricultural policy, advanced production economics, econometrics, research methodology, welfare economics, and agricultural development. His research has focused on agricultural labor markets, agricultural development, agricultural policy, technical change, the new household economics, and international trade and exchange rates. He has published four books on Brazilian agriculture and more than 100 research bulletins, journal articles, and technical papers.

Schuh was research associate at Harvard's Development Advisory Service in 1968–69. He was senior staff economist with the president's Council of Economic Advisors in 1974–75, and deputy secretary for International Affairs and Commodity Programs, USDA, 1978–79. He was also the first director of the Center for Public Policy and Public Administration at Purdue University in 1977–78.

Schuh has received four professional awards from the American Agricultural Economics Association. He won the Ph.D. Dissertation Award in 1961, the Published Research Award in 1970, Journal Article Award in 1974, and the Distinguished Policy Award in 1979. Schuh was elected professor *Honoris Causis* at the Federal University of Vicosa in 1965, elected Fellow of the American Academy of Arts and Sciences in 1977, and was elected Fellow of the American Agricultural Economics Association in 1984. He also has served two terms as director of the National Bureau for Economic Research at Cambridge, Massachusetts; director of the Economics Institute at Boulder, Colorado; director of the Minnesota Economics Association; director of the American Agricultural Economics Association; and served as president of the AAEA in 1982.

He is married to the former Maria Ignez Angeli, and they have three children.

# A List
# of Reference
# Books
# for the
# Beginning
# Student*

---

\* You can locate these in your campus library.

1. Barkley, Paul W. *Economics: The Way We Choose.* New York: Harcourt Brace Jovanovich, 1977.
2. Barlowe, Raleigh, *Land Resource Economics*, 4th ed., Englewood Cliffs, N.J.: Prentice-Hall, 1986.
3. Barry, Peter J., John A. Hopkin, and C. B. Baker. *Financial Management in Agriculture*, 3rd ed., Danville, Ill.: The Interstate Printers and Publishers, 1988.
4. Bishop, C. E., and W. D. Toussaint. *Introduction to Agricultural Economic Analysis*, New York: John Wiley & Sons, 1958.
5. Boehlje, Michael D., and Vernon Eidman. *Farm Management*, New York: John Wiley and Sons, 1984.
6. Bradford, Lawrence A., and Glenn L. Johnson. *Farm Management Analysis.* New York: John Wiley & Sons, 1953.
7. Brown, Lester R., and Erik P. Eckholm. *By Bread Alone.* New York: Praeger Publishers, 1981.
8. Buse, Rueben C., and Daniel W. Bromley. *Applied Economics: Resource Allocation in Rural America.* Ames, Iowa: Iowa State University Press, 1975.
9. Casavant, Kenneth L., and Craig Infanger. *Economics and Agricultural Management*, Reston, Va.: Reston Publishing, 1984.
10. Castle, Emery N., Manning H. Becker, and A. Gene Nelson. *Farm Business Management*, 3rd ed., New York: Macmillan, 1987.
11. Dahl, Dale C., and Jerome W. Hammond. *Market and Price Analysis: The Agricultural Industries.* New York: McGraw-Hill, 1977.
12. Duft, Kenneth D. *Principles of Management in Agribusiness.* Reston, Va.: Reston Publishing, 1979.
13. Eckert, Ross D., and Richard H. Leftwich. *The Price System and Resource Allocation*, 10th ed., Fort Worth: Dryden Press, 1988.
14. Epp, Donald J., and John W. Malone, Jr. *Introduction to Agricultural Economics.* New York: Macmillan, 1981.
15. Goodwin, John W. *Agricultural Economics*, 2nd ed., Reston, Va.: Reston Publishing, 1982.
16. Goodwin, John W. *Agricultural Price Analysis and Forecasting*, New York: John Wiley & Sons, 1994.
17. Gwartney, James D., and Richard Stroup. *Economics: Private and Public Choice*, 6th ed., Fort Worth: Dryden Press, 1992.
18. Halcrow, Harold G. *Economics of Agriculture*, New York: McGraw-Hill, 1980.
19. Halcrow, Harold G. *Food Policy for America*, New York: McGraw-Hill, 1977.
20. Heyne, Paul T. *Economic Way of Thinking*, 7th ed., New York: Macmillan, 1993.
21. Heyne Paul, and Thomas Johnson. *Toward Economic Understanding*, Chicago: Science Research Associates, 1976.

22. Ingram, James C., and Robert M. Dunn. *International Economics*, 3rd ed., New York: John Wiley & Sons, 1992.
23. Kadlec, John E. *Farm Management*, Englewood Cliffs, N.J.: Prentice-Hall, 1985.
24. Kohls, Richard L., and Joseph N. Uhl. *Marketing of Agricultural Products*, 7th ed., New York: Macmillan, 1990.
25. Lee, Warren F., Michael Boehlje, Aaron G. Nelson, and William G. Murray. *Agricultural Finance*, 8th ed., Ames, Iowa: Iowa State University Press, 1988.
26. Penson, John, Oral Capps Jr., and C. Parr Rosson, *Introduction to Agricultural Economics*, Englewood Cliffs, N.J.: Prentice-Hall, 1996.
27. Peterson, Willis L. *Principles of Economics: Micro*, 6th ed., Homewood, Ill.: Richard D. Irwin, 1987.
28. Roy, Ewell P. *Economics: Applications to Agriculture and Agribusiness*, 4th ed., Danville, Ill.: The Interstate Printers and Publishers, 1994.
29. Samuelson, Paul A., and William D. Nordhaus. *Economics*, 13th ed., New York: McGraw-Hill, 1989.
30. Seitz, Wesley D., Gerald C. Nelson and Harold G. Halcrow. *Economics of Resources, Agriculture, and Food*, New York: McGraw-Hill, 1994.
31. Shepherd, Geoffrey S. *Agricultural Price Analysis*, 5th ed., Ames, Iowa: Iowa State University Press, 1966.
32. Shepherd, Geoffrey S., Gene A. Futrell, and J. Robert Strain. *Marketing Farm Products—Economic Analysis*, Ames, Iowa: Iowa State University Press, 1976.
33. Snodgrass, Milton M., and L. T. Wallace. *Agriculture, Economics, and Resource Management*, 2nd ed., Englewood Cliffs, N.J.: Prentice-Hall, 1980.
34. Tomek, William G., and Kenneth L. Robinson. *Agricultural Product Prices*, 3rd ed., Ithaca, N.Y.: Cornell University Press, 1990.
35. Tweeten, Luther G. *Agricultural Policy Analysis*, Boulder, Col.: Westview Press, 1989.
36. Wilcox, Walter W., Willard W. Cochrane, and Robert W. Herdt. *Economics of American Agriculture*, 3rd ed., Englewood Cliffs, N.J.: Prentice-Hall, 1974.
37. Wills, Walter J. *An Introduction to Agribusiness Management*, 2nd ed., Danville, Ill.: The Interstate Printers and Publishers, 1979.

# Basic Sources of Agricultural Statistics*

* You can locate these in your campus library.

Economic Research Service, USDA, *African Food Needs Assessment* (annual), Washington, D.C.

Economic Research Service, USDA, *Agricultural Exports* (quarterly), Washington, D.C.

Economic Research Service, USDA, *Agricultural Income & Finance* (quarterly), Washington, D.C.

Economic Research Service, USDA, *Agricultural Land Values & Markets* (semiannual), Washington, D.C.

Economic Research Service, USDA, *Agricultural Outlook* (monthly Feb.–Dec.), Washington, D.C.

Economic Research Service, USDA, *Agricultural Resources: Inputs* (semiannual), Washington, D.C.

Economic Research Service, USDA, *Agriculture and Trade: Asia* (annual), Washington, D.C.

Economic Research Service, USDA, *Agriculture and Trade: China* (annual), Washington, D.C.

Economic Research Service, USDA, *Agriculture and Trade: Pacific Rim* (annual), Washington, D.C.

Economic Research Service, USDA, *Agriculture and Trade: Former USSR* (annual), Washington, D.C.

Economic Research Service, USDA, *Agriculture and Trade: Western Europe* (annual), Washington, D.C.

Economic Research Service, USDA, *Aquaculture* (semiannual), Washington, D.C.

Economic Research Service, USDA, *Cost of Production—Livestock and Dairy* (Apr., May), Washington, D.C.

Economic Research Service, USDA, *Cost of Production—Major Field Crops* (Apr., May), Washington, D.C.

Economic Research Service, USDA, *Cotton & Wool* (quarterly), Washington, D.C.

Economic Research Service, USDA, *Cropland, Water, & Conservation* (annual), Washington, D.C.

Economic Research Service, USDA, *Dairy* (quarterly), Washington, D.C.

Economic Research Service, USDA, *Farmline* (monthly), Washington, D.C.

Economic Research Service, USDA, *Feed* (quarterly), Washington, D.C.

Economic Research Service, USDA, *Food Consumption, Prices, and Expenditures* (annual), Washington, D.C.

Economic Research Service, USDA, *Food Cost Review* (annual), Washington, D.C.

Economic Research Service, USDA, *Food Marketing Review* (annual), Washington, D.C.

Economic Research Service, USDA, *Food Review* (quarterly), Washington, D.C.

Economic Research Service, USDA, *Food Spending in American Households* (annual), Washington, D.C.

Economic Research Service, USDA, *Foreign Agricultural Trade of the United States* (quarterly), Washington, D.C.

Economic Research Service, USDA, *Foreign Ownership of U.S. Agricultural Land* (annual), Washington, D.C.

Economic Research Service, USDA, *Fruit & Tree Nuts* (quarterly), Washington, D.C.

Economic Research Service, USDA, *Journal of Agricultural Economics Research* (quarterly), Washington, D.C.

Economic Research Service, USDA, *Livestock & Poultry* (quarterly), Washington, D.C.

Economic Research Service, USDA, *Livestock & Poultry Updates* (monthly), Washington, D.C.

Economic Research Service, USDA, *National Financial Summary* (Oct., Nov.), Washington, D.C.

Economic Research Service, USDA, *New Reports* (weekly), Washington, D.C.

Economic Research Service, USDA, *Oil Crops* (quarterly), Washington, D.C.

Economic Research Service, USDA, *Outlook Conference Proceedings* (annual), Washington, D.C.

Economic Research Service, USDA, *Outlook Conference Chartbook* (annual), Washington, D.C.

Economic Research Service, USDA, *Production & Efficiency Statistics* (annual), Washington, D.C.

Economic Research Service, USDA, *Reports Catalog* (quarterly), Washington, D.C.

Economic Research Service, USDA, *Rice* (triannual), Washington, D.C.

Economic Research Service, USDA, *Rural Conditions & Trends* (quarterly), Washington, D.C.

Economic Research Service, USDA, *Rural Development Perspectives* (triannual), Washington, D.C.

Economic Research Service, USDA, *State Financial Summary* (Feb., Mar.), Washington, D.C.

Economic Research Service, USDA, *Sugar & Sweeteners* (quarterly), Washington, D.C.

Economic Research Service, USDA, *Tobacco* (quarterly), Washington, D.C.

Economic Research Service, USDA, *U.S. Agricultural Trade Updates* (monthly), Washington, D.C.

Economic Research Service, USDA, *Vegetables & Specialties* (triannual), Washington, D.C.

Economic Research Service, USDA, *Weekly Weather & Crop Bulletin* (weekly), Washington, D.C.

Economic Research Service, USDA, *Wheat* (quarterly), Washington, D.C.

Economic Research Service, USDA, *World Agricultural Supply & Demand Estimates* (monthly), Washington, D.C.

Foreign Agricultural Service, USDA, *Agricultural Trade Highlights* (monthly), Washington, D.C.

Foreign Agricultural Service, USDA, *World Cotton Situation* (monthly), Washington, D.C.

Foreign Agricultural Service, USDA, *Dairy, Livestock & Poultry: U.S. Trade & Prospects* (monthly), Washington, D.C.

Foreign Agricultural Service, USDA, *Dairy Monthly Imports* (monthly), Washington, D.C.

Foreign Agricultural Service, USDA, *Export Markets for U.S. Grain & Products* (monthly), Washington, D.C.

Foreign Agricultural Service, USDA, *Horticultural Products Review* (monthly), Washington, D.C.

Foreign Agricultural Service, USDA, *U.S. Essential Oil Trade* (annual), Washington, D.C.

Foreign Agricultural Service, USDA, *U.S. Export Sales* (weekly), Washington, D.C.

Foreign Agricultural Service, USDA, *U.S. Seed Exports* (quarterly), Washington, D.C.

Foreign Agricultural Service, USDA, *U.S. Spice Trade* (annual), Washington, D.C.

Foreign Agricultural Service, USDA, *Wood Products: International Trade and Foreign Markets* (5 issues per year), Washington, D.C.

Foreign Agricultural Service, USDA, *World Agricultural Production* (monthly), Washington, D.C.

Foreign Agricultural Service, USDA, *World Cocoa Situation* (semiannual), Washington, D.C.

Foreign Agricultural Service, USDA, *World Coffee Situation* (semiannual), Washington, D.C.

Foreign Agricultural Service, USDA, *World Dairy Situation* (annual), Washington, D.C.

Foreign Agricultural Service, USDA, *World Grain Situation & Outlook* (monthly), Washington, D.C.

Foreign Agricultural Service, USDA, *World Honey Situation* (annual), Washington, D.C.

Foreign Agricultural Service, USDA, *World Livestock Situation* (semiannual), Washington, D.C.

Foreign Agricultural Service, USDA, *World Oilseed Situation & Market Highlights* (monthly), Washington, D.C.

Foreign Agricultural Service, USDA, *World Poultry Situation* (semiannual), Washington, D.C.

Foreign Agricultural Service, USDA, *World Sugar Situation & Outlook* (annual), Washington, D.C.

Foreign Agricultural Service, USDA, *World Tea Situation* (annual), Washington, D.C.

Foreign Agricultural Service, USDA, *World Tobacco Situation* (monthly), Washington, D.C.

National Agricultural Statistics Service, USDA, *Agricultural Prices Monthly* (monthly), Washington, D.C.

National Agricultural Statistics Service, USDA, *Agricultural Chemical Usage* (Mar. and June), Washington, D.C.

National Agricultural Statistics Service, USDA, *Agricultural Prices Annual* (annual), Washington, D.C.

National Agricultural Statistics Service, USDA, *Almond Production* (annual), Washington, D.C.

National Agricultural Statistics Service, USDA, *Broiler Hatchery* (weekly), Washington, D.C.

National Agricultural Statistics Service, USDA, *Catfish Processing* (monthly), Washington, D.C.

National Agricultural Statistics Service, USDA, *Catfish Production* (periodic), Washington, D.C.

National Agricultural Statistics Service, USDA, *Cattle* (semiannual), Washington, D.C.

National Agricultural Statistics Service, USDA, *Cattle on Feed* (monthly), Washington, D.C.

National Agricultural Statistics Service, USDA, *Cherry Production* (annual), Washington, D.C.

National Agricultural Statistics Service, USDA, *Citrus Fruits* (annual), Washington, D.C.

National Agricultural Statistics Service, USDA, *Cold Storage* (monthly plus annual), Washington, D.C.

National Agricultural Statistics Service, USDA, *Cotton Ginnings* (13 issues plus annual), Washington, D.C.

National Agricultural Statistics Service, USDA, *Cranberries* (annual), Washington, D.C.

National Agricultural Statistics Service, USDA, *Crop Production* (monthly), Washington, D.C.

National Agricultural Statistics Service, USDA, *Crop Progress* (weekly, Apr.–Nov.), Washington, D.C.

National Agricultural Statistics Service, USDA, *Crop Values* (annual), Washington, D.C.

National Agricultural Statistics Service, USDA, *Dairy Products* (monthly plus annual), Washington, D.C.

National Agricultural Statistics Service, USDA, *Egg Products* (monthly), Washington, D.C.

National Agricultural Statistics Service, USDA, *Eggs, Chickens and Turkeys* (monthly), Washington, D.C.

National Agricultural Statistics Service, USDA, *Farm Labor* (quarterly), Washington, D.C.

National Agricultural Statistics Service, USDA, *Farm Numbers and Land in Farms* (annual), Washington, D.C.

National Agricultural Statistics Service, USDA, *Farm Production Expenditures* (July and Aug.), Washington, D.C.

National Agricultural Statistics Service, USDA, *Floriculture Crops* (annual), Washington, D.C.

National Agricultural Statistics Service, USDA, *Grain Stocks* (quarterly), Washington, D.C.

National Agricultural Statistics Service, USDA, *Hazelnut Production* (annual), Washington, D.C.

National Agricultural Statistics Service, USDA, *Hogs and Pigs* (quarterly), Washington, D.C.

National Agricultural Statistics Service, USDA, *Honey* (annual), Washington, D.C.

National Agricultural Statistics Service, USDA, *Hop Stocks* (semiannual), Washington, D.C.

National Agricultural Statistics Service, USDA, *Livestock Slaughter* (monthly plus annual), Washington, D.C.

National Agricultural Statistics Service, USDA, *Meat Animals: Production, Disposition, and Income* (annual), Washington, D.C.

National Agricultural Statistics Service, USDA, *Milk Production* (monthly), Washington, D.C.

National Agricultural Statistics Service, USDA, *Milk Production, Disposition and Income* (annual), Washington, D.C.

National Agricultural Statistics Service, USDA, *Noncitrus Fruits and Nuts* (semiannual), Washington, D.C.

National Agricultural Statistics Service, USDA, *Peanut Stocks and Processing* (monthly), Washington, D.C.

National Agricultural Statistics Service, USDA, *Pistachio Production* (annual), Washington, D.C.

National Agricultural Statistics Service, USDA, *Potato Stocks* (6 issues plus annual), Washington, D.C.

National Agricultural Statistics Service, USDA, *Poultry Slaughter* (monthly), Washington, D.C.

National Agricultural Statistics Service, USDA, *Prices Received, Minnesota-Wisconsin Manufacturing Grade Milk* (annual), Washington, D.C.

National Agricultural Statistics Service, USDA, *Rice Stocks* (quarterly), Washington, D.C.

National Agricultural Statistics Service, USDA, *Sheep and Goats* (annual), Washington, D.C.

National Agricultural Statistics Service, USDA, *Sheep and Lambs on Feed* (3 issues per year), Washington, D.C.

National Agricultural Statistics Service, USDA, *Trout Production* (annual), Washington, D.C.

National Agricultural Statistics Service, USDA, *Turkey Hatchery* (monthly), Washington, D.C.

National Agricultural Statistics Service, USDA, *Vegetables* (5 issues plus annual), Washington, D.C.

National Agricultural Statistics Service, USDA, *Walnut Production* (annual), Washington, D.C.

National Agricultural Statistics Service, USDA, *Wool and Mohair* (annual), Washington, D.C.

World Agricultural Outlook Board, USDA, *World Agricultural Supply and Demand* (monthly), Washington, D.C.

# Glossary

**Absolute advantage** The ability to produce a greater physical output with a given set of resources than a similar set of resources can produce elsewhere.

**Administrative regulation** Rules and regulations established by administrative decision.

**Agent middlemen** Negotiate the transfer of goods from seller to buyer without themselves taking title to those goods.

**Aggregate demand curve** A curve showing the negative relationship between the price level and the quantities of real goods and services demanded during a given period of time.

**Aggregate supply curve** A curve showing the positive relationship between the quantities of real goods and services supplied by businesses and the price level during a given period of time.

**Agribusiness** Involves the manufacture and distribution of farm supplies; production operations on the farm; and the storage, processing, and distribution of farm commodities and items made from them.

**Agricultural bargaining** A group of producers organized to gain greater power in the market for its members.

**Agricultural economics** An applied social science dealing with how humankind uses technical knowledge and scarce productive resources to produce food and fiber and to distribute them to society for consumption over time.

**Agricultural fundamentalism** The school of thought holding that there is something special and unique about the farm way of life.

**Agricultural options** A type of financial instrument that gives the holder the right to buy or sell futures contracts.

**Agricultural policy** Public policies developed to achieve specific objectives desired for agriculture.

**Arable land** Land currently being cropped or capable of being cropped without requiring additional costs of development.

**Arbitrage** The simultaneous purchase and sale of a commodity in two different markets to take advantage of differences in the prices of that commodity in the markets.

**Assets** Items of money value owned by a business, including such items as land, buildings, tractors, and combines.

**Average fixed cost (AFC)** Fixed cost per unit of output, determined by dividing total fixed cost by output at each level of output.

**Average physical product (APP)** Units of output produced per unit of input for each level of variable input use.

**Average total cost (ATC)** Total cost per unit of output, determined by dividing total cost by output at each level of output.

**Average value product (AVP)** The value of output per unit of input for each level of variable input use.

**Average variable cost (AVC)** Variable cost per unit of output, determined by dividing total variable cost by output at each level of output.

**Balance of trade** The value of merchandise exports minus the value of merchandise imports.

**Bank reserves** The amount that a bank has on deposit with the Federal Reserve Bank or held as cash in its vault.

**Basis** The difference between a futures price and the cash price of a commodity at a particular time and place.

**Biological resources** Natural resources, such as forests and fisheries, that produce a harvestable yield from the resource stock.

**Budget line** Shows all possible combinations of goods that a consumer can buy with a given money income.

**Capital** Productive resources (goods) that are available, as a result of past human decisions, to produce other want-satisfying goods.

**Capitalism** A type of economic organization in which private individuals or groups own and manage all resources.

**Capitalized value** The present value of a resource obtained by discounting the value of its future net income stream.

*Ceteris paribus* Holding some variables constant, while letting specific variables change.

**Change in demand** A shift in the entire demand schedule.

**Change in quantity demanded** A movement along a given demand curve in response to a price change.

**Change in quantity supplied** A movement along a given supply curve in response to a price change.

**Change in supply** A shift in the entire supply schedule.

**Choice** Deciding on one action over another, or one good in preference to another, according to certain criteria.

**Collective farm** Typically, a farm in the FSU owned by the government, but operated by a number of families, which shared in the farm-derived revenue.

**Common market** A customs union that also provides for the free mobility of factors of production among the member nations.

**Common property resources** Resources that are owned by the government in the name of the public.

**Commonwealth of Independent States** A group of eleven republics of the former Soviet Union. These republics are Azerbaijan, Armenia, Byelarus, Kazakhstan, Kyrgyzstan, Moldova, Russia, Tajikistan, Turkmenistan, Uzbekistan, and Ukraine.

**Comparative advantage** A situation in which an individual, region, or nation is relatively superior at producing some goods because of lower opportunity costs, and gains by trading for goods that another is relatively more proficient at producing.

**Complements** Using two goods in combination because the presence of one good enhances the benefit obtained from the other.

**Conglomerate merger** The merging of two or more firms that operate in unrelated industries.

**Conservation** Preserving or extending the productive life of a resource.

**Constant returns** A constant input-output ratio.

**Consumer Price Index** The average price of a market basket of goods and services in index form to permit comparisons of the prices of goods and services over a period of years.

**Consumption component** The utility or satisfaction an individual derives from an education.

**Contract production** Contractual agreements between producers and their input suppliers or product marketing firms.

**Cooperative** An association of member-owners operating a business that provides services at cost to its patrons.

**Cost** The value of a sacrificed alternative.

**Cross price elasticity** The responsiveness of the quantity demanded of one commodity to a 1 percent change in the price of another commodity, *ceteris paribus*.

**Customs union** Group of nations that abolishes trade restrictions among themselves and imposes a common and uniform tariff on other nations.

**Demand curve** Shows the quantities of a good that a consumer will buy at different prices for that good, everything else unchanged.

**Dependent variable** One whose quantity changes as a result of a change in another (independent) variable.

**Derived demand** Demand schedules for resources that are used in producing final products. The demand for land depends on the intensity of demand for its products.

**Diminishing marginal rate of substitution** The rate at which one good substitutes for another diminishes because each added unit of that good replaces smaller and smaller amounts of the other good, *ceteris paribus*.

**Discount** The amount that is subtracted from a loan as an interest charge in advance. The process of determining the present value of future goods.

**Discount rate** The rate of interest charged by the Federal Reserve Bank on loans made to its member banks.

**Disposable personal income** The after-tax income of individuals available for spending on goods and services.

**Economic efficiency** Allocating resources (or consumption expenditures) in different possible uses so as to maximize net economic benefits (or satisfaction).

**Economic profit** The amount by which a firm's revenue exceeds its total (explicit plus implicit) costs. Also referred to as *pure profit*.

**Economic rent** Return in excess of opportunity cost.

**Economic supply** The part of the physical supply of resources that is used to satisfy human wants.

**Embargo** A complete prohibition against the imports or exports of a commodity.

**Engel curve** The relationship between changes in a consumer's income and the quantity of an item purchased by that person.

**Entrepreneurship** Organizing resources to produce and market goods and services.

**Equilibrium** A condition in which opposing forces within a system just offset one another.

**Equilibrium price** The price in a market at which quantity supplied and quantity demanded are equal.

**European Union (EU)** A group of countries that have reduced or abolished tariffs among themselves and established a common and uniform tariff to outsiders. The EU now includes 15 countries: West Germany, France, Italy, Belgium, the Netherlands, Luxembourg, Great Britain, Denmark, Ireland, Greece, Portugal, Spain, Austria, Finland, and Sweden.

**European Free Trade Association (EFTA)** A group of countries that have reduced or abolished tariffs among themselves, but have not established a common tariff to outsiders. The EFTA consists of Sweden, Norway, Switzerland, Austria, Liechtenstein, Iceland, and Finland.

**Exchange function** Includes all the activities throughout the market channels as buyers and sellers make their transactions.

**Expansion path** A line connecting the least cost combination points along the production surface. Also, a line connecting the most profitable combination points for each of a number of production possibilities curves.

**Expansionary fiscal policy** Increasing government expenditures or reducing taxes to cause economic activity to expand and increase the nation's GDP.

**Expansionary monetary policy** Government action to increase the money supply and lower the interest rate so as to expand economic activity and increase the nation's GDP.

**Expenditure method** Measuring GDP by totaling the value of all final goods produced in the economy during an accounting period.

**Explicit cost** An expenditure made for the use of a resource.

**Exports** Products and services sold to foreign countries.

**Externalities** Costs or benefits that are external to the decision maker and are imposed on others.

**External growth** Expanding the size of a firm by adding another firm through its purchase or other means of merger.

**Facilitating functions** Those activities throughout the marketing system that improve its operational and pricing efficiency.

**Facilitative organizations** Institutions that provide information or facilities for marketing commodities.

**Factor-factor relationship** The relationship between two factors or inputs used in production.

**Factor markets** Markets in which resources are bought and sold. One loop in a circular flow diagram that measures the earnings of all resources used in producing a nation's goods and services.

**Factor-product relationship** The functional relationship between a factor of production and its product.

**Family farm** A farm in which the family provides most of the labor required in its operation.

**Farm (1974 census of agriculture)** Any establishment which, during the census year, had or normally would have had agricultural product sales of $1000 or more.

**Final product** The finished product ready for sale to the ultimate user or consumer.

**Final product value** The market value of a final product.

**Firm** A decision-making business entity that uses resources hired from households to produce goods and services for sale to households or other consuming units.

**Fiscal policy** The use of government policy to achieve specific economic goals by manipulating expenditures or the tax rate.

**Fixed costs** Those costs incurred for resources that do not change as output is increased or decreased.

**Flow resources** Resources whose available quantities are constantly being replenished. For example, using the wind's power or the heat energy from the sun, does not reduce the supply of these resources to others.

**Form utility** The utility or satisfaction generated by the processing function in marketing as it delivers products in the desired forms.

**Free market** A market system in which buyers and sellers decide and act at their own initiative and in their own economic interests.

**Free trade area** Group of nations that abolishes trade restrictions among themselves without also imposing common tariffs on other nations.

**Full employment** A rate of unemployment no greater than that due to the normal functioning of the labor market.

**Fund or stock resources** Resources whose quantities are fixed in their natural state. Coal, oil, and iron ore are examples of these nonrenewable resources.

**Futures contracts** Standardized contracts for future delivery of commodities that are traded on organized exchanges.

**Futures markets** Places in which the activities of buyers and sellers determine the prices of commodities traded in those markets.

**GDP Price Deflator Index** An index used to remove the effects of inflation or deflation to determine changes in real GDP over time.

**Gosplan** The central planning organization of the Soviet Union.

**Gross domestic product (GDP)** The total market value of all finished goods and services produced within the domestic economy whether by foreign or American resources in a given period of time.

**Gross national product (GNP)** The total market value of all final goods and services produced in an economy, which is equal to gross domestic product adjusted to include the net income Americans earned overseas.

**Hectare** A metric measure of land area (one hectare equals 2.4710 acres).

**Hedging** Establishing opposite positions in the futures and cash markets as a protection against adverse price changes.

**Heterogeneous** Units within a group have dissimilar characteristics.

**Homogeneous** All units within a group are alike in their characteristics.

**Horizontal merger** A combination of two or more firms operating in the same industry.

**Household** The basic economic entity in society that consumes the final goods and services produced by the system.

**Household responsibility system (HRS)** China's restructured system for production and marketing decision making by individual households rather than at the commune (production team) level.

**Human capital** The educational investment that improves the knowledge and productivity of people.

**Imperfect competition** Any market structure in which firms do not exhibit the characteristics of perfect competition.

**Imperfect substitutes** Resources whose marginal rate of substitution changes as their proportions are changed, *ceteris paribus*.

**Implicit cost** A cost for which there is no cash outlay at the time a resource is being used, or for which no cash payment is required.

**Import quota** A maximum limit on the quantity of a commodity that can be imported.

**Imports** Goods and services purchased from foreign countries.

**Income effect** The change in a consumer's real income that occurs as the price of a good changes.

**Income elasticity of demand** The responsiveness of quantity purchased to a 1 percent change in income, *ceteris paribus*.

**Income method** Measuring GDP by totaling all labor and other resource earnings throughout the economy during an accounting period.

**Independent variable** A variable whose changes cause the quantity of another variable to be changed.

**Indifference curve** Shows all the combinations of goods that yield an individual the same amount of satisfaction or utility.

**Indifference map** The utility surface shown by a family of indifference curves.

**Input** A resource used to produce a valued product or service.

**Insurance hedging (or hedging)** Establishing an equal but opposite position in the futures market from the one taken in the cash market.

**Interdependent demand curves** A market in which each firm's demand curve is influenced by the pricing decisions of other firms in that market.

**Interest** A charge made for borrowed money. The rate at which we discount future economic goods.

**Interest rate** The price of borrowed money.

**Internal growth** Expanding a firm's capital and facilities as the market for that firm's product grows.

**International trade** Transactions with foreign nations in which goods and services are bought or sold.

**Investment component** The part of an education that is regarded as an investment in an individual's increased productivity that yields increased future earnings.

**Isocost line** Shows the various combinations of resources that can be purchased with a given dollar outlay.

**Isoquant** Shows the different combinations of inputs that can produce a given amount of output or product.

**Isorevenue line** Shows all the combinations of products sold that will yield the same total revenue.

**Labor** The physical effort of humans that is combined with other resources to produce valued goods and services.

**Land** All the productive attributes of the earth's surface, including space and the natural environment.

**Law of comparative advantage** The principle that each person, area, or nation will gain economically if each specializes in producing those goods and services where they have the greatest relative advantage, then trades with other similariy specialized individuals, areas, or nations.

**Law of diminishing marginal utility** As an individual consumes additional units of a specific good, holding everything else constant, the amount of satisfaction from each additional unit of that good decreases.

**Law of diminishing returns** As successive amounts of a variable input are combined with a fixed input, the total product will increase, reach a maximum, and eventually decline.

**Least cost combination** A combination of two or more resources in such a way that the resource cost of producing a given level of output is a minimum.

**Length-of-run** A planning concept relating to the proportion of a firm's inputs that may be fixed or varied in a given time period.

**Liabilities** The total claims of creditors against the value of assets owned by a business.

**Long position** Purchases of futures contracts not offset by sales.

**Long run** A time period long enough that all factors of production can be varied.

**Macroeconomic equilibrium** A situation where aggregate supply equals aggregate demand and there are no forces in the economy acting either to increase or decrease the nation's real GDP.

**Macroeconomics** The area of economics that deals with output, employment, incomes, or other activities in the aggregate.

**Management** The process of controlling or directing a situation. Also one of the four factors of production, with responsibility for decision making.

**Marginal** Additional or extra, either positive or negative.

**Marginal cost (MC)** The change in total cost when output is changed by one unit.

**Marginal factor cost (MFC)** The amount added to total cost when one more unit of the variable input is used in production.

**Marginal physical product (MPP)** The amount added to total physical product when another unit of the variable input is used in production.

**Marginal rate of product substitution (MRPS)** The rate at which one product substitutes for another, measured along the production possibilities curve.

**Marginal rate of substitution (MRS)** The rate at which one good (or resource) can be substituted for another at the margin without changing satisfaction (or output).

**Marginal revenue (MR)** The amount added to total revenue when an additional unit of output is produced and sold.

**Marginal value product (MVP)** The amount added to total value product when another unit of the variable input is used.

**Market** Consists of buyers and sellers with facilities to communicate with one another.

**Market demand curve** A horizontal summation of individual demand curves for a product.

**Market equilibrium** That point where the market demand curve intersects the market supply curve. The market clearing price where quantity demanded equals quantity supplied.

**Market orders and agreements** Government authorized programs that producers can use to determine the terms and conditions under which a commodity can be marketed.

**Market power** A firm's ability to influence product prices either on the buying or selling side of the market.

**Market supply curve** A horizontal summation of individual supply curves for a product.

**Marketing** Business activities that direct the flow of goods and services from producer to consumers.

**Marketing agreement** A voluntary agreement among handlers of a commodity with respect to the quality, quantity, and price of that good.

**Marketing efficiency** A comparison of the value of output to the value of inputs used in the marketing process.

**Marketing margin** The difference between the price that consumers pay for the final product and the price received by producers for the raw product.

**Marketing order** A price agreement that, once issued, is binding on all handlers of the commodity covered by that order.

**Merchant middlemen** Purchasers of goods from wholesalers for later sale to buyers.

**Merger** Combining two or more formerly independent firms under a single management entity.

**Metric ton** A metric measure of weight (one metric ton equals 2204.622 pounds).

**Microeconomics** The area of economics that deals with individual decision units—people, firms, or markets—within the economy.

**Middleman** Functions performed in moving goods and services from producer to consumer.

**Monetary policy** Manipulation of the money supply and the interest rate to achieve specific economic goals.

**Monopoly** Single seller in an industry.

**Monopsony** Single buyer in an industry.

**National income** Net national income minus indirect taxes, after accounting for depreciation; it is thus labor, management, and proprietary earnings plus net interest, net rent, and corporate profits.

**National income accounts** Dollar-term measures of the nation's economic performance.

**Natural resource** A factor of production provided by nature. Examples are land, water, minerals, wind, and tides.

**Net national product** The net value of all goods and services produced in the economy after deducting depreciation from GNP, to account for resources used up during the accounting period.

**Net worth** The excess of assets over liabilities, representing the owner's residual claim to assets.

**Nominal value** The dollar values of goods and services at their current market prices.

**Nonprice competition** Competing by product differentiation (advertising, services, or quality) rather than by price.

**Nonrecourse loan** A loan in which the lender, such as the CCC, has no recourse beyond the physical commodity in payment for that loan.

**Normal profit** A situation in which each of the firm's resources is earning a return just equal to its opportunity cost.

**Normative economics** A scientifically untestable belief or conclusion of what *should* be.

**Oligopolistic market** A market characterized by the small number of firms producing for that market.

**Oligopoly** A firm in an oligopolistic market.

**Open market operations** The purchase and sale in the open market of government securities by the Open Market Committee.

**Opportunity cost** The value of other opportunities given up in order to produce or consume any good.

**Optimum** A condition in which the economic outcome cannot be improved upon.

**Options** Contracts that give the holder the right to buy or sell a property.

**Parity** That price which gives a unit of agricultural commodity the same purchasing power as it had in a specified base period.

**Payment-in-kind (PIK) program** A program in which farmers are paid "in kind" (commodities, rather than money) for participating in the Conservation Reserve Program.

**Perfect complements** Resources that must be used in a given ratio in order to produce a product.

**Perfect substitutes** Resources that substitute for one another in production at a constant rate.

**Personal income** The total income of people before the payment of income taxes.

**Physical function** The processing, storage, and transportation of commodities in the marketing system.

**Physical supply** The total available quantity of a good or resource.

**Place utility** The utility or satisfaction derived from the transportation function in marketing that delivers products in the place desired by consumers.

**Positive economics** A scientifically testable conclusion of what is or what can be.

**Potentially arable land** Land that can be brought into production after undergoing sufficient physical development.

**Poverty level** Official designation of an income below that which is needed to provide the kind of living our society considers to be a basic right.

**Present value** The present value of money or an economic good to be received sometime in the future.

**Price** The dollar value per unit of a resource, good, or service as determined in the market or by other means.

**Price ceiling** A government-decreed maximum price for a commodity or service.

**Price elasticity of demand** The percentage change in quantity demanded due to a 1 percent change in price, *ceteris paribus*.

**Price elasticity of supply** The percentage change in quantity supplied due to a 1 percent change in price, *ceteris paribus*.

**Price floor** A government-decreed minimum price for a commodity.

**Price level** The prices of all currently produced goods and services taken as an average.

**Price searcher** A buyer or seller in an imperfectly competitive market structure who has market power and can determine the price or quantity sold.

**Price-support** A government-determined price that is above the equilibrium price.

**Price surface** A geographic map of concentric circles connecting all equal-price points around a central market.

**Price taker** A buyer or seller without market power in a purely competitive market.

**Pricing efficiency** The accuracy, precision, and speed with which prices reflect consumer demands and are passed back through the market channels to producers.

**Prime rate** The rate of interest that large banks charge their best customers.

**Principle of population** Malthus' general conclusion regarding the potential of population to grow at a more rapid rate than the food supply might grow.

**Privatization** The establishment and protection of private property rights by law. Government or centrally-held property is transferred to private ownership.

**Processors** Transform raw agricultural products into final products.

**Product markets** Markets in which commodities are bought and sold. One loop in a circular flow diagram measuring the value flow of final goods and services in an economy.

**Product-product relationship** The relationship between a firm's enterprises or products.

**Production** Any activity or process that satisfies a human desire directly or indirectly, presently or in the future.

**Production function** The input-output relationship between resources and their product.

**Production possibilities curve** Shows all combinations of products that can be produced with a given set of resources.

**Production surface** The production surface projected by a family of isoquants showing all the different output quantities resulting from using different quantities and proportions of two variable inputs.

**Productivity** A rate of output, such as the ratio of output to input services.

**Profit** A surplus over all opportunity costs.

**Public policy** Group action designed to achieve certain aspirations held by members of society.

**Pure competition** A market organization of many firms in an industry, a homogeneous product, and the freedom of firms to enter or leave the industry. No firm in this type of industry can influence the market price of its product.

**Pure monopoly** A market organization of one firm in an industry. Competition in the product market is absent.

**Pure profit** The amount by which a firm's total revenue exceeds its total (explicit plus implicit) costs. Also referred to as *economic profit*.

**Quality of life** A term used to reflect such general social goals as peace, security, freedom, and justice.

**Real value** Economic values expressed in terms of a base year index that eliminates the effect of price changes.

**Resources** Inputs with which goods and services are produced.

**Resource reserve** An inventorying concept related to the physical and economic availability of certain natural resources.

**Resource substitution** The ability of resources to substitute for one another in producing goods and services.

**Ridge line** Line on a production surface that connects all points of zero marginal physical product for each of two variable inputs as the other is held constant at different levels.

**Rural development** All the developmental activities that result in making rural America a better place in which to live and work.

**Sacrifice** Giving up one good or service in order to obtain another good or service.

**Scarcity** A basic economic condition in which our wants exceed the resources available to satisfy those wants.

**Scientific method** A systematic, logical approach to problem solving.

**Selective controls** Federal Reserve System controls over interest rates, margins, and installment credit in order to achieve federal policy objectives.

**Short position** Sales of futures contracts not offset by purchases.

**Short run** A time period where one or more of the factors of production are fixed.

**Social costs** The sum total of internal and external costs of an economic activity.

**Socialism** A type of economic organization in which the government owns all ~~d directs~~ all economic activity.

**Speculation** The purchase or sale of title to goods or financial obligations in the expectation of favorable price movements.

**Speculative middlemen** Own and hold commodities in the expectation of a price change in their favor.

**Spillover effect** Benefits or costs that are external to the decision maker and "spill over" to others in the economy.

**Stagflation** A situation in which a high rate of inflation and high unemployment occur simultaneously.

**State farm** Large farms owned and operated by the government (*sovkoz* in the FSU).

**Substitutes** Two different goods (or resources) between which a choice is made to satisfy human wants (or to produce a product).

**Substitution effect** The substitution of one good for another as their relative prices change.

**Supply curve** The amount of a good or service a producer is willing to offer for sale at different prices, holding everything else constant.

**Tariff** A tax on imported goods.

**Technical efficiency** The physical input-output relationships with which marketing functions are performed.

**Time preference** Our desire for economic goods now rather than in the future.

**Time utility** The utility or satisfaction derived from the storage function in marketing that delivers a product at the desired time.

**Total cost (TC)** The sum total of all the firm's fixed and variable costs at each level of output.

**Total fixed cost (TFC)** The implicit cost of fixed resources used to produce a good or service.

**Total physical product (TPP)** The total quantity of a product that is produced at each level of variable input use.

**Total revenue (TR)** A firm's total value of output, obtained by multiplying the units of product produced by the price of that product.

**Total value product (TVP)** The total value of output produced at each level of variable input use.

**Total variable cost (TVC)** Total of the costs of all the variable resources used in production.

**Transfer costs** The costs arising from moving commodities from one market to another.

**Utility** The satisfaction an individual gets from consuming goods and services.

**Value** The worth, in the market or to a person, of a good or service.

**Value added** The contribution to final product value by each stage in the production process.

**Value judgment** A conclusion that is based on the values that one holds.

**Values** Group or personally held principles regarding qualities that are worthwhile or desirable.

**Variable costs** Those costs that increase or decrease as varying amounts of the resources are used.

**Vertical coordination** The linkage of successive stages in producing or marketing a product in a single decision entity.

**Vertical integration** The linkage of firms in different stages of production or marketing under the ownership of a single firm.

**Vertical merger** Combining two or more firms in different stages of production or marketing into one firm.

**World Trade Organization (WTO)** Created as a single organization to handle international trade agreements, international trade disputes, and as an international trade policy review mechanism.

# Photo Credits

**CHAPTER 13**    *Page 346:* Ric Ergenbright/Tony Stone Images/New York, Inc.    *Page 379:* Courtesy Daniel W. Bromley.

**CHAPTER 14**    *Page 382:* J. Clark/Courtesy U.S. Department of Agriculture.    *Page 408:* Courtesy University of Minnesota.

**CHAPTER 15**    *Page 410:* Philip H. Coblentz/Tony Stone Images/New York, Inc.    *Page 441:* Courtesy D. Gale Johnson.

**CHAPTER 16**    *Page 444:* Courtesy of Lakehead Harbour Commission. *Page 473:* Courtesy Andrew Schmitz.

**CHAPTER 17**    *Page 476:* Clive Rowat/Tony Stone Images/New York, Inc.    *Page 508:* Courtesy G. Edward Schuh.

# Index

Page numbers followed by *n* refer to footnotes.